An Essential Guide to Electronic Material Surfaces and Interfaces

An Essential Guide to Electronic Material Surfaces and Interfaces

LEONARD J. BRILLSON

Ohio State University, USA

WILEY

Library of Congress Cataloging-in-Publication Data

Names: Brillson, L. J., author.
Title: An essential guide to electronic material surfaces and interfaces / by
    Leonard J. Brillson, Ohio State University.
Description: Hoboken, New Jersey : John Wiley & Sons, Inc., [2016] | ?2016 |
    Includes bibliographical references and index.
Identifiers: LCCN 2016006828 | ISBN 9781119027119 (cloth) | ISBN 111902711X
    (cloth)
Subjects: LCSH: Electronics – Materials. | Surfaces (Technology) – Analysis. |
    Spectrum analysis. | Semiconductors – Materials.
Classification: LCC TK7871 .B748 2016 | DDC 621.381 – dc23 LC record available at http://lccn.loc.gov/2016006828

A catalogue record for this book is available from the British Library.

ISBN: 9781119027119

Typeset in 10/12pt TimesLTStd by SPi Global, Chennai, India.

Printed in Singapore by C.O.S. Printers Pte Ltd

1   2016

Contents

Preface

This book is intended to introduce scientists, engineers, and students to surfaces and interfaces of electronic materials. It is designed to be a concise but comprehensive guide to the essential information needed to understand the physical properties of surfaces and interfaces, the techniques used to measure them, and the methods now available to control them. The book is organized to provide readers first with the basic parameters that describe semiconductors, then the key features of their interfaces with vacuum, metal, and other semiconductors, followed by the many experimental and theoretical techniques available to characterize electronic material properties, semiconductor surfaces, and their device applications, and finally our current understanding of semiconductor interfaces with metals and with other semiconductors – understanding that includes the methods, both macroscopic and atomic-scale, now available to control their properties.

Electronic material surfaces and interfaces has become an enormous field that spans physics, chemistry, materials science, and electrical engineering. Its development in terms of new materials, analytic tools, measurement and control techniques, and atomic-scale understanding is the result of nearly 70 years of activity worldwide. This book is intended to provide researchers new to this field with the primary results and a framework for new work that has developed since then. More in-depth information is provided in a Track II, accessed online, that provides advanced examples of concepts discussed in the text as well as selected derivations of important relationships. The problem sets following each chapter provide readers with exercises to strengthen their understanding of the material presented. Figures for use by instructors are available from the accompanying website. For more advanced and detailed information, the reader is referred to *Surfaces and Interfaces of Electronic Materials* (Wiley-VCH, Weinheim, 2010) as well as the lists of Further Reading following each chapter.

On a personal note, the author wishes to acknowledge his Ph.D. thesis advisor, Prof. Eli Burstein at the University of Pennsylvania and Dr. Charles B. Duke at Xerox Corporation in Webster, New York for their inspiration and mentoring. Here at The Ohio State University, he thanks his many Electrical & Computer Engineering and Physics colleagues who have provided such a stimulating environment for both teaching and research. Most of all, the author's deepest thanks are to his wife, Janice Brillson, for her patience, love, and support while this book was written.

Columbus, Ohio July 31, 2015

About the Companion Websites

This book is accompanied by Instructor and Student companion websites:

www.wiley.com\go\brillson\electronic

The Instructor website includes

- Supplementary materials
- PPT containing figures
- Advanced topics
- Solutions to problems

The student website includes

- Supplementary materials
- Advanced topics

1

Why Surfaces and Interfaces of Electronic Materials

1.1 The Impact of Electronic Materials

This is the age of materials – we now have the ability to design and create new materials with properties not found in nature. Electronic materials are one of the most exciting classes of these new materials. Historically, electronic materials have meant semiconductors, the substances that can emit and react to light, generate and control current, as well as respond to temperature, pressure, and a host of other physical stimuli. These materials and their coupling with insulators and metals have formed the basis of modern electronics and are the key ingredient in computers, lasers, cell phones, displays, communication networks, and many other devices. The ability of these electronic materials to perform these functions depends not only on their inherent properties in bulk form but also, and increasingly so, on the properties of their surfaces and interfaces. Indeed, the evolution of these materials over the past 70 years has been for ever decreasing size and increasing complexity, features that have driven ever increasing speeds and the ability to manage ever larger bodies of information. In turn, these are enabling our modern day quality of life.

1.2 Surface and Interface Importance as Electronics Shrink

Surfaces and interfaces are central to microelectronics. One of the most common micro-electronic devices is the transistor, whose function can illustrate the interfaces involved and their increasing importance as the scale of electronics shrinks. Figure 1.1 shows the basic structure of a transistor and the functions of its interfaces. Current passes from a *source* metal to a *drain* metal through a semiconductor, in this case, Si. Voltage applied to a *gate* metal positioned between the source and drain serves to attract or repel charge

An Essential Guide to Electronic Material Surfaces and Interfaces, First Edition. Leonard J. Brillson.
© 2016 John Wiley & Sons, Ltd. Published 2016 by John Wiley & Sons, Ltd.
Companion Website: www.wiley.com/go/Brillson/

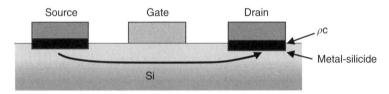

Figure 1.1   Source–gate–drain structure of a silicon transistor. (Brillson 2010. Reproduced with permission of Wiley.)

from the semiconductor region, the "*channel*", through which current travels. The result is to control or "gate" the current flow by this third electrode. This device is at the core of the microelectronics industry.

The interfaces between the semiconductor, metal, and insulator of this device play a central role in its operation. Barriers to charge transport across the metal–semiconductor interface are a major concern for all electronic devices. Low resistance contacts at Si source and drain contacts typically involve metals that produce interface reactions, for example, Ti contacts that react with Si with high temperature annealing to form $TiSi_2$ between Ti and Si, as shown in Figure 1.2a. These reacted layers reduce such barriers to charge transport and the contact resistivity ρ_C. Such interfacial silicide layers form low resistance, planar junctions that can be integrated into the manufacturing process for integrated circuits and whose penetration into the semiconductor can be controlled on a nanometer scale.

The gate–semiconductor interface is another important interface. This junction may involve either: (i) a metal in direct contact with the semiconductor where the interfacial barrier inhibits charge flow or (ii) as widely used in Si microelectronics, a metal–insulator–semiconductor stack to apply voltage without current leakage to the semiconductor's channel region (Figure 1.2b). Lattice sites within the insulator and at the insulator–semiconductor interface can trap charges and introduce electric dipoles that oppose the voltage applied to the gate metal and its control of the channel current. A major goal of the microelectronics industry since the 1950s has been to minimize the formation of these localized charge states.

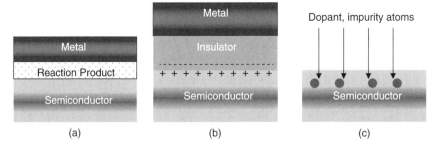

(a)   (b)   (c)

Figure 1.2   Interfaces involved in forming the transistor structure including: (a) the source or drain metal–semiconductor contact with a reacted interface, (b) the gate metal–semiconductor contact separated by an insulator in and near which charges are trapped, and (c) dopant impurity atoms implanted below the surface of a semiconductor to control its carrier concentration. (Brillson 2010. Reproduced with permission of Wiley.)

A third important interface involves the implantation of impurity atoms into the semi-conductor to add donor or acceptor "dopants" that control the transistor's n- or p-type carrier type and density within specific regions of the device. This process involves acceleration and penetration of ionized atoms at well-defined depths, nanometers to microns, into the semiconductor, as Figure 1.2c illustrates. This ion implantation also produces lattice damage that high temperature annealing can heal. However, such annealing can introduce diffusion and unintentional doping in other regions of the semiconductor that can change their electronic activity. Outdiffusion of semiconductor constituents due to annealing is also possible, leaving behind electronically active lattice sites. Precise design of materials, surface and interface preparation, thermal treatments, and device architectures are required in order to balance these competing effects.

At the circuit level, there are multiple interfaces between semiconductors, oxides, and metals. For Si transistors, the manufacturing process involves (i) growing a Si boule in a molten bath that can be sectioned into wafers, (ii) oxidizing, diffusing, and implanting with dopants, (iii) overcoating with various metal and organic layers, (iv) photolithographically patterning and etching the wafers into monolithic arrays of devices, and (v) dicing the wafer into individual circuits that can be mounted, wire bonded, and packaged into chips. Within individual circuit elements, there can be many layers of interconnected conductors, insulators, and their interfaces. Figure 1.3 illustrates the different materials and interfaces associated with a 0.18 μm transistor at the bottom of a multilayer Al–W–Si-oxide dielectric assembly [1]. Reaction, adhesion, interdiffusion, and the formation of localized electronic states must be carefully controlled at all of these interfaces during the many patterning, etching, and annealing steps involved in assembling the full structure. The materials used in this multilayer device architecture have continued to change over decades to compensate for the otherwise increasing electrical resistance as interconnects between layers shrink into the nanometer regime. These microelectronic materials and architectures continue to evolve in order to achieve higher speeds, reliability, and packing density. This continuing evolution highlights the importance of interfaces since they are an increasing proportion of the entire structure as circuit sizes decrease and become ever more complex.

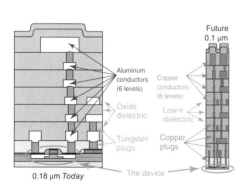
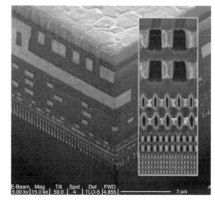

Figure 1.3 *Multilayer, multi-material interconnect architectures at the nanoscale. Feature size of interconnects at right is 45 nm [1].*

Besides transistors, many other electronic devices rely on interfaces for their operation. For example, solar cells operate by converting incident light into free electrons and holes that separate and generate current or voltage in an external circuit. The charge separation requires built-in electric fields that occur at metal–semiconductor or semiconductor–semiconductor interfaces. These will be discussed in later chapters. Transistors without metal gates are another such device. Channel current is controlled instead by molecules that adsorb on this otherwise free surface, exchanging charge and inducing electric fields analogous to that of a gate. Interfaces are also important in devices that generate photons, microwaves, and acoustic waves. These devices require low resistance contacts to inject current or apply voltage to the layer that generates the radiation. Otherwise, power is lost at these contacts, reducing or totally blocking power conversion inside the semiconductor. Semiconductor cathodes that emit electrons when excited by incident photons are also sensitive to surface conditions. Surface chemical treatments of specific semiconductors are required in order to promote the emission of multiple electrons when struck by single photons, useful for electron pulse generation or photomultipliers.

For devices on the quantum scale, surfaces and interfaces have an even larger impact. A prime example is the quantum well, one of the workhorses of optoelectronics. Here a semiconductor is sandwiched between two larger gap semiconductors that localize both electrons and holes in the smaller gap semiconductor. This spatial overlap between electrons and holes increases electron–hole recombination and light emission. Since the quantum well formed by this sandwich is only a few monolayers thick, the allowed energies of electrons and holes inside the well are quantized at discrete energies, which promotes efficient carrier population inversion and laser light emission. Imperfections at the interfaces of these quantum wells introduce alternative pathways for recombination that reduce the desired emission involving the quantized states. Such recombination is even more serious for three-dimensional quantum wells, termed *quantum dots*. Another such example is the *two-dimensional electron gas* (2DEG) formed when charges accumulate in narrow interfacial layers, only a few tens of nanometers thick, between two semiconductors. The high carrier concentration, high mobility 2DEG is the basis of *high electron mobility transistors* (HEMTs). As with quantum wells, lattice defects at the interface can produce local electric fields that scatter charges, reduce mobility, and alter or even destroy the 2DEG region. Yet another quantum-scale structure whose interfaces play a key role is the *cascade laser*, based on alternating high- and low-bandgap semiconductors. In this structure, charge must tunnel through monolayer thin barrier layers, formed by the higher band gap semiconductors, into quantized energy levels. Here again, efficient tunneling depends on interfaces without significant imperfections that could otherwise introduce carrier scattering.

In addition to carrier scattering and recombination, the alignment of energy levels between constituents is a major feature of interfaces. How energy levels align between a semiconductor and a metal, as well as between two semiconductors, is a fundamental issue that is still not well understood. The interface band alignment determines the barriers to charge transport between constituents and the carrier confinement between them. There are several factors that can play a role: (i) the band structure of the constituents at the junction, (ii) the conditions under which the interface forms, and (iii) any subsequent thermal or chemical processing. Hence, it is important to measure and understand the properties of these interfaces at the microscopic, indeed atomic, level in order to learn and control how surfaces and interfaces impact electronics. The future of electronics depends on our ability to design, measure, and control the physical properties of these surfaces and interfaces. This book is intended to describe the key properties of these surfaces and interfaces along with the tools and techniques used to measure them.

1.3   Historical Background

1.3.1   Contact Electrification and the Development of Solid State Concepts

The recognition that contacts to materials could be electrically active dates back to the Greco-Roman times and the discovery of *triboelectricity*, the charge transfer between solids generated by friction – for example, the generation of static electricity by rubbing cat's fur with amber. Two millennia later, Braun discovered *contact rectification*, the unequal flow of charge in two directions, by applying positive and negative voltage to metal contacts with selenium [2]. In 1905, Einstein explained the *photoelectric effect* at metal surfaces based on light as wave packets with discrete energies [3]. This established both the particle behavior of light and the concept of work function, an essential component in the description of energy bands in solids. These energy bands will be discussed in Chapter 2 and form the basis for discussing all the physical phenomena associated with surfaces and interfaces of electronic materials.

The theory of energy bands in solids was first developed by Wilson [4,5] and forms the basis for interpreting semiconductor phenomena. Siemens *et al.* [6] and Schottky *et al.* [7] showed that contact rectification at the interface exhibited behavior that was distinct from that of the semiconductor interior. Subsequently, Mott [8], Davidov [9,10], and Schottky [11–13] each published theories of contact rectification that involved an energy band bending region near the interface. Mott proposed an insulating region between the metal and semiconductor to account for current–voltage features of copper oxide rectifiers. Davidov showed the importance of thermionic work function differences in forming the band bending region. Schottky introduced the concept of a stable space charge in the semiconductor instead of a chemically distinct interfacial layer. Appropriately, this interface concept became the *Schottky barrier*. Bethe's theory of thermionic emission of carriers over an energy barrier [14], based partly on Richardson's earlier thermionic cathode work [15] provided a picture of charge transport across the band bending region that accounted for the rectification process in most experimental situations. These developments laid the foundation for describing barrier formation and charge transfer at semiconductor interfaces.

1.3.2   Crystal Growth and Refinement

The understanding and control of semiconductor properties began with advances in the growth and refinement of semiconductor crystals. Once researchers learned to create crystals with ultrahigh purity (as low as one impurity atom in ten billion) and crystalline perfection, they could unmask the inherent properties of semiconductors, free of effects due to lattice impurities and imperfections, and begin to control these properties in electronic devices. Beginning in the 1940s, crystal pulling and zone-refining techniques became available that could produce large, high-purity semiconductor single crystals.

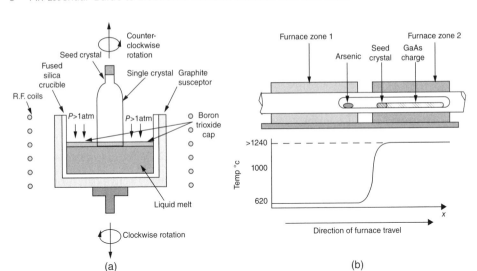

(a)   (b)

Figure 1.4   (a) LEC growth of GaAs crystal pulled from a molten bath inside a heated cru-cible. (Lammers, D www.semiconductor.net/article/CA6513618.html. Used under CC-BY 3.0 https://creativecommons.org/licenses/by-sa/3.0/deed.en.) (b) Horizontal Bridgman growth of GaAs traveling through the temperature gradient of a furnace.

Figure 1.4a illustrates the *liquid encapsulated Czochralski* (LEC) method of growing large single crystals from a molten bath in which a seed crystal nucleates larger boules of crystal as the seed is gradually raised out of the melt. Figure 1.4b shows how molten semiconductor constituents are nucleated at a seed over which a furnace is gradually pulled from high to low temperature.

The crystal thus produced is then passed again through a melt zone along its length, causing impurities to segregate from high to low temperature regions in a process termed *zone-refining* [16]. Numerous other crystal growing techniques such as liquid phase epitaxy (LPE), pulsed laser deposition (PLD), and molecular beam epitaxy (MBE) are now available that are capable of producing multilayer crystal structures with atomic-scale precision. Figure 1.5 illustrates the LPE and MBE methods. These and other techniques are able to create materials not found in nature and having unique properties that enable a myriad of devices.

1.3.3   Transistor Development and the Birth of Semiconductor Devices

Semiconductor surfaces and interfaces have been an integral part of the transistor's development. Motivated by the need to replace vacuum tubes as amplifiers and switches in telecommunications equipment, researchers at Bell Labs succeeded in demonstrating transistor action in 1947 using closely spaced point contacts on germanium crystals [18]. Figure 1.6 shows this early transistor configuration, which relied on high purity crystals supplied by Purdue University [19] to obtain carrier diffusion lengths long enough to reach between contacts. Contrast this early transistor with the nanoscale device structures pictured in Figure 1.3. Developing this first transistor, Bardeen, Shockley, and Brattain found that gate modulation of the charge and current inside the semiconductor was much weaker than

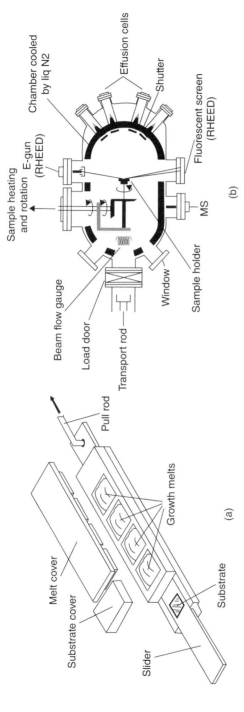

Figure 1.5   (a) LPE method in which a growth substrate slides under pockets of melted constituents that condense to form a stack of semi-conductor layers on the substrate. (b) MBE method in which constituents evaporated from individual crucibles of pure elements deposit layer-by-atomic layer monitored by reflection high energy diffraction (RHEED) on a heated substrate in vacuum. (Barron, W. D http://cnx.org/content/m25712/latest. Used under Creative Commons Attribution 4.0 License.)

Figure 1.6   *A point contact transistor with three gold contacts on a germanium crystal [18].*

expected. This led to the discovery that electric charges immobilized on the semiconductor surface were responsible, which Bardeen correctly interpreted as electronic states localized at the semiconductor interface [20]. By reducing the density of these states and their influence on the field effect pictured in Figure 1.1, Bardeen *et al.* could demonstrate the transistor action clearly, for which they received the 1956 Nobel Prize. The interface concepts developed with the transistor continue to drive the evolution of future electronics technology. By the early 1950s, semiconductor electronics began to develop rapidly. First came the switch from germanium to silicon, since the native oxide of silicon was much more resistant to water vapor than that of germanium [21]. By 1958, the first integrated circuits appeared [22], based on arrays of transistor circuits on silicon wafers. The 1950s and early 1960s saw the development of many other devices, including photodiodes, sensors, photomultipliers, and light emitting diodes. All required high purity semiconductor as well as refined techniques to form their electrical contacts. This led to the development of high-vacuum technology, including the pumps, chambers, and pressure gauges to be discussed in Chapter 5. These tools could be used to control both the localized states at the transistor's semiconductor–insulator interface and those at Schottky barriers of the metal–semiconductor junctions found in most of these devices. Early Schottky barrier experiments involved semiconductor crystals cleaved in medium vacuum, $\sim 10^{-6}$ Torr and in a stream of evaporating metal atoms to minimize contamination [23]. This work presaged the use of *ultrahigh vacuum* (UHV) to obtain atomically-clean surfaces and interfaces.

1.3.4   Surface Science and Microelectronics

The use of ultrahigh vacuum technology with chamber pressures of 10^{-10} Torr or less began in the late 1960s and marked the birth of *surface science* as a field that spans multiple disciplines and that continues to grow. With the ability to prepare clean surfaces and maintain their cleanliness for long periods of time, researchers were now able to study a wide range of electronic, chemical, and atomic structural phenomena. Beginning in the 1970s, much of this work aimed to understand the nature of surface and interface states that could account for the insensitivity of transistors to applied bias voltage. For the oxide interface with silicon, researchers found that careful surface cleaning and precise high temperature gas ambient annealing – processes developed using surface science techniques – could reduce interface state densities by orders of magnitude. By the mid-1970s, the materials science or microelectronic metallurgy of the various electronic material interfaces helped refine the monolithic (extended surface) electronic circuitry on silicon wafers. This work was instrumental in understanding and predicting the interfacial reactions at silicon–metal interfaces, for example, Figure 1.2a [24,25].

Another major challenge was the insensitivity of band bending and Schottky barriers to different metals at the metal–semiconductor interface. This was attributed to surface or interface states that "pinned" the semiconductor Fermi level in a narrow range of energies within the band gap. Initial studies of metal contacts to compound semiconductors suggested that the ionicity of the semiconductor played a role in creating such states [26], and theoretical calculations of the abrupt surface termination of the bulk crystal structure supported this. However, surfaces of many compound semiconductors, particularly those prepared without steps or other defects, exhibited an absence of such states in the semiconductor band gap. Instead, measurements of the bonding structure at many compound semiconductor surfaces revealed atomic rearrangements that served to remove surface states from within the semiconductor band gap. Theoretical studies confirmed that these surface atomic rearrangements or reconstructions in general minimized bond energies and removed such *intrinsic* surface states from the semiconductor band gap.

Other models to account for this *Fermi level pinning* were: (i) adsorbate bonding that forms electrically-active defects at the semiconductor surface; (ii) wave function tunneling that forms localized states close to the interface; (iii) chemical reactions that produce interfacial layers with new dielectric properties and/or defects. During this period, literally thousands of studies, both experimental and theoretical, were devoted to testing these models of Fermi level pinning. While metal deposition on one type of III-V semiconductor exhibited barrier insensitivity for many adsorbates, researchers found that the same compounds grown with higher crystalline perfection could exhibit wide barrier height ranges. Wave function tunneling based on intrinsic semiconductor properties provided a mechanism for theoretical predictions of barriers heights and their systematics for different semiconductors with an inert metal [27–30]. However, experimental results show serious inconsistences with predicted barrier height values for both ionic [31] and covalent semiconductors [32].

Starting in the late 1980s, researchers found that metals deposited on semiconductors could produce chemical reactions, even near room temperature [33,34]. Surface science techniques were uniquely capable of detecting such reactions even though such reactions and atomic rearrangement often took place on a monolayer scale. The chemical composition of these reactions, the diffusion of atoms out of the semiconductor lattice, and the correlation of their reactivity with Schottky barrier heights suggested that such reactions produced native point defects within the semiconductor where atoms had outdiffused [35,36]. Such defects are electrically active and mobile and can segregate near surfaces and interfaces [35,37]. They also have energy levels that can account for reported Fermi level pinning energies of different semiconductors [36].

The 1980s saw the invention and development of the *scanning tunneling microscope* (STM). With this instrument, researchers were now able to image the arrangement of individual atoms on semiconductor surfaces, resolving many questions previously unanswered by less direct techniques. Derived from STM in the 1990s, *atomic force microscopy* (AFM) provided a new way to measure surface morphology on a nanometer scale and to probe semiconductors and insulators lacking the conductivity required for STM. Also related to the STM was *ballistic electron energy microscopy* (BEEM), which revealed heterogeneous electronic features across semiconductor surfaces that reflected multiple local Schottky barrier heights within the same macroscopic contact. These insulator–semiconductor and metal–semiconductor studies also provided insights into the mechanisms controlling heterojunction band offsets, which establish the basic properties of many micro- and optoelectronic devices.

1.4   Next Generation Electronics

New electronic materials and device architectures have emerged over the past two decades where surfaces and interfaces play an integral role. These include: high dielectric constant (hi-K) insulators, complex oxides, magnetic semiconductors, spin conductors and insulators, nano-scale and quantum-scale structures, two-dimensional (2D) semiconductors, and topological insulators. All of these new materials and device architectures pose challenges for researchers and technologies because of their ultra-small dimensions. In turn, these dimensions magnify the effects of diffusion and reaction that can occur with the very high temperatures and chemically reactive environments that such materials may encounter. As with previous generations of electronic development, we expect a synergy between the science and technology of electronic materials. This dual track of progress will again involve an interplay between growth, characterization, device design and testing that will require many of the techniques described in this book.

This chapter aimed to motivate the study of electronic materials' surfaces and interfaces. Their properties influence band structure, charge densities and electronic states on a scale ranging from microns down to atomic layers. Although these phenomena exist on a microscopic scale, they manifest themselves electrically on a macroscopic scale. Indeed they already dominate electronic and optical properties of current electronics. Coupled with this "race to the bottom" are atomic and nanoscale techniques to characterize the electronic, chemical, and geometric properties of these electronic material structures. This introductory book aims to provide the essential framework to understand the physical properties of electronic material surfaces and interfaces and the techniques developed to measure them.

1.5   Problems

1. Give one example each of how surface and interface techniques have enabled modern CMOS (*complementary metal-oxide semiconductor*) technology (a) structurally, (b) chemically, and (c) electronically.
2. Name three desirable solar cell properties that could be affected by interface defects.
3. Name three interface effects that could degrade quantum well operation.

4. As electronics shrinks into the nanoscale regime, actual numbers of atoms become significant in determining the semiconductor's physical properties. Consider a 0.1 μm Si field effect transistor with a 0.1×0.1 μm^2 cross-section and a doping concentration of 10^{17} cm^{-3}. How many dopant atoms are in the channel region? How many dopant atoms are there altogether?

5. Assume the top channel surface has 0.01 trapped electrons per unit cell. How many surface charges are present? How much do they affect the channel's bulk charge density?

6. Even small numbers of interface states inside electronic devices can have large effects. For a GaAs field effect transistor with gate length 0.2 μm with a 0.1 μm thick $\times$ 0.5 μm wide cross-section and doping density of 5×10^{17} cm^{-3}, how many dopant atoms are in the channel region? How many atoms are there altogether? If there are 0.02 electrons per unit cell at the gate–channel interface, how many electrons are there altogether? How large an effect can these interface states have on the transistor properties? Explain.

7. What fraction of atoms is within one lattice constant of the channel surfaces?

8. Figure 1.2a indicates that reaction products can form between metal contacts and Si. Describe the advantages of depositing multilayers of Ti rather than Au on this surface if a submonolayer of air molecules is present at the Si surface prior to deposition.

9. Figure 1.4 illustrates the complexity of the transistor chip structure. Describe three interface effects that could occur in these structures that could degrade electrical properties during microfabrication.

10. Si emerged as the material of choice for microelectronics. (a) Give five reasons why. (b) What feature of the Si/SiO$_2$ is the most important for device operation and why?

11. Give two reasons and two examples why GaAs, InP and many other III-V compound semiconductors are used for quantum-scale optoelectronics.

12. Which would you choose for high power, high-frequency transistors, GaAs or GaN?

References

1. http://it-material.de/category/wirtschaft/computerindustrie/chip-herstellung/ (accessed 1 Feb.2016).
2. Braun, F. (1874) Über die stromleitungdurchschwefelmetalle. *Ann. Phys. Chem.*, **53**, 556.
3. Einstein, A. (1905) Concerning an heuristic point of view toward the emission and transformation of light. *Ann. Phys.*, **17**, 132.
4. Wilson, A.H. (1931) The theory of electronic semi-conductors. *Proc. R. Soc. Lond.*, **A133**, 458.
5. Wilson, A.H. (1931) The theory of electronic semi-conductors II. *Proc. R. Soc. Lond.*, **A134**, 277.
6. Siemens, G. and Demberg, W. (1931) Über detektoren. *Z. Phys.*, **67**, 375.
7. Schottky, W., Störmer, R., and Waibel, F. (1931) On the rectifying action of cuprous oxide in contact with other metals. *Z. Hochfrequenztztechnik*, **37**, 162.
8. Mott, N.F. (1939) The theory of crystal rectifiers. *Proc. R. Soc. Lond. A*, **171**, 27.

9. Davidov, B. (1938) On the photo-electromotive force in semiconductors. *J. Tech. Phys. USSR*, **5**, 87.

10. Davidov, B. (1939) On the contact resistance of semiconductors. *Sov. J. Phys.*, **1**, 167.

11. Schottky, W. (1939) Zurhalbleitertheorie der sperrschict-und spitzengleichrichter. *Z. Phys.*, **113**, 367.

12. Schottky, W. and Spenke, E. (1939) Zur quantitativen Durchführung der Raumladungs- und Randschichttheorie der Kristallgleichricht. *Wiss. Veröffentl. Siemens-Werken*, **18**, 225.

13. Schottky, W. (1942) Vereinfachte und erweitertetheorie der randschichtgleichrichter. *Z. Phys.*, **118**, 539.

14. Bethe, H.A. (1942) Radiation Laboratory Report No. 43-12, Massachusetts Institute of Technology, November.

15. Richardson, O.W. (1921) *The Emission of Electricity from Hot Bodies*, Longmans-Green, Harlow, Essex.

16. Pfann, W.G. (1958) *Zone Melting*, Wiley, New York.

17. Barron, A. (2009) *Molecular Beam Epitaxy. OpenStax QA, July 13, 2009.* http://legacy-textbook-qa.cnx.org/content/m25712/1.2/.

18. Courtesy of AT&T Archives and History Center.

19. Bray, R. (2014) *The Origin of Semiconductor Research at Purdue*, Purdue University, Department of Physics http://www.physics.purdue.edu/about_us/history/semi_conductor_research.shtml

20. Bardeen, J. (1947) Surface states and rectification at a metal semi-conductor contact. *Phys. Rev.*, **71**, 717.

21. Teal, G.K. (1976) Single crystals of germanium and silicon – basic to the transistor and integrated circuit. *IEEE Trans. Electron Devices*, **ED-23**, 621 and references therein.

22. http://www.ti.com/corp/docs/kilbyctr/kilby.shtml

23. Mead, C.A. and Spitzer, W.G. (1964) Fermi level position at metal-semiconductor interfaces. *Phys. Rev.*, **134**, A713.

24. Poate, J.M., Tu, K.N., and Mayer, J.W. (1978) *Thin Films – Interdiffusion and Reactions*, John Wiley & Sons, Inc.

25. Mayer, J.W. and Lau, S.S. (1990) *Electronic Materials Science: For Integrated Circuits in Si and GaAs*, Macmillan, New York.

26. Kurtin, W., McGill, T.C., and Mead, C.A. (1970) Fundamental transition in the electronic nature of solids. *Phys. Rev. Lett.*, **22**, 1433.

27. Flores, F. and Tejedor, C. (1979) Energy barriers and interface states at heterojunctions. *J. Phys. C*, **12**, 731.

28. Cohen, M.L. (1980) Electrons at interfaces. *Adv. Electron. Electron Phys.*, **51**, 1.

29. (a) Tersoff J. (1984) Schottky barrier heights and the continuum of gap states. *Phys. Rev. Lett.*, **52**, 465; (b) Tersoff, J. (1984) Theory of semiconductor heterojunctions: The role of quantum dipoles. *Phys. Rev. B*, **30**, 4874(R).

30. Mönch, W. (1990) On the physics of metal-semiconductor interfaces. *Rep. Prog. Phys.*, **53**, 221.

31. Allen, M.W. and Durbin S.M. (2010) Role of a universal branch-point energy at ZnO interfaces. *Phys. Rev. B*, **82**, 165310.

32. (a) Mönch, W. (1989) Mechanisms of barrier formation in Schottky contacts: Metal-induced surface and interface states. *Appl. Surf. Sci.*, **41/42**, 128; (b) Mönch, W.(1988) Chemical trends in Schottky barriers: Charge transfer into adsorbate-induced gap states and defects. *Phys. Rev. B*, **37**, 7129.
33. Brillson, L.J. (1978) Transition in Schottky barrier formation with chemical reactivity. *Phys. Rev. Lett.*, **40**, 260.
34. Brillson, L.J. (1978) Chemical reactions and local charge redistribution at metal-CdS and CdSe interfaces. *Phys. Rev. B*, **18**, 2431.
35. Brillson, L.J., Brucker, C.F., Katnani, A.D. *et al.* (1981) Chemical basis for InP-metal Schottky-barrier formation. *Appl. Phys. Lett.*, **38**, 784.
36. Brillson, L.J., Dong, Y., Doutt, D. *et al.* (2009) Massive point defect redistribution near semiconductor surfaces and interfaces and its impact on Schottky barrier formation. *Physica B*, **404**, 4768.
37. Mosbacker, H.L., Strzhemechny, Y.M., White, B.D., *et al.* (2005) Role of near-surface states in ohmic-Schottky conversion of Au contacts to ZnO. *Appl. Phys. Lett.*, **87**, 012102.

Further Reading

Orton, J. (2004) *The Story of Semiconductors*, Oxford University Press, Oxford.
Turton, R. (1995) *The Quantum Dot, A Journey into the Future of Microelectronics*, Oxford University Press, Oxford.
Roulston, D.J. (1999) *An Introduction to the Physics of Semiconductor Devices*, Oxford, New York.

2

Semiconductor Electronic and Optical Properties

The two most important features of semiconductors are their electronic and optical properties. These properties can be understood on the most basic level – that of atoms either arranged randomly in a solid, termed *amorphous*, or arranged in regular geometric lattices, termed *crystalline*. While both semiconductor types exhibit interesting features, crystalline semiconductors can be designed and assembled into multicomponent structures on an atomic scale that possess advanced electronic and optical properties not found in nature.

2.1   The Semiconductor Band Gap

The fundamental characteristic of a semiconductor is its band gap – the energy separating the highest energy level of a solid that is filled with electrons from the next highest energy level that is empty. Materials for which this energy is zero, for example, partially filled bands, are termed *metals*, which readily conduct electricity with low resistance. Materials whose band gaps are very large, exceeding several electron volts, are termed *insulators*, which exhibit very high resistance to conducting electricity. In between these two material classes lie semiconductors, whose electrical resistance can be controlled over many orders of magnitude and which are capable of absorbing or emitting light extending from the infrared to the ultraviolet. It is these properties and the ability to control them with external stimuli that make semiconductors so useful for advanced microelectronic and optoelectronic applications.

   The semiconductor band gap arises from the energy levels associated with the atoms in a solid. Each of these atoms has electrons that arrange themselves into discrete levels according to specific atomic orbitals. When these atoms bond to one another and form a solid, the atomic orbitals least bound to the atomic nuclei interact with each other, forming

An Essential Guide to Electronic Material Surfaces and Interfaces, First Edition. Leonard J. Brillson.
© 2016 John Wiley & Sons, Ltd. Published 2016 by John Wiley & Sons, Ltd.
Companion Website: www.wiley.com/go/Brillson/

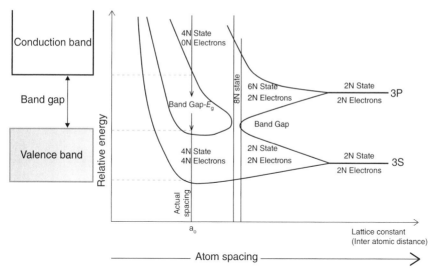

Figure 2.1   *Evolution of atomic orbitals from discrete levels (right) to energy bands (center) and a band gap (left) as atomic separation decreases inside a semiconductor, in this case silicon with two electrons each in its outer 3s and 3p orbitals [1].*

chemical bonds and "bands" of closely spaced levels at whose energies electrons can move between atoms in the solid. The atoms' electrons fill these levels and bands from the deepest atomic orbitals up to a maximum energy in the bands, leaving empty levels at energies above. Figure 2.1 illustrates how the electronic orbitals of electron of the atoms comprising the solid are discrete at large separation but broaden as the atomic distances become shorter. This broadening is a quantum mechanical effect due to the *Pauli Exclusion Principle* that no two electrons can occupy the same space, that is, their orbitals overlap, with the same four *quantum numbers*, parameters that fully describe an electron orbiting an atom.

As this atomic separation decreases further, these bands split, leaving an energy gap in between. Depending on the spacings between atoms in the solid, termed crystal *lattice constants*,this energy gap increases as the lattice constants decrease. The energy separation between filled bands, termed *valence bands* and empty bands, termed *conduction bands*, can be on the order of fractions of a volt to several volts. This energy gap can be determined experimentally by measuring the threshold energy for optical absorption of the semiconductor as a function of incident photon energy. Figure 2.2 illustrates these thresholds for semiconductor alloys of $Al_xGa_{1-x}As$, which increase with increasing Al content.

2.2   The Fermi Level and Energy Band Parameters

For all three types of electronic materials – metals, insulators, and semiconductors – there is a characteristic energy at which the probability of finding an electron is exactly 50%. This energy is termed the *Fermi level* and denoted as E_F, a very useful concept to describe the nature of charge conduction in solids. Figure 2.3 illustrates the most important energy levels in a semiconductor. The position of E_F depends on the number of free electrons or

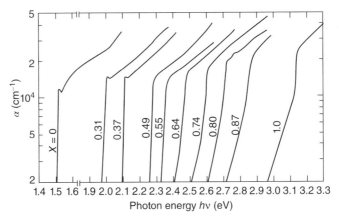

Figure 2.2   *Absorption coefficient α versus photon energy hv spectra for Al$_x$Ga$_{1-x}$As. Absorption threshold increases as Al content x and band gap increase. (Monemar 1976 [2]. Reproduced with permission of American Institute of Physics.)*

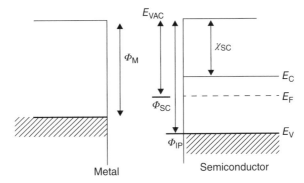

Figure 2.3   *Schematic energy band diagram to illustrate semiconductor vacuum level E$_{VAC}$, conduction band E$_C$, Fermi level E$_F$, valence band E$_V$, and ionization potential Φ$_{IP}$ along with metal work function Φ$_M$, semiconductor work function Φ$_{SC}$, and electron affinity χ$_{SC}$. (Brillson 2010. Reproduced with permission of Wiley.)*

holes within the semiconductor. It is closer to the conduction band for high densities of electrons and closer to the valence band for high densities of holes (missing electrons that act as positively charged carriers). Furthermore, its energy level is constant at equilibrium throughout the semiconductor and across its interfaces with other materials.

Along with Fermi level E_F, the conduction band E_C and valence band E_V comprise the basic semiconductor energy levels used to describe most electronic and optical properties. However, other energy levels can play an important role in semiconductor properties. First, there are the atomic orbital levels below and extending up to the valence band. These levels and their separations are at energies characteristic of particular atoms. These levels can shift because of chemical bonding between atoms, particularly those closest to the valence and conduction bands.

The minimum energy at which an electron can escape the solid defines the *vacuum level* E_{VAC}. For metals, the energy difference between its E_F, the highest states filled with electrons, and E_{VAC} is termed the *work function Φ_M*. In other words, the metal work function is the energy required to raise an electron from its highest filled state to an energy at which it can escape the solid. A table of element work functions measured experimentally appears in Appendix 5, Track II. In a semiconductor or an insulator, the highest filled state is at the top of the valence band, for example, at E_V. The energy difference between E_V and E_{VAC} is termed the *ionization potential Φ_{IP}*. Finally, the energy difference between E_C and E_{VAC} is termed the *electron affinity χ_{SC}*. All of these physical properties are important for the design and control of electronic device properties and measurable by surface and interface techniques.

2.3   Band Bending at Semiconductor Surfaces and Interfaces

The energy bands pictured in Figure 2.3 are flat within the semiconductor bulk at thermal equilibrium and in the absence of applied forces such as external electric fields and optical illumination. However, at a surface or interface, changes in atomic bonding or the presence of new atomic species can add electric charges that introduce local electric fields that in turn repel or attract charges inside the semiconductor. As a result, the carrier density in the semiconductor near that surface or interface changes to offset or *screen* this electric field. This screening extends over a characteristic distance determined by the semiconductor's *dielectric permittivity ε* and bulk *carrier density n or p*, depending on whether the majority free carriers are electrons or holes. Figure 2.4 illustrates the resultant band bending that occurs near the surface or interface. For *n-type band bending*, free holes are swept toward the surface while free electrons are swept into the bulk. As a result, the region in which electric fields are present is depleted of free charge. Neglecting any free charge within this depletion region, the negative charges present at the surface are equal and opposite to those in the region depleted of free charge, the *depletion region*. The ability to control this electric field near a surface or interface enables many electronic applications.

2.4   Surfaces and Interfaces in Electronic Devices

Surfaces and interfaces play a key role in the operation of many if not most electronic devices. Beside the transistor pictured in Figure 1.1 with its ohmic source and drain contacts, metal–insulator and insulator–semiconductor interfaces, many of today's advanced electronic applications depend sensitively on surfaces and interfaces. Figure 2.5a illustrates a band bending region similar to Figure 2.4 with incident light creating electrons and holes that separate into the semiconductor bulk and the metal, respectively, due to the built-in electric field. This charge separation produces current and voltage and is the basis for solar cell operation.

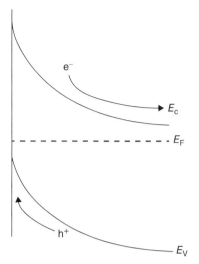

Figure 2.4 *Conduction and valence band bending at an n-type semiconductor surface. The slope of band bending corresponds to the electric field within the depletion region that moves electrons toward the bulk and holes toward the surface. E_F remains constant relative to the bulk conduction and valence bands. (Brillson 2010. Reproduced with permission of Wiley.)*

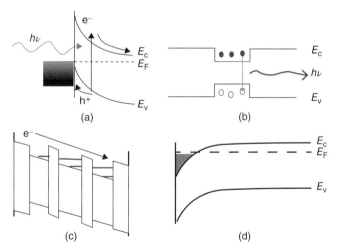

Figure 2.5 *Representative electronic device structures whose operation depends on interfaces. (a) Solar cell based on Schottky barrier ; (b) quantum well based on carrier confinement by a double heterojunction; (c) resonant tunnel diode based on multilayer heterojunctions; (d) high electron mobility transistor based on surface inversion layer. (Brillson 2010. Reproduced with permission of Wiley.)*

Figure 2.5b illustrates a quantum well engineered to confine electrons and holes within a small volume in order to promote efficient electron–hole recombination. This charge confinement provides the mechanism to promote efficient recombination and light between conduction and valence band states. This carrier confinement is the basis for all solid state lasers and light emitting diodes. Figure 2.5c illustrates a resonant tunnel diode structure across which an applied electric field can vary the alignment of energy levels of confined carriers inside each well. The resultant *negative differential resistance* is useful for very high frequency oscillators. Figure 2.5d illustrates the *accumulation layer* formed by E_F above E_C in an n-type semiconductor due to downward or p-type band bending. Such *two-dimensional electron gas* (2DEG) layers formed either at a surface or a heterojunction with a larger band gap semiconductor can have very high charge density without an otherwise accompanying dopant density. Such structures are the basis for *high electron mobility* transistor (HEMTs).

2.5   Effects of Localized States: Traps, Dipoles, and Barriers

Localized electronic states can strongly affect electronic properties for all electronic material structures. At the semiconductor interface with a metal, an insulator, or another semiconductor, these localized states have electronic levels within the band gap that can trap charge. This trapped charge can introduce dipoles that alter the voltage drop across the semiconductor's band bending region from what would otherwise be determined by classical charge transfer between the two constituents. For the transistor, such dipoles reduce the control of band bending by applied voltages, essential for device switching and amplification. For the solar cell, these states not only alter the semiconductor band bending required to separate opposite charges, but they also introduce new avenues of charge recombination that reduce the efficiency of charge carrier collection. For Schottky diodes based on such band bending, the presence of such states within the depletion region can alter the carrier density, changing the depletion width and hence the effective barrier height while introducing sites for carrier hopping that increase leakage through the diode. Similarly, in the quantum well, states within the well and at its heterojunction reduce the band-to-band recombination efficiency by creating alternate, lower energy, recombination pathways. Localized states within the resonant tunnel diode also trap charge and introduce new pathways for carrier recombination as well as altering the heterojunction band alignments. Finally, trapped charge states near the 2DEG interface introduce scattering centers that decrease carrier mobility. They also introduce dipoles that can shift the band alignment to reduce the 2DEG well width and hence the carrier density contained within.

2.6   Summary

This chapter aimed to introduce the most important features of semiconductors, their surfaces and interfaces. Besides the energy levels that form the semiconductor band gap and describe the carrier type and density, the positions of these levels relative to the vacuum level are important to understand how charge distributes at semiconductor interfaces. The band bending that occurs due to this charge transfer plays a central role in the operation of a wide variety of microelectronic and optoelectronic devices. Localized electronic states can form at these interfaces that affect this band bending, the ability to control charge transport through these bent band regions with applied voltage, the transport of charge across heterojunctions, the charge density within 2DEG layers, and the efficiency of light emission at confined quantum well structures.

The chapters to follow describe all the basic features of surfaces and interfaces needed to understand their effects on electronic material behavior. First, Chapter 3 examines quantitatively the effect of these localized states on band bending and Schottky barriers and introduces the macroscopic techniques used conventionally to measure Schottky barrier heights. Next, Chapter 4 describes the features of localized states on a microscopic scale, and Chapter 5 introduces the ultrahigh vacuum techniques needed to measure these micro- to atomic-scale properties. The following chapters describe the various techniques used to study electronic material surfaces and interfaces. The final chapters describe the most important properties of electronic material surfaces, heterojunctions, and metal–semiconductor interfaces discovered over several decades of research, ending with the new frontiers of electronic materials surface and interface research.

2.7   Problems

1. Explain why Ge was the first semiconductor considered for a transistor rather than Si. Hint: Use Appendix D: Semiconductor Properties.
2. Which semiconductors would be useful as phosphors for light displays among the following: CdSe, InP, GaAs, GaP, and CdS. Explain why.
3. Under mechanical compression, would you expect a band gap to increase or decrease? Why? Provide a quantum mechanical analogy.
4. Given the ionization potential of a known semiconductor, how could you calculate the electron affinity?
5. List at least four semiconductor properties necessary to grow a quantum well such as pictured in Figure 2.5b and to produce visible light emission. Give an example of materials that would satisfy these conditions.
6. Consider a semiconductor with ionization potential $\Phi_{IP} = 5.7$ eV, band gap $E_G = 1.7$ eV, and work function $\Phi_{SC} = 4.2$ eV. Assuming no band bending, what is χ? Assuming n-type (upward) band bending of 0.3 eV, what are Φ_{SC} and Φ_{IP} now?
7. In a resonant tunnel diode such as in Figure 2.5c, an applied voltage changes the tilt of the conduction and valence bands. Explain why this could lead to negative differential resistance. How would interface roughness affect the NDR?

References

1. Streetman, B.G. and Banerjee, S.K. (2006) *Solid State Electronic Devices*, Pearson Prentice Hall, Upper Saddle River, NJ, Ch. 3.
2. Monemar, B., Shih, K.K., and Pettit, G.D. (1976) Some optical properties of the $Al_xGa_{1-x}As$ alloy system. *J. Appl. Phys.*, **47**, 2604.

Further Reading

Halliday, D., Resnick, R. and Walker, J. (1997) *Fundamentals of Physics Extended*, 5th edn, John Wiley & Sons, Inc., New York, Ch. 42.

3

Electrical Measurements of Surfaces and Interfaces

This chapter describes the techniques available to characterize the electrical properties of surfaces and interfaces at the macroscopic level. Included are methods to determine: (i) sheet resistance across electronic material surfaces, (ii) contact resistance of non-rectifying metal contacts to these surfaces, (iii) Schottky barrier heights of rectifying contacts, and (iv) heterojunction band offsets. Later chapters will describe the range of techniques available to measure these and related properties at the microscopic and atomic scale where the physical mechanisms that determine these properties take place.

3.1 Sheet Resistance and Contact Resistivity

Macroscopic techniques to measure properties of surfaces and interfaces begin by taking into account the semiconductor's sheet resistance and its metal contact resistivity. Two-terminal electrical measurements of Schottky barriers and heterojunction band offsets involve applied voltages that fall not only across the interface under test but also across the series resistances between the voltage terminals and the interface. A standard technique to obtain both sheet resistance and contact resistivity is the *transfer length method* (TLM). Here current–voltage (I–V) curves between metal electrodes of width W on the surface spaced apart by increasing distances d_i provide a plot of resistance values versus electrode spacing [1]. Figure 3.1 illustrates the geometry of such a set of electrodes and the resultant plot of total resistance R_T versus distance d with d values on a typical scale of 5 to 50 microns. This is one of several linear as well as circular TLM geometries. The slope $\Delta(R_T)/(\Delta d)$ of this plot yields *sheet resistance* ρ_S divided by contact width Z determined independently. The intercept of this line at $d = 0$ corresponds to $R_T = 2\rho_C$ where ρ_C is the *contact resistance*. The intercept at $R_T = 0$ corresponds to *transfer length* L_T where

$$L_T = \sqrt{(\rho_C/\rho_S)} \tag{3.1}$$

An Essential Guide to Electronic Material Surfaces and Interfaces, First Edition. Leonard J. Brillson.
© 2016 John Wiley & Sons, Ltd. Published 2016 by John Wiley & Sons, Ltd.
Companion Website: www.wiley.com/go/Brillson/

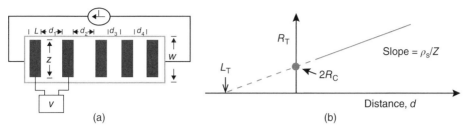

Figure 3.1 *(a) Representative TLM test structure and (b) plot of total resistance versus distance. (Brillson 2010. Reproduced with permission of Wiley.)*

and corresponds to the "1/e" distance over which electric potential decreases from the metal edge nearest the current flow (where resistance to the semiconductor is least) to interior points under the contact. Thus such TLM measurements of semiconductor surfaces provide three useful parameters for interface measurements – *sheet resistance* ρ_S, *contact resistance* ρ_C, and *specific contact resistivity* $\rho_C(\Omega \text{ cm}^2) = \rho_C \cdot A$ where A is the active area of the contact.

3.2 Contact Measurements: Schottky Barrier Overview

The potential barrier set up by band bending at the semiconductor interface is at the heart of all modern electronics. This barrier equals the energy difference between the Fermi level of a metal $E_F{}^M$ and the band edge of the semiconductor's majority charge carrier, that is, the conduction band for electrons and the valence band for holes. This barrier determines the "ohmic" or "rectifying" behavior of the electrical contact. Figure 3.2a illustrates the *I–V* behavior of an "ohmic" metal–semiconductor contact. Here the resistance $R = V/I$ is constant for both forward and reverse bias. By definition, forward bias corresponds to majority carriers moving from semiconductor to metal. The steep linear slope in Figure 3.2a corresponds to a low resistance, which is needed for circuits in which the contact introduces negligible voltage drop. Ohmic metal–semiconductor contacts are essential in delivering all the applied voltage to the intended circuit element without loss of voltage or energy to intermediate components. Examples include the field effect transistor (FET) pictured in Figure 1.1 used to perform computer functions with the minimum possible energy loss, the solar cell pictured in Figure 2.5a used to generate power without voltage loss away from the Schottky barrier, the quantum well light emitter in Figure 2.5b used to generate

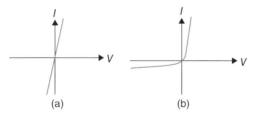

Figure 3.2 *I – V plot of (a) ohmic contact and (b) rectifying contact. (Brillson 2010. Reproduced with permission of Wiley.)*

optical power with minimum energy loss at the metal contacts, and the high power HEMT for RF power wave generation with minimal resistive contact heating. Subsequent chapters describe many of the techniques used to obtain low-resistance contacts.

Figure 3.2b illustrates the *I–V* behavior of a "rectifying" metal–semiconductor contact. Here resistance is low in the forward direction and high in the reverse direction. The asymmetric behavior is useful to control circuit current and voltage in many ways. Examples include current rectifiers that block alternating current in one direction, solar cells that supply voltage proportional to the built-in voltage of the band bending region, and photodetectors whose bent bands separate photo-induced electron–hole pairs to increase detection efficiency. The forward bias current in Figure 3.2b also has an exponential dependence of carrier density on voltage so that small applied voltages can produce disproportionately large current changes. Later chapters discuss the wide range of results measured for Schottky barriers on different semiconductors and their sensitivity to growth methods, surface chemical treatment and thermal processing. The understanding, prediction, and control of these Schottky barriers represent in fact a major challenge for next generation electronics, one that surface and interface techniques are well suited to address. This chapter introduces the basic concepts of Schottky barriers and the conventional techniques used to measure them in the laboratory. A case study serves to illustrate differences between results expected from this classical framework versus what is measured experimentally. Such differences between classical theory and experiment extend across most semiconductor contacts. They provide the main motivation for understanding and ideally controlling charge transfer at surfaces and interfaces on an atomic scale.

3.2.1 Ideal Schottky Barriers

The ideal or "classical" model for Schottky barrier formation between a metal and a semiconductor involves charge transfer at the interface to equalize the Fermi levels of the two materials. Thus electrons from a semiconductor with an equilibrium E_F^{SC} higher in energy (closer to E_{VAC}) than E_F^M in the contacting metal will transfer to the metal. Since the electron density in the metal is orders of magnitude greater than that of the transferred charge, E_F^M remains essentially unchanged. In the semiconductor, however, the transferred charge leaves behind a depleted surface region with an equal and opposite charge density over a screening distance, typically 10^{-4} to 10^{-6} cm thick, determined by its dielectric permittivity ε and bulk charge density n. With the two Fermi levels E_F^M and E_F^{SC} aligned, a double layer forms consisting of transferred electrons at the metal surface (since excess carriers in a metal reside on its surface) and an equal and opposite charge induced density distributed through an extended depletion region inside the semiconductor. The double layer produces a band bending in the semiconductor since $E_C - E_F^{SC}$ at the surface and in the bulk are different. E_C at the surface remains unchanged while E_F^{SC} at the surface now aligns with E_F^M at the interface, and $E_C - E_F^{SC}$ beyond the semiconductor depletion region remains unchanged. The resultant band diagram before and after charge transfer appears as Figure 3.3a [2]. The n-type depletion region in Figure 3.3a is a layer of high resistance. Thus, a voltage applied to this junction falls mostly across this *surface space charge region*. In this simple model, band bending qV_B depends on the thermionic work function difference between the metal and the semiconductor so that

$$qV_B = \Phi_M - \Phi_{SC} \tag{3.2}$$

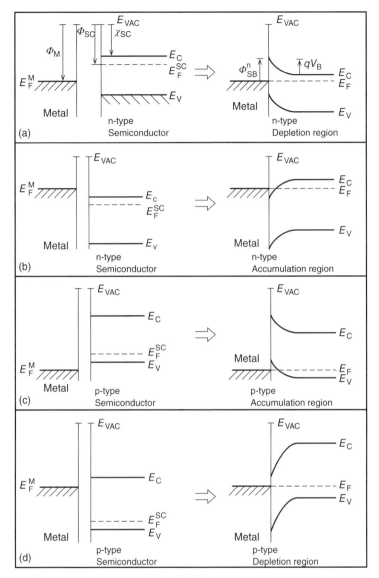

Figure 3.3 *Ideal Schottky barrier formation for (a) high work function metal on an n-type semiconductor, (b) low work function metal on an n-type semiconductor, (c) high work function metal on a p-type semiconductor, and (d) low work function metal on a p-type semiconductor. (Brillson, L. J. 1982 [2]. Reproduced with permission of Elsevier.)*

where Φ_M and Φ_{SC} are the metal and semiconductor work functions, respectively, and

$$\Phi_{SC} = \chi_{SC} + E_C{}^{bulk} - E_F \qquad (3.3)$$

For typically doped n-type semiconductors, $E_C{}^{bulk} - E_F < 0.1$ eV.

The Schottky barrier at this interface is defined as the energy difference between $E_F{}^M$ and the conduction band edge at the interface. Hence the n-type Schottky barrier is

$$\Phi^n{}_{SB} = \Phi_M - \chi_{SC} \tag{3.4}$$

so that $\Phi^n{}_{SB}$ for different metals on the same semiconductor should increase with increasing Φ_M. Figure 3.3b illustrates the case for $\Phi_M < \chi_{SC}$. With the two Fermi levels $E_F{}^M$ and $E_F{}^{SC}$ aligned, $E_F{}^{SC}$ now resides above E_C, forming an *accumulation layer* with excess electrons in the semiconductor near the interface that presents no barrier for electrons to move across the junction. In contrast to the *rectifying* contact in Figure 3.3a, this figure represents an *ohmic* contact.

Equation (3.2) also holds for p-type semiconductors, but now the barrier height is defined as the energy difference between $E_F{}^M$ and the valence band edge at the interface. Thus

$$\Phi^P{}_{SB} = \Phi_M - \chi_{SC} - E_G \tag{3.5}$$

for a semiconductor with band gap E_G. Figure 3.3c illustrates the case for $\Phi_M > \chi_{SC} + E_G$ for a p-type semiconductor. With the two Fermi levels aligned, $E_F{}^{SC}$ now resides below E_V, now forming an accumulation layer with excess holes near the interface and no barrier for holes to move across the junction. This case represents an *ohmic* contact for a p-type semiconductor.

Finally, Figure 3.3d illustrates the case of $\Phi_M < \chi_{SC} + E_G$ for a p-type semiconductor. Analogous to Figure 3.3a, a *rectifying* contact forms but with bands that bend down and a space charge region that is depleted of holes.

For all four cases of this band bending model, the potential Φ within the semiconductor satisfies Poisson's equation[1]

$$\nabla^2\Phi(x) = -\rho(x)/\varepsilon_s \tag{3.6}$$

In Equation (3.6), ρ is the charge density in the *surface space charge region* of width W, x is the coordinate axis normal to the metal–semiconductor interfaces, and ε_s is the *static dielectric constant* of the semiconductor. Within the abrupt approximation for bulk concentration N of ionized impurities within the surface space region, $\rho \simeq qN$, for $x < W$ versus $\rho \simeq 0$ and $d\Phi/dx = 0$ for $x > W$. Integrating Equation (3.6), one obtains a voltage

$$V = -qN(x - W)^2/2\varepsilon_S + V_0 \text{ for } 0 \leq x \leq W$$

$$= V_0 \text{ for } x > W \tag{3.7}$$

and a depletion width of

$$W = [2\varepsilon_S(V_0 - V)/qN]^{1/2} \tag{3.8}$$

Equations (3.7)–(3.8) show that the abrupt metal–semiconductor junction contains a parabolic band bending region whose width increases with ε_S and decreases with N.

3.2.2   Real Schottky Barriers: Role of Interface States

The classical model of ideal Schottky barriers described by Equations (3.2)–(3.8) and illustrated in Figure 3.3 provides a useful starting point to understand metal–semiconductor

---

[1] Throughout this book, we use the rationalized MKS system with centimeters as a convenient unit of length.

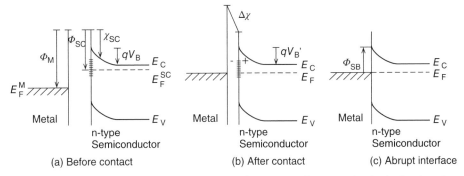

Figure 3.4 *Energy band diagram of a metal and a semiconductor with a high density of surface states (a) before contact, (b) after contact, and (c) as shown conventionally with interface dipole not shown. (Brillson, L. J 1982 [2]. Reproduced with permission of Elsevier.)*

interface formation. However, experimental measurements of real Schottky barriers often show major disagreements with the barrier heights expected. For example, Φ^n_{SB} for different metals on Si deviate from those predicted according to Equation (3.4) by several tenths of volts for both chemically prepared and aged as well as clean cleaved Si surfaces [3–5]. Similarly, different metals on bulk GaAs produce Φ^n_{SB} that vary only by 0.1–0.2 eV for metals with Φ_M varying over a volt [6]. These deviations can be understood in terms of localized states at the interface that can accumulate charge and generate dipoles that take up much of the metal–semiconductor difference in potential [7]. These states can have several different physical origins that surface and interface analysis techniques can measure.

Figure 3.4 illustrates the effect of such interface states. Before contact, Figure 3.4a shows that band bending is already present, even without the metal contact due to the presence of surface states on the semiconductor surface. This is due to negative charge that fills states localized at or near the surface at energies below the Fermi level E_F^{SC}. Upon contact with the metal, the higher E_F^{SC} relative to E_F^M induces charge transfer toward the metal. However, the presence of interface states induces a dipole with a voltage equal to part or all of the initial $\Phi_M - \Phi_{SC}$ difference. If the surface state density is high enough, for example, $\sim 10^{14}\,\text{cm}^{-3}$, the presence of the metal does not alter the band bending appreciably, as shown in Figure 3.4b. Here most of the contact potential difference $\Phi_M - \Phi_{SC}$ falls across an atomically thin interface dipole region instead of the semiconductor space charge region. The voltage drop $\Delta\chi$ across this dipole depends on the density of localized states that are filled with electrons as well as the dielectric constant of the layer across which this voltage drops. Assuming that this dipole layer is sufficiently thin that charge tunnels freely across it, the resultant abrupt interface is typically pictured without the dipole as in Figure 3.4c. Here $\Phi_{SB} = qV_B + E_C^{bulk} - E_F^{SC}$ is not equal to $\Phi_M - \chi_{SC}$. Instead

$$\Phi_{SB} = \Phi_M - \chi_{SC} - \Delta\chi \tag{3.9}$$

For high enough densities of interface states, small E_F movements within this energy range of states produce large changes of localized state occupancy. As a result, most of the potential difference between metal and semiconductor before contact produces changes in the interface dipole rather than in the surface space charge region after contact. The barrier height Φ_{SB} and band bending qV_B are then relatively independent of the metal work function Φ_M. In this case, the Fermi level at the surface is then termed *pinned* by surface states within a narrow range of energy in the semiconductor band gap.

The earliest direct evidence of this *Fermi level pinning* was the field effect measurement associated with the development of the transistor [7]. The presence of states at the gate–semiconductor interface of the transistor structure pictured in Figure 1.1 reduced the effect of an applied gate voltage on the Fermi level movement within the surface space charge region, thereby preventing the otherwise large changes in carrier concentration intended to move within the transistor's channel between source and drain. Ideally, an applied gate voltage increases or decreases charge density in the space charge region, changing the conductivity $\sigma = ne\mu$ and current density $J = \sigma\mathcal{E}$, where μ is carrier mobility, e is electron charge, and $\mathcal{E}$ is applied electric field between source and drain. With surface states present as in Figure 3.4b, control of this conductivity is reduced. Thus interface states are very important since the control of charge carriers in the surface space charge region by an applied voltage is essential to the operation of all active electronics.

3.2.3   Schottky Barrier Measurements

Several methods are available to measure barrier heights between metals and semiconductors. They differ in how difficult they are to perform and the factors that can complicate their interpretation. The simplest to perform is also the most challenging to interpret.

3.2.3.1   *Current–Voltage Technique*

The most direct method is the current–voltage (*J–V*) technique. Thermionic emission–diffusion theory yields a forward *J–V* characteristic given as

$$J = A^{**}T^2 \ \exp(-q\Phi_{SB}/k_BT)[\exp(qV/k_BT) - 1] \tag{3.10}$$

where $A^{**} = f_p f_q A^* /(1 + f_p f_q v_R/v_D)$, $A^* = m^*/m_0$ (120 A cm^{-2} K^{-2}), f_p = probability of electron emission over the semiconductor potential maximum into the metal without electron – optical phonon scattering, f_q = ratio of total J with versus without tunneling and quantum-mechanical reflection, v_R = *recombination velocity*, v_D = effective *diffusion velocity* with thermionic emission, m^* is the carrier's *effective mass*, m_0 is the free electron mass, T is *temperature*, and V is the *applied voltage*.

For $V > 2 \ k_BT/q$ in the forward direction, A^{**} can be approximated by A^* and Φ_{SB} consists of a barrier height extrapolated to zero field, Φ_{SB0}, minus a term $\Delta\Phi$ due to the combined effects of applied electric field and image force [8]. The effect is given by

$$\Phi_{SB} = \Phi_{SB0} - \Delta\Phi = \Phi_{SB0} - (q^3\mathcal{E}/\varepsilon_S)^{1/2} \tag{3.11}$$

and is illustrated in Figure 3.5. Here the *applied electric field* $\mathcal{E} = V/W$, where W is the width of the surface space charge region and any dipole layer. Depending on N and ε_S, the contact's depletion region width W is typically <1000 Å so that voltages of only 1V can produce field gradients $\mathcal{E}$ of 10^5 V cm^{-1} or higher. As an example, $\Delta\Phi = 0.03$ eV with $\varepsilon_S = 16\varepsilon_0$, for example, Ge, and $\mathcal{E} = 10^5$ V cm^{-1}. While the $\Delta\Phi$ term is only several tens of millivolts, it can nevertheless affect the *J–V* behavior for non-zero voltages.

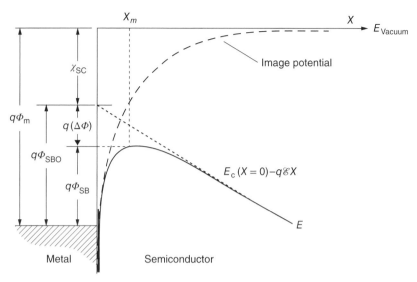

Figure 3.5 *Energy band diagram with image force lowering at the metal–semiconductor interface. The effective barrier is lowered from $q\Phi_{SB0}$ to Φ_{SB} with an applied electric field. (Sze, S.M. 2007 [8]. Reproduced with permission of Wiley.)*

The forward current density in Equation (3.10) extrapolated logarithmically to zero applied forward bias has an intercept at

$$J_S = A^{**}T^2 \exp(-q\Phi_{SB0}/k_BT) \tag{3.12}$$

so that the barrier Φ_{SB0} can be extracted from a plot of $\ln J$ versus applied forward voltage. As shown in Figure 3.6, the slope of the J–V plot also provides the "*ideality factor*" n of the contact, defined by [8]:

$$n \equiv (q/k_BT)\partial V/\partial(\ln\ J) = [1 + \Delta\partial\Phi/\partial V + (k_BT/q)\ \partial(\ln A^{**})/\partial V]^{-1} \tag{3.13}$$

In general, the second and third terms of Equation (3.13) are much less than one so that $n \approx 1$. However, a variety of physical processes can increase n. These include tunneling through the barrier [9,10], intermediate layers with new dielectric and transport properties [10,11], and recombination or trapping at states near the interface and within the semiconductor band gap [11]. Later chapters will discuss the chemical and electronic interactions responsible for these processes.

An approximation to Equation (3.10) that contains n explicitly is:

$$J = A^{**}\ T^2 \exp(-q\Phi_{SB0}/k_BT)[\exp(qV/nk_BT) - 1] \tag{3.14}$$

However, extrapolating $\ln J$ to $V = 0$ in order to obtain Φ_{SB} is not a reliable procedure for n values that are significantly different from unity [11]. Nevertheless, Equation (3.14) can be used to extract Φ_{SB0} from the reverse current, provided that generating-recombination currents within the depletion region are small compared with the Schottky emission current.

Sheet resistance and contact resistivity both affect J. Finite voltage drops across the semiconductor bulk or its otherwise low resistance back contact will reduce $\ln J$ in Figure 3.6

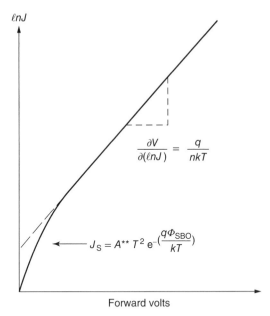

Figure 3.6 *Forward current density versus applied voltage for a metal–semiconductor contact.*

with increasing forward voltage. Thus, $V_{applied} = V_{interface} + IR^{SC}_{bulk} + IR^{SC}_{back\ contact}$, where R^{SC}_{bulk} and $R^{SC}_{back\ contact}$ are parasitic semiconductor bulk and back contact resistances, respectively, so that $V_{interface} < V_{applied}$. If the deviation of ln J versus $V_{Forward}$ is linear, its slope yields the total parasitic resistance $R^{SC}_{bulk} + R^{SC}_{back\ contact}$ and the correction term $\Delta V_{applied} = I_{measured}(R^{SC}_{bulk} + R^{SC}_{back\ contact})$. Finally, the field lowering effect depicted in Figure 3.5 will also produce a gradual increase in current with reverse bias.

3.2.3.2   Capacitance–Voltage Technique

The second most common method to measure carrier heights is the capacitance – voltage (*C–V*) technique. Capacitance *C* of the surface space charge region is

$$C = \varepsilon_S A/W \tag{3.15}$$

where $A =$ capacitor area and depletion width $W = [2\varepsilon_S(V_0 - V)/qN]^{1/2}$ from Equation (3.8) so that *C* changes with applied bias as

$$C = A[Nq\varepsilon_S/2(V_0 - V)]^{1/2} \tag{3.16}$$

As reverse bias increases, *W* increases and *C* decreases. Since *W* decreases with decreasing reverse voltage and vanishes at $V = V_0$, *C* becomes infinite and $1/C^2$ decreases to zero. Thus the intercept of a $1/C^2$ versus *V* plot yields the band bending V_B, as illustrated by Figure 3.7, and the barrier height is given by

$$\Phi_{SB}{}^n = qV_B + qV_n - \Delta\Phi \tag{3.17}$$

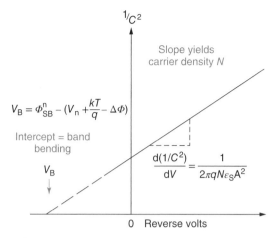

Figure 3.7 *Plot of $1/C^2$ versus V for an ideal metal–semiconductor contact. Voltage intercept V_B equals the band bending within the surface space charge region of the semiconductor and relates to Schottky barrier height Φ_{SB} as indicated.*

where $\Delta\Phi$ is the *image force correction* and the *diffusion potential* $V_n = E_C{}^{bulk} - E_F$ is known from the bulk doping and the effective mass m^*. Specifically,

$$n_0 = N_C \ \exp[-(E_C - E_F)/k_B T] \tag{3.18}$$

where n_0 is the *equilibrium carrier concentration* in the bulk, $N_C = 2(2\pi m_e{}^* k_B T/h^2)^{3/2}$ is the *conduction band density of states*, and h is *Planck's constant*. Furthermore, the slope of a $1/C^2$ plot versus reverse voltage

$$d(1/C^2)/dV = 2/(N q \varepsilon_S A^2) \tag{3.19}$$

yields a straight line for constant semiconductor carrier concentration N. Hence one can use $d(1/C^2)/dV$ and $W(V)$ to obtain doping profiles of the surface space charge region.

As with *J–V* measurements, the C–V technique has several sources of possible error, including (i) an insulating layer between the metal and semiconductor, (ii) variation of surface charge (interface states) with voltage, (iii) series resistance of the junction, (iv) traps with the depletion width, and (v) variations in effective contact width with depletion layer width. The capacitance can also exhibit a frequency dependence due to trapped carriers with different capture and emission rates.

This frequency dependence is in fact useful to identify trapped charge levels and cross sections, for example, by *deep level transient spectroscopy* (DLTS) [12], *thermally stimulated capacitance* (TSCAP) [13], admittance spectroscopy [14] and photocapacitance spectroscopy [15]. These techniques yield information on electronic properties of impurity and defect centers in semiconductor surface space charge regions. These properties include: (i) energy levels, (ii) charge state multiplicity, (iii) thermal and optical emission rates, (iv) thermal and optical cross sections, and (v) the dependences of the cross sections on temperature, static electric field, and photon energy [16,17].

3.2.3.3   Internal Photoemission Technique

The most direct method of determining the Schottky barrier height is *internal photoemission spectroscopy* (IPS). This technique uses photoexcitation to measure the energy barrier over which carriers must move in order to cross the metal–semiconductor interface. The IPS technique employs monochromatic light to excite transitions that produce free carriers and a photocurrent in an external circuit. Figure 3.8a shows photoexcitation either through the front contact or the back, while Figure 3.8b shows photo-induced excitation of transitions from the Fermi level of the metal to the conduction band edge (1) as well as band-to-band transitions (2). Light passing through the semiconductor with energy $h\nu \geq E_G(\equiv E_C - E_V)$ will be strongly attenuated by absorption before it reaches the metal–semiconductor interface. The free carriers so generated lie in a field-free region, recombine and do not contribute to the photoresponse. Hence the technique requires either: (a) that the metal film be thin enough for front illumination to be absorbed at the metal–semiconductor interface or (b) back illumination through the semiconductor. For example, in back illumination,

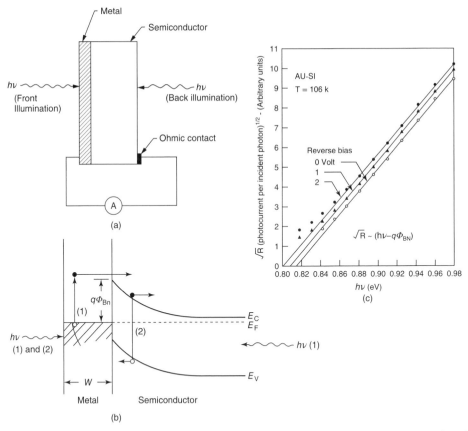

Figure 3.8   *(a) Schematic diagram of internal photoemission measurement. (b) Energy band diagram for photoexcitation process. (c) Au-Si internal photoemission spectra dependence on applied bias. (Sze, S.M. 2007 [8]. Reproduced with permission of Wiley.)*

transition (1) will dominate a spectrum of photoexcited current versus incident photon energy. The photocurrent per absorbed photon per unit area is given by the Fowler theory [18] as

$$J_R \sim [T^2/(E_S - h\nu]^{1/2}]\{(x^2/2) + \pi^2/6 - [e^{-x} - (e^{-2x}/4) + (e^{-3x}/9) - \dots]\} \text{ for } x > 0 \tag{3.20}$$

where $x \equiv h(\nu - \nu_0)/k_B T$, $\nu =$ photon frequency, $h\nu_0 = q\Phi_{SB}$, the barrier height, and E_S is the sum of $q\Phi_{SB}$ and the metal Fermi energy. For $h\nu > \Phi_{SB} + 3k_B T$, photocurrent density J_R varies quadratically as a function of incident photon energies above the barrier height.

$$J_R \sim (h\nu - h\nu_0)^2 \tag{3.21}$$

Thus a plot of $\sqrt{(J_R)}$ versus $h\nu$ yields a straight line that intersects the $h\nu$ axis at a value equal to the barrier height Φ_{SB}. Furthermore, the bias dependence of this $\sqrt{(J_R)}$ versus $h\nu$ plot yields the image force lowering term in Equation (3.11), from which one obtains $\varepsilon_S/\varepsilon_0$ near the interface [19]. Figure 3.8(c) illustrates $\sqrt{(J_R)}$ versus $h\nu$ plots for three bias conditions of Au–Si Schottky barriers. For zero bias, the barrier height equals 0.815 eV. From the applied voltage and bulk carrier concentration, one can plot $\Delta\Phi$ versus electric field and obtain $\varepsilon_S/\varepsilon_0$ [19]. The presence of localized states at the metal–semiconductor interface will increase the image force dielectric constant $\varepsilon_S/\varepsilon_0$ and the image force lowering of the barrier [20–22]. In addition, the temperature dependence of the $\sqrt{(J_R)}$ intercept can reflect changes in the semiconductor electronic structure. In general, E_G increases with decreasing temperature so that changes in Φ_{SB} versus E_G can associate E_F at the interface with either E_C or E_V, which typically change at different rates.

Case Study: IPS Measurement of GaAs Schottky Barriers

Figure 3.9 illustrates IPS measurements of $q\Phi_{SB}$ for different metals deposited on clean MBE-grown GaAs [23]. The extrapolated leading edge of (current/photon)$^{1/2}$ versus $h\nu$ for each IPS plot yields $q\Phi_{SB} = 0.41, 0.935,$ and 1.01 eV for Al, Cu, and Au overlayers, respectively, on MBE-grown GaAs. How do these barrier heights compare with $q\Phi_{SB}$ calculated from the ideal Schottky barrier model? How do they compare with conventional GaAs $q\Phi_{SB}$?

Solution: The preferred work functions for Al, Cu, and Au are 4.28, 4.65, and 5.1 eV, respectively. From Appendix 4, the electron affinity of GaAs is 4.07 eV. From Equation (3.4), $\Phi^n_{SB} = \Phi_M - \chi_{SC}$. Substituting for Φ_M and χ_{SC}, one obtains: $\Phi^n_{SB}(Al) = 4.28 - 4.07 = 0.21$ eV, $\Phi^n_{SB}(Cu) = 4.65 - 4.07 = 0.58$ eV, and $\Phi^n_{SB}(Au) = 5.1 - 4.07 = 1.03$ eV. The IPS-measured barrier height range $\Delta\Phi^n_{SB} = 1.01 - 0.41 = 0.6$ eV versus $\Delta\Phi_M = 5.1 - 4.28 = 0.82$ eV for barriers that vary with metal work function. Thus the measured $\Delta\Phi^n_{SB}$ exhibits a somewhat weaker than predicted dependence on $\Delta\Phi_M$. The Φ^n_{SB} measured by IPS deviates from the ideal model by 0.2 eV for Al, 0.337 eV for Cu, and −0.02 eV for Au. The close agreement between Φ^n_{SB} for Au measured by IPS and from Equation (3.4)

(continued)

(*continued*)

may be due to the chemically inert nature of the Au–GaAs interface, whereas both Al and Cu can react on a monolayer scale with GaAs and introduce localized states. Chapter 4 addresses such chemical reactions. Conventional GaAs $q\Phi_{SB}$ span a much narrower range, typically 0.1 eV or less. See Chapter 15, Section 15.5.2. Such barrier differences are much larger than those commonly reported for melt-grown GaAs and can be explained by differences in crystal quality, deposition methods, and chemically-induced defects to be discussed in later chapters.

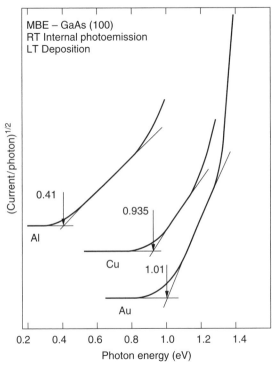

Figure 3.9 *Internal photoemission spectroscopy plots of MBE-grown GaAs barrier heights for Al, Cu, and Au. (Chang et al. 1992 [23]. Reproduced with permission of American Institute of Physics.)*

3.2.4 Schottky Barrier Conclusions

Each of the measurement techniques presented in this section is based on a different physical mechanism. Each is susceptible to experimental complications, and features within the interface can introduce different types of errors. For example, *J–V* measurements are particularly sensitive to tunneling through the barrier region and to trapping and recombination by deep levels within the band gap. Artifacts of the measurement technique and deep level trapping affect the *C–V* technique. Image force lowering of the Schottky barrier affects all three techniques. In addition, measurements obtained with these techniques become difficult to interpret if the barrier is more complex than a region of parabolic band bending. Such non-parabolicity can result, for example, if electrically-active sites such as dopants or traps are not uniformly distributed throughout the surface space charge region. Subsequent chapters provide Schottky barriers measured by these different methods. Systematic differences between Φ_{SB} measured by different techniques typically reflect these experimental complications. Hence, it is advisable to use more than one technique to interpret complex Schottky barrier features.

3.3   Heterojunction Band Offsets: Electrical Measurements

The semiconductor heterojunction has been integral to state-of-the-art electronic device technology for the past several decades and has enabled a wide range of fundamental physical phenomena to be explored. Engineered crystal structures created by epitaxial growth techniques have led to novel band structures with features such as quantum confinement and record breaking free carrier mobilities. In turn, these features have led to high speed, high power transistors, ultra-sensitive photodetectors, and lasers emitting from the far infrared to the ultraviolet.

The heterojunction band offsets are primary factors determining these physical phenomena and device applications. As with Schottky barriers, these offsets along with the band bending within each semiconductor's surface space charge region are determined not only by the bulk properties of the constituents but by dipoles and localized electronic states. Here we present the macroscopic electrical technique for measuring these offsets. Later chapters will describe advanced techniques to measure the physical properties that determine the heterojunction dipoles and band bending on an atomic scale.

Electrical measurements of heterojunction band offsets are based on the $C–V$ technique. Figure 3.10 illustrates a schematic energy band diagram of a heterojunction composed of two semiconductors with different band gaps in a *nesting* or *straddling* energy band line-up. Band bending contributes to the variation of E_C and E_V across the interface. Here, semiconductor A has an n-type (upward) band bending due to ionized impurities within the depletion region while semiconductor B has a p-type (downward) band bending due to charge accumulated in an inversion region. The single-headed arrows denote the sign convention for the ΔE_C and ΔE_V discontinuities. In this figure, ΔE_C is positive when the conduction-band edge of semiconductor A is above that of B. ΔE_V is positive when the valence band of A is below that of B. Hence ΔE_C and ΔE_V in Figure 3.10 are both positive. The convention used here is that $E_G(A) > E_G(B)$ so that

$$\Delta E_G = E_G(A) - E_G(B) \tag{3.22}$$

is >0 and

$$\Delta E_G = \Delta E_C + \Delta E_V \tag{3.23}$$

The vacuum levels of the two semiconductors are in general not aligned after they join. Instead they are shifted by the built-in potential V_{bi}, which depends on the difference in semiconductor surface and interface dipoles. This built-in potential is equal to the sum of

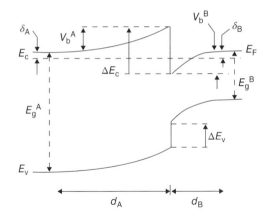

Figure 3.10   Schematic heterojunction band diagram for two semiconductors with nested band gaps.

the individual band bendings V_b^A and V_b^B, analogous to the built-in potential of a p–n homojunction.

$$V_{bi} = V_b^A + V_b^B \qquad (3.24)$$

The band edge energies as a function of position in Figure 3.10 follow from the individual band bendings V_b^A and V_b^B and their depletion widths as calculated from Poisson's equation, Equation (3.6). A complete calculation of heterojunction band diagrams involves ΔE_C and ΔE_V as well as V_b^A and V_b^B, δ_A and δ_B from the semiconductor doping levels, and the dielectric constants ε_A and ε_B. See, for example, Reference [24]. Approximate band diagrams can be drawn based on values of the band offsets, the semiconductor doping levels and the dielectric constants [25].

Macroscopic measurements involve charge transport, optical excitation, or both. In the case of transport methods, $C–V$ measurements can extract the built-in potential V_{bi}, assuming a square-root dependence for C on $V_{bi}-V$, as with Equation (3.16). One can obtain ΔE_C and ΔE_V by combining V_{bi} with δ_A and δ_B.

Thus, equating energies on either side of Figure 3.10, one obtains

$$\Delta E_C - V_b^A - \delta_A = V_b^B - \delta_B \qquad (3.25)$$

Since the built-in potential $V_{bi} = V_b^A - V_b^B$,

$$\Delta E_C = V_{bi} + \delta_A - \delta_B \qquad (3.26)$$

Within each semiconductor, the free carrier concentration $n = |N_d - N_a|$, where N_D and N_A are donor and acceptor densities, and N_C is the conduction band density of states. Then,

$$\delta_A = kT \, \ln(n_A/N_C^A) \qquad (3.27a)$$

$$\delta_B = kT \, \ln(n_B/N_C^B) \qquad (3.27b)$$

and

$$\delta_A - \delta_B = kT \, [\ln(n_A/n_B) + 1.5 \, \ln(m_B^*/m_A^*)] \qquad (3.28)$$

where m_B^* and m_A^* are the effective electron masses in the two semiconductors. Since the free carrier concentrations and effective masses are known, ΔE_C follows in a straightforward way from the measurement of V_{bi} and Equation (3.26). For a heterojunction with B, a p-type semiconductor, Equation (3.26) becomes

$$\Delta E_C = V_{bi} - E_G{}^B + \delta_A + \delta_B \qquad (3.29)$$

The built-in voltage V_{bi} follows from the capacitance, which is obtained from Poisson's equation in one dimension. Requiring that the electric displacement be continuous across the heterojunction boundary, one obtains

$$\varepsilon_A \mathscr{E}_A = \varepsilon_B \mathscr{E}_B \qquad (3.30)$$

for dielectric permittivity ε and electric field $\mathscr{E}$. Considering the heterojunction as two capacitors in series and with the depletion approximation, the capacitance per unit area is [26]

$$C^2 = [qN_A N_B \varepsilon_A \varepsilon_B]/[2(\varepsilon_A N_A + \varepsilon_B N_B)(V_{bi}-V)] \qquad (3.31)$$

for an applied bias V. ΔE_C follows from Equation (3.26) using the intercept value of V_{bi} in a plot of $1/C^2$ versus V.

This analysis is complicated by any impurity gradients or near-interface charge [26]. Among such capacitance techniques, the most reliable are measurements of interface charge at the heterojunction with known doping profiles [27]. Likewise, carrier recombination, tunneling, and shunt currents can seriously affect the interpretation of *I–V* measurements.

Photoexcitation combined with electrical transport can also probe heterojunction band offsets. Analogous to the IPS technique for measuring Schottky barriers, the incident photons excite electrons from a band populated with carriers in one semiconductor to an empty band in the other, which produces measureable photocurrent. The threshold energy for conduction marks the minimum transition energy across the junction. Figure 3.11 illustrates the photoexcitation of electrons from the valence band of one semiconductor to the conduction band of the other. In this case, photon energies exceeding the smaller band gap excite the transitions. Internal photoemission is also possible from one conduction

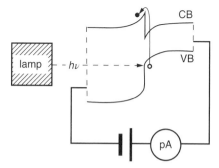

Figure 3.11 *Internal photoemission at semiconductor heterojunctions involving valence-to–conduction band transitions from the smaller to the larger band gap semiconductor. (Coluzza et al. 1993. [28]. Reproduced with permission of American Physical Society.)*

band to the other. If the conduction band of semiconductor B is accumulated in Figure 3.11, photoexcitation of these carriers into the conduction band of semiconductor A measures ΔE_C. Such measurements typically involve intense, monochromatic infrared photons with energies $h\nu \sim 0.1 - 0.5$ eV and special material structures, for example, superlattices, so that optical excitation across the interface is the dominant contribution to photoconduction. As with IPS of Schottky barriers, interface-specific phenomena such as chemical reactions and localized electronic states can complicate band offset measurements by IPS. However, it is the most direct of the transport techniques.

3.4   Summary

This chapter introduced the conventional macroscopic techniques, I–V, C–V, and IPS, used to measure the Schottky barriers and heterojunction band offsets at semiconductor interfaces. Each technique provides useful information but different sources of error are inherent in each. Thus more than one technique is advisable to characterize Schottky barriers. We have also seen that interface states can "pin" the Fermi level within the semiconductor band gap in a narrow energy range such that the barrier height becomes insensitive to the metal work function. Such Fermi level "pinning" is due to additional electronic states near or at the semiconductor surface or interface that introduce dipoles that alter the macroscopic barriers. In general, the classical picture described here is complicated by localized charge states that arise due to both intrinsic and extrinsic phenomena. Fermi level pinning underscores the importance of controlling chemical, electronic and geometric structure on an atomic scale. Later chapters will present methods by which such control can be achieved.

3.5   Problems

1. A TLM structure with 10 μm wide electrodes and area 20μm^2 on a semiconductor gives the following values:

$d(\mu m)$	1	3	7	10
$R_T(\Omega)$	9.2	14.41	24.83	32.65

   Determine the sheet resistance ρ_S, the contact resistance $\rho_C(\Omega)$, and the specific contact resistance $\rho_C(\Omega\ cm^2)$.

2. Assume that an ideal Schottky barrier is formed by a metal with work function 4.8 eV on n-type Si with $N_d = 10^{17}$ cm^{-3}. The Fermi level position in the band gap is obtained from the relation $n_0 = n_i e^{(E_F - E_i)/kT}$, where the intrinsic carrier concentration n_i for Si is 1.5×10^{10} and E_i is approximately at the midpoint of the band gap. Draw the intrinsic Fermi level position E_i and the equilibrium semiconductor Fermi level position E_{FS} in the band gap. Calculate the Schottky barrier height. What is the band bending? Draw an equilibrium band diagram as in Figure 3.3.

3. Calculate the depletion width for the Schottky diode in Problem 2. Use MKS units so that $W = (2\varepsilon_s(V - V_0)/Nq)^{1/2}$. What is the depletion width with a reverse bias of 0.5 V? Draw the corresponding energy band diagram. How does it change?

4. Assume that a dipole forms at the Si surface due to negative surface charges with a density $n_s = 2 \times 10^{13}$ cm^{-2} separated by a distance $d = 10$ Å. What is the corresponding change in barrier height and band bending for the junction in Problem 2.

5. Calculate the capacitance for the diode in (2) at zero bias with a diode diameter of 400 μm.

6. Calculate the image force lowering of the barrier for a metal–Si diode with an applied electric field of 10^4 V cm^{-1}.

7. Calculate the forward current for a Si barrier height 0.5 eV at room temperature and $V_F = 0.3$ V forward bias for a diode with 200 μm circular diameter. Neglect image force lowering. For Si, use $m_e{}^*/m_0 = 0.26$.

8. An n-type semiconductor with $\chi = 4$ eV, $N_D = 10^{16}$cm^{-3}, and $n_i = 10^{10}$ cm^{-3} has a room temperature band gap $E_G = 1.2$ eV. For metals with work function $\Phi_M = 4, 4.25, 4.5, 4.75$, and 5 eV, calculate and draw Φ_{SB} versus Φ_M. Calculate and draw qV_0 versus Φ_M. For $\varepsilon/\varepsilon_0 = 15$, determine a numerical equation for the depletion region width versus Φ_M for applied voltage $V_A = 0$.

9. Consider a Si Schottky diode of area 1 cm^2 and doping N_D with a capacitance such that $1/C^2$ versus applied voltage V_A varies with slope -4.4×10^{13} F^{-2} V^{-1}. (a) Determine the semiconductor doping. (b) If $C = 168$ nF at $V = 0$, what is the band bending qV_0?

10. The bias dependent internal photoemission curves in Figure 3.8 show a decrease in the metal–Si barrier height with increasing reverse bias. For n-type doping of 3×10^{16} cm^{-3} and a barrier height of 0.825 eV, calculate the maximum electric fields $-qN_Dd/\varepsilon$ at the interface for 0 and 2 V reverse bias. Use $n_i = 1.5 \times 10^{10}$ cm^{-3} for the intrinsic carrier concentration. What image force dielectric constant is consistent with the barrier height change?

11. From Figure 1.6 showing the first transistor, speculate on how the inventors achieved low resistance contacts.

12. Solar cells operate by absorbing light, generating free electron–hole pairs that separate inside a built-in electric field, generating a voltage. Give two reasons why the external contacts to the solar cell must be both Ohmic and low resistance.

13. Among the I–V, C–V, and IPS techniques to measure Schottky barrier heights, which would be easiest to interpret for: (a) diodes with two different Φ_{SB} under the same metal contact, (b) diodes on an indirect band gap semiconductor, and (c) diodes on a semiconductor with a voltage-dependent dielectric permittivity?

References

1. Schroder, D.K. (1998) *Semiconductor Material and Device Characterization*, 2nd edn, Wiley-Interscience, New York, Chapter 3 and reference therein.
2. Brillson, L.J. (1982) The structure and properties of metal-semiconductor interfaces. *Surf. Sci. Rep.*, **2**, 123.

3. Turner, M.J. and Rhoderick, E.H. (1968) Metal-silicon Schottky barriers. *Solid-State Electron.*, **11**, 291.

4. Archer, R.J. and Atalla, M.M. (1963) Metal contacts on cleaved silicon surfaces. *Am. Acad Sci. NY*, **101**, 697.

5. Crowell, C.R., Spitzer, W.G., Howarth, I.E., and LaBate, E.E. (1962) Attenuation length measurements of hot electrons in metal films. *Phys. Rev.*, **127**, 2006.

6. Spicer, W.E., Lindau, I., Skeath, P., and Su, C.Y. (1980) Unified mechanism for Schottky – barrier formation and III-V oxide interface states. *Phys. Rev. Lett.*, **44**, 420.

7. Bardeen, J. (1947) Surface states and rectification at a metal semi-conductor contact. *Phys. Rev.*, **71**, 717.

8. Sze, S.M. (2007) *Physics of Semiconductor Devices*, 3rd edn, Wiley-Interscience, New York, Chapter 5.

9. Duke, C.B. (1969) *Tunneling in Solids*, Academic, New York, pp. 102–110.

10. Crowell, C.R. (1965) The Richardson constant for thermionic emission in Schottky barrier diodes. *Solid-State Electron.*, **8**, 979.

11. Shaw, M.P. (1981) in *Handbook on Semiconductors, Device Physics*, Vol. 4, (ed. C. Hilsum), North-Holland, Amsterdam, Chapter 1.

12. Lang, D.V. (1974) Deep-level transient spectroscopy: A new method to characterize traps in semiconductors. *J. Appl. Phys.*, **45**, 3023.

13. (a) Carballes, J.C., Varon, J., and Ceva, T. (1971) Capacitives methods of determination of the energy distribution of electron traps in semiconductors. *Solid State Commun.*, **9**, 1627; (b) Sah, C.T., Chan, W.W., Fu, H.S., and Walker, J.W. (1972) Thermally stimulated capacitance (TSAP) in p-n junctions. *Appl. Phys. Lett.*, **20**, 193 (1972); (c) Buehler, M.G. (1972) Impurity centers in PN junctions determined from shifts in the thermally stimulated current and capacitance response with heating rate. Impurity centers in PN junctions determined from shifts in the thermally stimulated current and capacitance response with heating rate. *Solid-State Electron.*, **15**, 69.

14. (a) Sah, C.T. and Walker, J.T. (1973) Thermally stimulated capacitance for shallow majority-carrier traps in the edge region of semiconductor junctions. *Appl. Phys. Lett.*, **22**, 384; (b) Losee, D.L. (1972) Admittance spectroscopy of deep impurity levels: ZnTe Schottky barriers. *Appl. Phys. Lett.*, **21**, 54.

15. (a) Kukimoto, H., Henry, C.H., and F.R. Merritt, F.R. (1973) Photocapacitance studies of the oxygen donor in GaP:. I. Optical cross sections, energy levels, and concentration. *Phys. Rev. B.*, **7**, 2486; (b) White, A.M., Dean, P.J., and Porteous, P. (1976) Photocapacitance effects of deep traps in epitaxial GaAs. *J. Appl. Phys.*, **47**, 3230.

16. Sah, C.T. (1976) Bulk and interface imperfections in semiconductors. *Solid-State Electron.*, **19**, 975.

17. Sah, C.T., Forbes, L., Rosier, L.L., and Tasch, Jr., A.F. (1970) Thermal and optical emission and capture rates and cross sections of electrons and holes at imperfection centers in semiconductors from photo and dark junction current and capacitance experiments. *Solid-State Electron.*, **13**, 759.

18. Fowler, R.H. (1931) The analysis of photoelectric sensitivity curves for clean metals at various temperatures. *Phys. Rev.*, **38**, 45.

19. Sze, S.M., Crowell, C.R., and Kahng, D. (1964) Photoelectric determination of the image force dielectric constant for hot electrons in Schottky barriers. *J. Appl. Phys.*, **35**, 2534.

20. Crowell, C.R., Shore, H.B., and Labate, E.E. (1965) Surface-state and interface effects in Schottky barriers at n-type silicon surfaces. *J. Appl. Phys.*, **36**, 3843.
21. Parker, G.H., McGill, T.C., Mead, C.A., and D. Hoffman, D. (1968) Electric field dependence of GaAs Schottky barriers. *Solid-State Electron.*, **11**, 20.
22. Crowell, C.R., Sze, S.M., and Spitzer,W.G. (1964) Equality of the temperature dependence of the gold-silicon surface barrier and the silicon energy gap in Au n-type Si diodes. *Appl. Phys. Lett.*, **4**, 91.
23. Chang, S., Shaw, J.L., Brillson, L.J., *et al.* (1992) Inhomogeneous and wide range of barrier heights at metal/molecular-beam epitaxy GaAs(100) interfaces observed with electrical measurements. *J. Vac. Sci. Technol., B*, **10**, 1932.
24. Casey Jr., H.C. and Panish, M.B. (1978) *Heterojunction Lasers, Part A: Fundamental Principles; Part B: Materials and Operating Characteristics*, Academic Press, New York.
25. Streetman, B.G. and S. Banarjee, S. (2000) *Solid State Electronic Devices*, 5th edn, Prentice Hall, Upper Saddle River, NJ, p. 229.
26. Forrest, S.R. (1987) in: *Heterojunction Band Discontinuities: Physics and Device Applications*, (eds. F. Capasso and G. Margaritondo), North-Holland, Amsterdam, Chapter 8.
27. Kroemer, H., Polasko, J., and Wright, S.J. (1980) On the (110) orientation as the preferred orientation for the molecular beam epitaxial growth of GaAs on Ge, GaP on Si, and similar zincblende-on-diamond systems. *Appl. Phys. Lett.*, **36**, 763.
28. Coluzza, C., Tuncel, E., Stachli, J.-L., *et al.* (1992) Interface measurements of heterojunction band lineups with the Vanderbilt free-electron laser. *Phys. Rev. B*, **46**, 12834(R).

Further Reading

Streetman, B.J. and Banarjee, S.K. (2015) *Solid State Electronic Devices*, 7th edn, Pearson, Upper Saddle River, New Jersey.

4

Localized States at Surfaces and Interfaces

Each of the preceding chapters has emphasized the importance of localized electronic states at surfaces and interfaces. These states have been the subject of intense experimental and theoretical study for nearly 70 years, resulting in literally millions of scientific articles. In this chapter, we introduce the various classes of semiconductor localized states, describe the many different physical phenomena that can introduce such states, and present key results that have shaped our current understanding.

4.1   Interface State Models

Trapped charge at the surface and interfaces can be viewed in terms of either the lattice termination of the bulk crystal or localized states associated with the surface band structure. The crystal lattice termination at the free surface can produce a major change in electronic structure versus the bulk since atoms at the surface possess fewer atomic neighbors and hence more broken chemical bonds. States at the vacuum–semiconductor interface with energies in the band gap will be localized in the two-dimensional plane of the surface.

   Figure 4.1a illustrates how the wave function of such in-plane states decays into vacuum as well as into the semiconductor since no allowed states are present either in vacuum or in the semiconductor for energies in the band gap. The peak in wave function at the surface corresponds to a localized state with the ability to trap charge. These states are termed *intrinsic* since they are associated with the discontinuity of bulk lattice potential at the semiconductor–vacuum interface, that is, they are an intrinsic property of the semiconductor. A simple model calculation illustrates how quantum-mechanical boundary conditions on the semiconductor wave function in Figure 4.1a introduce localized states at the surface.

An Essential Guide to Electronic Material Surfaces and Interfaces, First Edition. Leonard J. Brillson.
© 2016 John Wiley & Sons, Ltd. Published 2016 by John Wiley & Sons, Ltd.
Companion Website: www.wiley.com/go/Brillson/

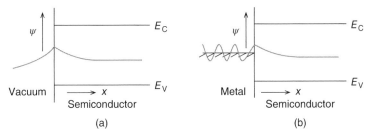

Figure 4.1   *Schematic energy band diagrams of wave function localization due to (a) intrinsic surface states at the semiconductor–vacuum interface and (b) a metal-induced state at the semiconductor–metal interface.*

Localized states can also arise due to wave function tunneling from a metal overlayer into the semiconductor. Figure 4.1b represents schematically a periodic wave function in the metal that decays into the semiconductor for energies in the semiconductor band gap. Again, there is a peak in the wave function at the metal–semiconductor interface that corresponds to a localized state that can trap charge. This tunneling state is also considered *intrinsic* since it involves only properties intrinsic to the two constituents.

A third type of localized state involves imperfections of the semiconductor surfaces, such as impurities or lattice defects common to both the surface and the bulk. These states are *extrinsic*, that is, their properties are different from those of the pure bulk semiconductor.

A particularly important type of *extrinsic* state is interface-specific, for example, interface chemical reactions, interdiffusion, or adsorbate-specific local bonding. This class of localized state is especially important since its presence depends on the way the interface is prepared and subsequently treated.

4.2   Intrinsic Surface States

4.2.1   Experimental Approaches

Researchers have developed a wide range of experimental techniques to measure localized states at surfaces and interfaces of electronic materials. The electrical techniques described in Chapter 3 provided direct measurements of Schottky barriers for various metals in contact with many different semiconductors. These macroscopic studies revealed striking differences between classes of semiconductors in terms of their metal interface behavior, that is, how strongly different metal work functions controlled a given semiconductor's Schottky barrier. This *index of interface behavior S* is defined as [1]

$$\Phi_{SB} = S(\Phi_M - \Phi_{SC}) + C \tag{4.1}$$

where C is a constant. According to Equation (4.1), semiconductors with little or no interface dipole formation should exhibit ideal Schottky barrier behavior, that is, $S \simeq 1$ as the metal work function varies whereas semiconductors that form dipoles with E_F pinning should exhibit $S \simeq 0$. With this definition for S, Figure 4.2 shows a striking transition in interface behavior with semiconductor *ionicity*. A measure of this ionicity is the *electronegativity difference* ΔX between anion and cation [2]. Figure 4.2 illustrates how S varies

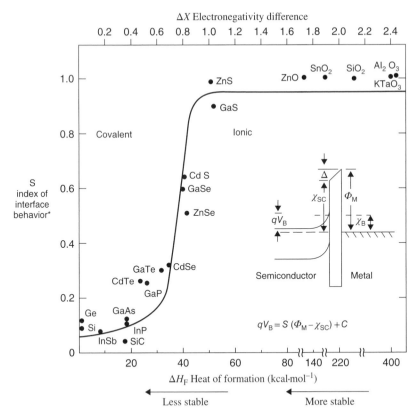

Figure 4.2 *Transition in the index of interface behavior S between covalent and ionic semiconductors plotted versus semiconductor electronegativity ΔX (upper scale) or chemical heat of formation ΔH_F (lower scale). The same curve illustrates the dependence of S on chemical bonding and stability against chemical reactions. After [1] and [3].*

with ΔX for various metals on a wide range of semiconductors. Semiconductors whose anion–cation bonding is more covalent, such as Si and GaAs, display S close to zero whereas more ionic semiconductors, such as ZnO and SiO_2, exhibit S values near unity. This behavior represents evidence for intrinsic surface states since lattice disruption at the surface of a crystal lattice is larger for covalent than for ionic semiconductors. Hence a lower lattice disruption would generate a lower density of surface states, smaller interface dipoles, and a larger S value.

Beyond these macroscopic methods are experimental techniques to probe the nature of surface and interface states at the atomic scale. Chief among them is *photoemission spectroscopy*, which use photons to excite emission of electrons from occupied states at the surface. Various forms of photoemission spectroscopy can provide information on chemical composition and bonding, E_F within the band gap, work function, valence band state density, unoccupied states above E_F, heterojunction band offsets, atomic bonding symmetry, and Brillouin zone dispersion. Electron beam techniques, such as *cathodoluminescence spectroscopy*, *Auger electron spectroscopy*, and *electron microscopy*, can provide chemical composition, bonding, and reaction, surface atomic geometry and interface lattice structure, energies and densities of states within the band gap, barrier heights, heterojunction band offsets, plasmons, interband transitions, and interface compound formation. Ion beam techniques, such as *Rutherford backscattering spectrometry*, can provide elemental composition and depth distribution, bonding and mapping. Optical techniques, such as *photoluminescence spectroscopy*, can provide absorption edges, direct and indirect band gaps, gap state energies and densities, as well as surface and interface dielectric response. Scanned probe techniques such as *scanning tunneling microscopy* and *atomic force microscopy* can provide atomic geometry and morphology, filled and empty state geometry, work function, band bending, energies and densities of states in the semiconductor band gap, heterojunction band offsets, as well as magnetic moment and polarization. The chapters to follow describe all these techniques and the methods used to obtain the physical properties listed.

4.2.2   Theoretical Approaches

In general, theoretical techniques to describe localized states at semiconductor surfaces require either: (i) a well-established surface atomic geometry or (ii) self-consistency between calculations with electronic and structural measurements. Well-established surface atomic geometries are challenging since there are very few available. On the other hand, there are many avenues to match theoretical calculations to electronic and structural measurements. These include calculations to: (i) predict lowest energy bonding/lattice configuration, (ii) fit near-surface electron diffraction profiles, and (iii) fit densities of states calculations to photoemission spectra.

Successful theoretical approaches that can account for the photoemission spectroscopy features of elemental semiconductor surfaces include: (a) self-consistent calculations of potential and charge density, (b) empirical tight-binding methods and (c) surface energy minimization. For Si, today's most important elemental semiconductor, the picture of its surface is incomplete, despite extensive experiments, due to the complexity of its surface structure. For compound semiconductors, theoretical methods can account for the atomic structure of non-polar surfaces, which are better understood [4].

As an example of self-consistent potential and charge density calculations, Figure 4.3 illustrates valence charge density contours, total valence-charge densities averaged in planes parallel to the interface plane, and local densities of states versus energy. Here, pseudopotentials approximate the potentials of the ion cores for the clean Si surface and a *jellium* model, corresponding to a "sea" of loosely bound electrons that approximate the Al metal overlayer at an Al–Si interface. Figure 4.3a shows the contours of valence-charge density on an atomic scale in a cross section of the Al–Si(111) interface. Figure 4.3b illustrates the total valence charge density $\rho_{total}(z)$ as a function of distance z perpendicular to the same interface. The constant $\rho_{total}(z)$ within the Al results from the jellium model assumed for the metal, while the peaks in $\rho_{total}(z)$ within the semiconductor correspond to contributions from the semiconductor bond charges [4]. Near the interface, $\rho_{total}(z)$ exhibits new features unlike those of either the bulk Al or the Si. Panels III and IV in Figure 4.3b show that some charge transfer takes place from the Al to the Si at the interface in this model, yielding an interface dipole extending over atomic dimensions. The corresponding densities of states are shown in Figure 4.3c [6]. Region I displays a

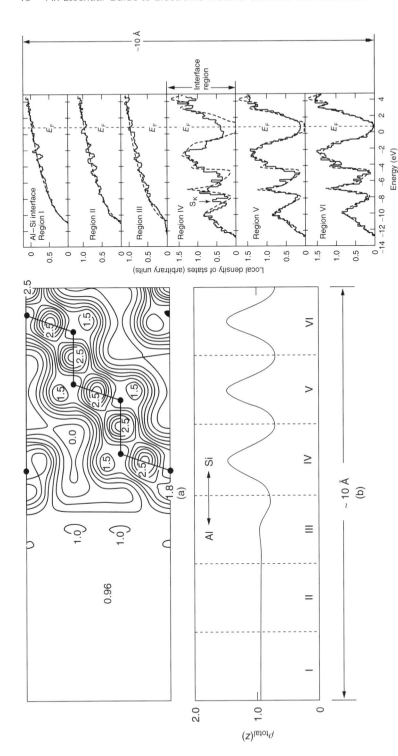

Figure 4.3 *(a) Total valence-charge-density contours in a (110) plane for the Al–Si(111) interface (out of the page). Dots indicate the Si atoms. (b) Total valence-charge density averaged parallel to the interface. Spatial regions I–VI lie in the direction normal to the interface. (c) Local density of states. Dotted lines indicate bulk densities of Al or Si states [6]. (Louie, S.G. and Cohen, M.I. 1976 [5]. Reproduced with permission of American Physical Society.)*

density of states just as in bulk Al, region VI just as in bulk Si, whereas new features appear near the plane of the Al–Si junction plane. These include interface states that are bulk-like in the Si but that decay rapidly in the Al, states that are bulk-like in the Al but that decay rapidly in the Si – as in Figure 4.1b, and truly localized interface states that decay away from the interface in both directions. The bulk-like Al states decay into the Si whenever the range of the semiconductor band gap is inside the metallic band. Louie and Cohen associated these "*metal-induced gap states*" (MIGS) with the Fermi level pinning because the energy of these states within the semiconductor band gap agrees with experimental values [7] for Si Schottky barriers. Other calculations have refined the potential used in these calculations [8,9]. These reconfirm the disappearance of the intrinsic Si surface states and the appearance of metal-induced gap states [4].

Empirical tight-binding methods treat surface atoms that are not fully coordinated with their neighbors as they are in the bulk in terms of the energy levels of the free atoms. Figure 4.4 shows two atomic levels A and B that form the bulk semiconductor's valence and conduction bands, respectively. Since the surface atoms have fewer bonds with their neighbors, their electronic energy levels are closer to those of the free atom, splitting off from the band edges into the band gap. The wave functions of these *split-off levels* are built up from conduction and valence band wave functions. Depending on their conduction or valence band nature, these states have acceptor- or donor-like character. Thus conduction band-derived surface states are negative when filled and act as acceptors, whereas valence band-derived surface states are positive when empty and act as donors [10].

An example of energy minimization is the determination of surface atomic rear-rangement, termed *reconstruction*, at the GaAs(110) (equal numbers of Ga and As atoms) surface. Here the minimum or ground state energy of the surface atomic geom-etry is expressed in terms of structural parameters, such as bond lengths and angles. Figure 4.5 shows densities of states for the ideal and reconstructed surface and a bond-length-conserving 27° rotation of outer layer atoms. This reconstruction lowers the

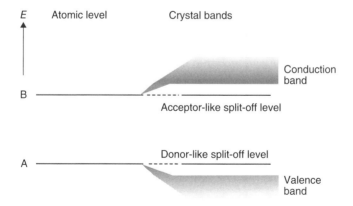

Figure 4.4 *Split-off surface-state levels in a tight-binding picture. Surface atoms have fewer bonding partners than bulk atoms so that their electronic energy levels shift toward those of the free atoms. Such states can accept or donate charge depending on their valence or conduction band origin [10].*

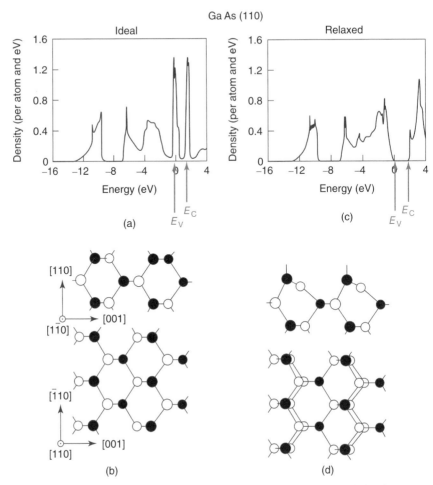

Figure 4.5 *Surface state densities and the corresponding structure models for the GaAs(110) surface. (a, c) Calculated surface state densities for the ideal, non-reconstructed (a) and the relaxed (c) surface. $E_V = 0$ corresponds to the valence band edge (Chadi, D.J. 1978 [11]. Reproduced with permission of American Physical Society.) (b, d) Structure models (side and top view, filled circles As, open circles Ga) for the ideal, non-reconstructed (b) and relaxed (d) surface (Lubinsky et al. 1976 [12]. Reproduced with permission of American Physical Society.) After Himpsel [13].*

ground state energy by moving the more (less) negatively charged As (Ga) atom outward (inward). The ideal GaAs(110) surface exhibits a density of states that includes states within the band gap, whereas the bond rotation in the reconstructed, more specifically in this case *relaxed*, surface sweeps these states out of the band gap. This relaxation is significant in that it accounts for the absence of intrinsic states in the band gap of this and analogous non-polar III-V compound semiconductor surfaces.

Overall, intrinsic surface states can be grouped into four theory categories: (i) *Shockley states* [14] corresponding to the nearly-free electron model used, (ii) *Tamm states* [15] consisting of a linear combination of atomic eigenstates (that is, quantum-mechanical wave functions of electrons orbiting a specific atom) as in Figure 4.4, (iii) *dangling bond states* associated with particular orbital lobes for chemical bonding, for example, sp^3 for As versus sp^2 for Ga in Figure 4.5, and (4) *back bond states* associated with surface-induced modification of chemical bonds between atomic layers.

4.2.3   Key Intrinsic Surface State Results

Experimental and theoretical work to date provides several general conclusions about intrinsic surface states in semiconductors. For elemental semiconductors such as Si and Ge: (i) intrinsic surface states exist near or within the band gap. (ii) The density of intrinsic surface states depends sensitively on the detailed surface reconstruction. (iii) Complete determination of the surface state eigenvalue spectrum (their energies and wave functions) requires a better determination of their respective surface atomic geometries.

For compound semiconductors such as GaAs and InP, non-polar surfaces obtained by UHV cleavage show: (i) There are *no* intrinsic surface states within the band gaps of III-V and II-VI compound semiconductors. (ii) Reconstruction sweeps states out of the band gap. (iii) Surface states are very sensitive to detailed atomic structure. For non-cleavage compound semiconductor surfaces: (i) States can exist inside the band gap. (ii) Surface preparation (for example, polishing or cleaning by ion bombardment) introduces damage that can extend many layers below the surface. The extensive literature devoted to intrinsic surface states supports these conclusions for a wide range of semiconductors [4].

In general, surface states and their associated energy band structure depend sensitively on surface atomic structure, which can differ substantially from the bulk lattice structure in multiple variants. Detailed comparisons of experimental and theoretical features are required to characterize these surface atomic structures.

4.3   Extrinsic Surface States

Extrinsic states at semiconductor surfaces fall into three general categories: (i) trapped charge inherent in the semiconductor bulk. These may include bulk defects, grain boundaries, and impurities. (ii) Imperfections created at the semiconductor surface. These may include cleavage steps, mechanically-induced point defects, dislocations, and chemisorbed impurity atoms such as adsorbed air molecules. (iii) Charge states created by interaction between the semiconductor and a contact metal. Beyond metal-induced gap states, the creation of electrically active sites may include interface chemical reaction, semiconductor atom outdiffusion, and metal indiffusion.

Both solid state research and electronic device applications typically seek to avoid semiconductors with crystal defects, impurities, and imperfections. Numerous microscopic methods are useful for detecting such features. More difficult to observe are charge states produced by chemical reaction and interdiffusion, which can occur on a nanometer or even atomic scale. Nevertheless, such interactions can produce localized electronic states that can dominate the macroscopic interface properties, such as Schottky barrier formation and heterojunction band offsets. Indeed there is considerable evidence that such chemical interactions take place and that they may dominate junction properties.

Early work showed that chemical reactions could occur at metal-compound semiconductor interfaces with thermal annealing [16,17]. Schottky barriers at Si interfaces with transition metal silicides – intentionally reacted to form contacts to Si microelectronics (See Figure 1.1) – exhibited a correlation between Φ_{SB} and the heats of formation of the transition metal silicides [18]. The interpretation of this systematic behavior was that the degree of hybridized bonding between the transition metal with Si atoms scaled with ΔH_F. Thus the localized charge redistribution associated with this atomic bonding introduced the dipole variations that contributed to Φ_{SB}, as in Figure 3.4. This correlation was the first to highlight the role of microscopic chemical bonding in forming macroscopic electronic properties. Later metal–Si Schottky barrier studies noted a correlation between Φ_{SB} and the lowest melting *eutectic temperature* (which determines the first metallurgical phase to form) of the transition metal silicide [19]. This correlation demonstrated that the physical properties of the interface related to the bond strength of the interfacial layer, in this case the reacted silicide. Also reflecting the electronic properties of the reacted transition metal–Si interface was a correlation between Φ_{SB} and the *effective work function* of the silicide according to the stoichiometry of its constituents [20].

The transition in the coefficient of interface behavior displayed in Figure 4.2 can be viewed in terms of the semiconductor's thermodynamic stability against chemical reaction with metals. Semiconductors with a high heat of formation ΔH_F are more resistant to chemical reaction with metals than low ΔH_F semiconductors. Here, covalent semiconductors such as Si, GaAs, and InP are less stable than ionic semiconductors such as ZnO, SiO_2, and Al_2O_3. Thus the same S values discussed with Figure 4.2 correlate with ΔH_F, implying that chemical thermodynamics dominates Schottky barrier formation for compound semiconductors. The Φ_{SB} values for various metals in contact with both ionic and covalent semiconductors provide further evidence for the electronic effects of chemical reaction. These values, which were used to obtain the S values in Figure 4.2, also display a Φ_{SB} transition in terms of chemical thermodynamics. The stability of a metal–semiconductor interface can be described quantitatively in terms of a chemical *heat of reaction* ΔH_R, defined as [3]:

$$\Delta H_R = (1/x)[H_F(CA) - H_F(M_X A)] \qquad (4.2)$$

for the reaction $M + (1/x)CA \Rightarrow (1/x)[M_X A + C]$, where M, C, and A signify metal, semiconductor cation and anion, respectively, X is the number of metal atoms, and $M_X A$ is the most stable metal-anion product, that is, having the most negative ΔH_R.

Figure 4.6 shows a systematic correlation between Φ_{SB} (measured by IPS) versus ΔH_R for the same semiconductor with different metals. All of the semiconductors display the same qualitative behavior, regardless of ionicity. A *critical heat of reaction* $\Delta H_R{}^C \sim 0.5$ eV per metal atom marks the transition between reactive and unreactive interfaces. Interface-specific measurements confirm that metals with $\Delta H_R < 0.5$ eV/metal atom reacted with compound semiconductors while metals with $\Delta H_R > 0.5$ eV/metal do not [3,22][1]. The p-type barrier height $\Phi_{SB}{}^P$ – that is, $E_F - E_V$ for the data in Figure 4.2 – also shows systematic correlations with ΔH_R for the same metal on many different semiconductors. This again emphasizes the metal–anion interaction since anion orbitals contribute most to the valence band properties [22]. Other compound

---

[1] Metal-cation alloying may also contribute to Equation (4.2) at elevated temperatures but the corrections are typically small.

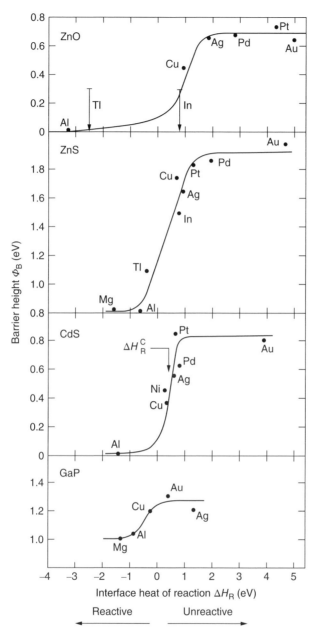

Figure 4.6 *Barrier heights correlated with heats of interface chemical reactions ΔH_R for metals on representative ionic and covalent semiconductors. A* critical *heat of reaction $\Delta H_R^C \sim 0.5$ eV per metal atom determined experimentally marks the center of the transition region between reactive and unreactive interfaces. (Brillson, L.J. 1978 [3]. Reproduced with permission of American Physical Society.)*

semiconductors exhibit similar behavior [4]. These results demonstrate that chemical stability has a systematic effect on Schottky barrier formation.

Case Study: Interface Heat of Reaction

Using Equation (4.2), calculate ΔH_R for the Al–GaAs interface. Is this a reactive interface? $H_F(\text{GaAs}) = 74.1 \text{ kJ mol}^{-1}$, $H_F(\text{AlAs}) = 116.3 \text{ kJ mol}^{-1}$. AlAs is the only stable product that can form at the Al–GaAs interface.

$$\Delta H_R = (1/1)\,[74.1 - 116.3] = -42.2 \text{ kJ mol}^{-1} = -0.44 \text{ eV/atom}$$

Since $\Delta H_R < 0.5$ eV, this interface is reactive.

In addition to models of localized charge redistribution with hybridized interface orbitals or effective work function changes, chemical reactions may also introduce native point defects. The metal – semiconductor reaction can draw atoms from the semiconductor, leaving behind vacancies and other defects. These defects can be electrically-active, mobile, and may accumulate at the metal-semiconductor interface. These defects can alter near-interface doping and localized states that affect Schottky barriers [23,24].

4.4 The Solid State Interface: Changing Perspectives

Perspectives on localized interface states have evolved since the late 1940s. Traditional models started with ideal Schottky barriers at atomically abrupt interfaces that states intrinsic to the clean semiconductor surface could modify. The advent of UHV surface science techniques led to models of local charge redistribution with interface chemical bonds, which introduced surface and interface formation at the monolayer level. Further studies at the monolayer level produced evidence for atomic interdiffusion and new chemically-reacted phases – even at or below room temperature. Depending on the nature of these microscopic interface reactions, interfacial layers could introduce new dielectric properties or native point defects that could alter Schottky barriers. Measurements of the initial stages of Schottky barrier formation show that interface electronic structure could not be extracted simply from the ideal properties of the abrupt metal–semiconductor interface. Instead, a more complex interface picture has emerged with chemically-reacted and/or diffused regions, local charge redistribution, built-in potential gradients, and new dielectric properties. The chapters to follow will describe the techniques used to obtain detailed information about these interfaces and the salient results thereof.

4.5 Problems

1. (a) For a localized state in GaAs with energy level at mid-gap, calculate the decay length of its wavefunction into vacuum. (b) Assuming a metal $\psi(z)$ decaying into the GaAs

with an effective potential barrier $= \frac{1}{2}E_G$ and $m = m^*$. Considering only this valence band to mid-gap energy barrier, what is the corresponding decay length?

2. Compare the heats of nitride interface reaction for In, Al, and Cr (-117.2 kJ mol^{-1}) on GaN.

3. (a) For Mg on GaN, Mg_3N_2 ($\Delta H_F = -461.5$ kJ mol^{-1}) can form. Does Mg react more strongly (per metal atom) with GaN or ZnO at room temperature? Ignore entropy contributions. (b) How strongly does Ti react with GaN to form TiN (-338.1 kJ mol^{-1}) versus ZnOto form TiO_2 (-944 kJ mol^{-1})?

4. Metals deposited on clean GaAs (110) surfaces can produce exchange reactions with the semiconductor if the reacted compound bonding is more stable than GaAs. (a) Of the metals In, Ti, Ni, and Al, which metal is most likely to produce such a reaction at room temperature? (b) Which is the least likely? Hint: the most thermodynamically stable metal-arsenide reaction products for Cu, In, Ni, Al, and Ti are: Cu_3As, InAs, NiAs, AlAs and TiAs with H_F values of -2.8, -57.7, -72, -116.3, and -149.7 kJ mol^{-1}, respectively.

5. (a) A metal with work function $\Phi_M = 5.4$ eV in contact with Si forms a Schottky barrier $\Phi_{SB} = 0.8$ eV. Calculate the density of interface states per unit area located at $E_C - E_{SS} = 0.6$ eV, assuming a dipole length of 10 Å and a bulk $E_C - E_F = 0.1$ eV. Hint: For a capacitor with charge Q, area A, and capacitance C, $Q = CV = \varepsilon AV/d$. (b) Suppose a new metal with work function $\Phi_M = 4.2$ eV forms the same density of interface states at this gap energy. What is the expected barrier height?

6. If a metal with work function $\Phi_M = 5.4$ eV forms a Schottky barrier with GaAs ($\chi = 4.07$ eV, $\varepsilon/\varepsilon_0 = 13.2$), what is the Schottky barrier height without any interface defects? If $E_C - E_F = 0.1$ eV in the semiconductor bulk, what is the band bending?

7. Calculate the dipole contribution to the Schottky barrier height of GaAs for a filled surface acceptor density $\sigma_a^- = 10^{13}$ cm^{-2} located 10 Å below the surface. How does the dipole shift the Fermi level – up or down? Draw a band diagram of the interface including the dipole.

References

1. Kurtin, S., McGill, T.C., and Mead, C.A. (1970) Fundamental transition in the electronic nature of solids. *Phys. Rev. Lett.*, **22**, 1433.

2. Pauling, L. (1960) *The Nature of the Chemical Bond*, 3rd edn, Cornell University Press, Ithaca.

3. Brillson, L.J. (1978) Transition in Schottky barrier formation with chemical reactivity. *Phys. Rev. Lett.*, **40**, 260.

4. Brillson, L.J. (2010) *Surfaces and Interfaces of Electronic Materials*, Wiley-VCH Verlag GmbH, Weinheim, Chapter 4.

5. Louie, S.G. and Cohen, M.I. (1976) Electronic structure of a metal-semiconductor interface. *Phys. Rev.*, **B13**, 2461.

6. Louie, S.G. and Cohen, M.I. (1975) Self-consistent pseudopotential calculation for a metal-semiconductor interface. *Phys. Rev. Lett.*, **35**, 866.

7. A.Thanalaikas, A. (1975) Contacts between simple metals and atomically clean silicon. *J. Phys. C (Solid State Phys.)*, **8**, 655.

8. Chelikowsky, J.R. (1977) Electronic structure of Al chemisorbed on the Si (111) surface. *Phys. Rev.*, **B16**, 3618.

9. Zhang, H.I. and Schlüter, M. (1978) Studies of the Si(111) surface with various Al overlayers. *Phys. Rev.*, **B15**, 1923.

10. Lüth, H. (2001) *Solid Surfaces, Interfaces, and Thin Films*, 4th edn, Springer, Berlin, pp. 266–270.

11. Chadi, D.J. (1978) Energy-minimization approach to the atomic geometry of semiconductor surfaces. *Phys. Rev. Lett.*, **41**, 1062.

12. Lubinsky, A.R., Duke, C.B., *et al.* (1976) Semiconductor surface reconstruction: The rippled geometry of GaAs(110). *Phys. Rev. Lett.*, **36**, 1058.

13. Himpsel, F.J. (1990) Inverse photoemission from semiconductors. *Surf. Sci. Rep.*, **12**, 1.

14. Shockley, W. (1939) On the surface states associated with a periodic potential. *Phys. Rev.*, **56**, 317.

15. Tamm, I. (1932) On the possible bound states of electrons on a crystal surface. *Phys. Z. Soviet Union*, **1**, 733.

16. Kipperman, A.H.M., and Leiden, H.F. (1973) Electrical properties of metal surface barriers on the layer structures of GaS and GaSe. *J. Phys. Chem. Solids*, **31**, 597.

17. Van Den Vries, J.G.A.M. (1974) Influence of chemical reactions on the barrier height of metal semiconductor junctions. *J. Phys. Chem. Solids*, **35**, 130.

18. Andrews, J.M. and Phillips, J.C. (1975) Chemical bonding and structure of metal-semiconductor interfaces. *Phys. Rev. Lett.*, **35**, 56 (1975).

19. Ottaviani, G., Tu, K.N. and J.W. Mayer, J.W. (1980) Interfacial reaction and Schottky barrier in metal-silicon systems. *Phys. Rev. Lett.*, **44**, 284.

20. Freeouf, J.L. (1980) Silicide Schottky barriers: An elemental description. *Solid State Commun.*, **33**, 1059.

21. Brillson, L.J. (1978) Chemical reaction and charge redistribution at metal-semiconductor interfaces. *J. Vac. Sci. Technol.*, **15**, 1378.

22. Brillson, L.J. (1982) The structure and properties of metal-semiconductor interfaces. *Surf. Sci. Rep.*, **2**, 123.

23. Brillson, L.J., Dong, Y., Doutt, D. *et al.* (2009) Massive point defect redistribution near semiconductor surfaces and interfaces and its impact on Schottky barrier formation. *Physica*, **B404**, 4768.

24. Brillson, L.J. (2013) *Semiconductors and Semimetals: Oxide Semiconductors*, Vol. 88 (eds Svensson, B.G., Pearton, S.J., and Jagadish C.) SEMSEM, Academic Press, Burlington, p. 87.

Further Reading

Brillson, L.J. (1982) The structure and properties of metal–semiconductor interfaces, *Surf. Sci. Rep.*, **2**, 123.

Poate, J.M., Tu, K.N., and Mayer, J.M. (1978) *Thin Films – Interdiffusion and Reactions*, The Electrochemical Society, Wiley-Interscience, New York.

5

Ultrahigh Vacuum Technology

Many of the techniques used to characterize electronic material surfaces and interfaces require *ultrahigh vacuum* (UHV) environments. These conditions allow measurements of surface atomic properties over periods of hours without air molecules or other contaminants contacting these surfaces and altering those properties. Furthermore, techniques involving electrons and ions need UHV conditions so that these particles can travel to and from a probed specimen without being scattered by gas molecules. Besides reducing the number of transiting particles, such scattering alters their energy and momenta, which often contain key physical information.

5.1 Ultrahigh Vacuum Chambers

5.1.1 Ultrahigh Vacuum Pressures

In order to maintain clean surfaces over the course of a few hours, UHV chambers must maintain pressures in the range of 10^{-10} Torr or below, where 760 Torr = 1 atmosphere (atm). Alternatively, 1 Newton (N) m^{-2}(= 1 Pascal(Pa)) = 1×10^{-5} bar = 7.5×10^{-3} Torr. For N$_2$ molecules, which comprise 78% of air, the momentum transfer associated with a pressure of 10^{-6} Torr yields an arrival rate of $\sim 3 \times 10^{14}$ cm^{-2} s^{-1} or 1 monolayer/second for a typical semiconductor (GaAs) surface atom density $1/a_0{}^2$ with lattice constant $a_o = 5.65 \times 10^{-8}$ cm and a square lattice mesh. The probability of these molecules sticking to the surface is termed the "*sticking coefficient*" S. For metals, $S \cong 1$ while for semiconductors, S can range from unity to 10^{-6} or less, depending on the adsorbate and its bonding to the substrate. For unity sticking coefficient $S = 1$, every impinging particle sticks to the surface. Maintaining a surface "clean" for one hour, that is, 3600 s, such that contamination is less than 10% of a monolayer requires a pressure $P \sim 1 \times 10^{-10}$ Torr. Gas exposure is measured in units of *Langmuir* (L), the dosage

An Essential Guide to Electronic Material Surfaces and Interfaces, First Edition. Leonard J. Brillson.
© 2016 John Wiley & Sons, Ltd. Published 2016 by John Wiley & Sons, Ltd.
Companion Website: www.wiley.com/go/Brillson/

corresponding to exposure of the surface to a gas pressure of 10^{-6} Torr for 1 s. Gas dosage is reciprocal so that 1 L is also equivalent to 100 s of exposure at 10^{-8} Torr, 1000 s at 10^{-9} Torr, and so on.

Modern equipment to enable preparation and analysis of specimens consists of (i) stainless steel chambers, (ii) pumps to remove air and other gaseous contaminants, (iii) pressure gauges to monitor the pressures, (iv) manipulators to move the sample remotely from one location to another inside the chamber, and (v) analytic equipment to measure the properties of the specimens before and after various processes. In order to prepare or modify surfaces, UHV systems may also include (i) evaporation sources, (ii) deposition monitors, (iii) sample heaters and temperature sensors, (iv) techniques to monitor surface crystallography and stoichiometry, and (v) gas transport and/or plasma processing for sample growth or modification.

5.1.2   Stainless Steel UHV Chambers

Sealed vessels that can reach pressures of $\leq 10^{-10}$ Torr are commonly constructed from stainless steel in order to support the weight of analytic equipment and to be assembled in modular fashion. Stainless steel components are joined together by copper gaskets sandwiched between circular flanges. Each flange has a circular "*knife edge*" that cuts into the gasket when the two surrounding flanges are drawn together by bolts. The resultant seal has negligible gas leakage into the chamber at UHV pressures. Inside the chamber should be only materials such as glass and ceramics with low vapor pressure at temperatures of several hundred degrees. Such modular systems are capable of manipulating, processing, and analyzing specimens inside the UHV environment.

To reach UHV pressures, the water molecules in air that adhere to the inside walls of the chamber must be pumped away. To accelerate this process, the system is "baked out" at temperatures of $\sim 150 - 180\,°C$ for periods of 24 h or longer. Without this "*bakeout*," pressures of only $\sim 10^{-8}$ Torr are readily attainable. The higher the *bakeout* temperature, the faster the water desorption. However, elastomer valve seals at the chamber degrade above $\sim 200\,°C$, limiting the maximum bakeout temperature.

A combination of pumps that operate in different ranges are used in order to achieve UHV pressures. Valves separate these pumps since they would otherwise act as leaks below their operating pressure for other pumps that operate down to lower pressures. Figure 5.1 provides a schematic view of a UHV system. A *scroll pump* (two interleaved scrolls or spirals with one orbiting eccentrically relative to the other) or a rotary vane oil pump reduces the chamber pressure from atmosphere to $\sim 10^{-3}$ Torr by removing gas through the roughing valve shown. *Sorption pumps* can also reduce the pressure to this range. Here zeolite pellets inside a stainless steel canister cooled by liquid nitrogen adsorb air molecules and reduce the pressure to the 10^{-3} Torr range due to their large surface area. Once this pressure is reached, the roughing valve is closed and the foreline valve is opened to the turbo pump, which can then be started. The high vacuum valve is then opened to the main chamber. Once the turbo pump reduces pressure to 10^{-6} Torr or lower, this high vacuum valve is closed and a valve from the main chamber to an *ion pump* and *titanium sublimation pump* (TSP) is opened. These pumps can reduce the pressure into the UHV range. Turbopumps with high enough pumping speeds can also reach UHV pressures.

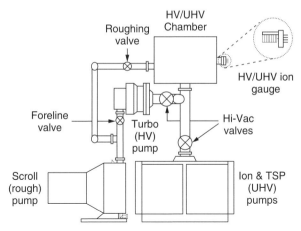

Figure 5.1 *Schematic view of a UHV system. (Reproduced with permission of Agilent Technologies.)*

5.2 Pumps

The turbopump operates as a turbine engine, pushing air through a series of vanes rotating at high speeds, typically 15 000–75 000 rpm. Depending on the pumping speed, turbopumps can reach pressures of 10^{-9} Torr or less. Figure 5.2a shows a schematic cutout of a turbopump that illustrates its operating principle [1,2]. In Figure 5.2b and c, rotating vanes are tilted at an angle to produce a higher mean free path away from the UHV side so that gas molecules are pushed away preferentially from the UHV side to the backing side of the turbopump, where a roughing pump draws them away. The pumping speed depends on the high-to-low pressure *compression ratio K*, which depends on the rotor velocity and the molecular weight of molecules passing through. Figure 5.2d shows how this compression ratio increases exponentially with speed and with the square root of *molecular weight M_W*. Thus K increases by a factor of nearly 10^6 for N_2 ($M_W = 28$) versus H_2 ($M_W = 2$), and K for N_2 increases by a factor of over 10^2 for a 50% increase in rotor velocity.

Once the turbopump backed by the roughing pump reduces the chamber pressure to the 10^{-6} range, an ion pump can reduce the pressure into the 10^{-10} Torr range. Figure 5.3a shows the interior of an ion pump schematically. A multicell array of tube-like anodes at high voltage, typically 5 kV, is sandwiched between two or more plates of Ti at ground potential that act as cathodes. Within an individual cell, Figure 5.3b shows the motion of electrons and molecules subject to both the high electric fields and a magnetic field B applied normal to the cathode plates. Electrons spiraling around the magnetic field hit residual gas molecules, ionizing them so they accelerate to the cathode, where they are trapped on the active cathode surface. The accelerated molecules may also impact the auxiliary cathode surface and knock off, that is, "*sputter*," Ti atoms, which in turn help to trap other gas ions [1].

Besides their capability to reach UHV pressures, ion pumps have the advantage of no moving mechanical parts so that they require little or no regular maintenance. However, a disadvantage of ion pumps is their limited ability to pump high gas loads for extended

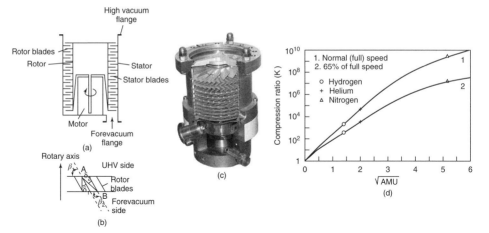

Figure 5.2   *Schematic cutout view of a turbomolecular pump: (a) rotor and stator arrangement, inclined to one another. (b) Qualitative view of the arrangement of the rotor blades with respect to the axis of rotation. (c) Cutout of turbopump showing tilted, rotating vanes and drive motor. The vane angles geometrically determine the possible paths of molecules from the UHV side to the backing side and vice versa. (d) Compression ratio of a turbomolecular pump as a function of molecular weight M at normal versus 65% of full speed. ((a) (b) Lüth, H. 2001 [1]. Reproduced with permission of Springer. (c) http://en.wikipedia.org/wiki/ Vacuum_pump. Used under Creative Commons Attribution-Share Alike 3.0 Unported. https:// creativecommons.org/licenses/by-sa/3.0/deed.en)*

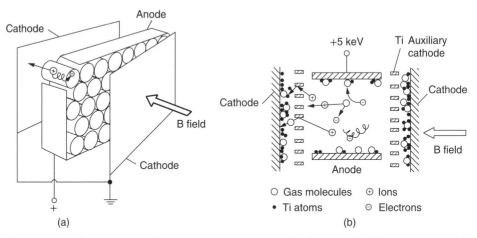

Figure 5.3   *Schematic view of an ion getter pump: (a) The basic multicell arrangement. (b) Electron and molecule motion within an individual cell. (Lüth, H. 2001 [1]. Reproduced with permission of Springer.)*

periods. To assist in reducing pressures in the 10^{-5} Torr or high range, titanium sublimation filaments can add pumping speed, operating in a similar way to ion pump cells, creating Ti atoms that combine with gas molecules and condense on the chamber walls but with higher Ti vapor formation over short (minutes) time periods. Turbo pumps can be used to pump large gas loads for extended periods. The high energy collisions produced by ion pumps also present problems for pumping large molecules, such as organics, causing fragmentation or *cracking* that can produce unwanted chemical reactions in surface chemical studies. Turbopumps can be used in tandem with ion pumps to alleviate this cracking – first, turbopumps reduce the initially high gas loads and any molecules prone to fragmentation, leaving the ion pump to maintain UHV pressures afterwards.

Prior to the advent of turbo and ion pumps, *vapor* or *diffusion pumps* were used to reach high vacuum ($10^{-8} - 10^{-9}$ Torr). Such pumps operate by heating a reservoir of organic fluid that vaporizes, rises into a series of baffles, and condenses on gas molecules, which then drop back into the reservoir, where they can be removed by a roughing pump. There is no molecular cracking, but the baffling reduces pumping speed. A *cold trap* cooled by liquid nitrogen above the baffles can minimize vapor diffusing out of the pump into the vacuum chamber but also reduces pumping speed. In general, such oil-based pumping should be avoided in growth chambers where even trace impurities can degrade the quality of electronic materials.

Cryopumps can augment turbo and ion pumping, particularly for noble and very low mass gases. As shown in Figure 5.4a, cryopumping involves the simple adsorption of gases on a condenser coil connected to a source of cryogenic gas generated by an expander module. The cryogenic fluid flows through a closed loop so there is no chemical interaction between the pump and the chamber gas. Temperatures can reach ~10 K at the expander module so that most common constituents of air and other small molecules reach saturation vapor pressures in the UHV range. Figure 5.4b shows that the main residual gases at UHV pressures are chemically-inert helium and molecular hydrogen, despite being 5×10^{-4} and 5×10^{-5} mole%, respectively, of air at atmospheric pressure. Cryopumps can accumulate

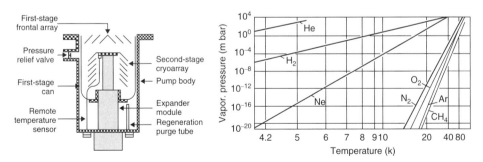

Figure 5.4 (a) Schematic cutout of cryopump showing condenser vanes and expander module used to condense residual gases inside the vacuum chamber. (Reproduced with permission of Agilent Technologies.) (b) Saturation vapor pressures of various gases versus temperature. (Lüth, H. 2001 [1]. Reproduced with permission of Springer.)

large amounts of gas before their pumping speed is reduced. At that point, one seals off the cryopump from the main chamber, letting it warm to room temperature, then pumps away the accumulated gases with a rough pump. This gas removal is critical to avoid reaching potentially explosive pressure if the cryopump warms up without a *pressure relief valve* to vent unsafe pressures.

The ultimate pressure that a pumping system can achieve depends on the volume of gas to be pumped, the desorption rate v_d of molecules per second and per unit area A from the chamber walls, the *pumping speed* S_p, and the *conductance* C_p of pipes from the pump to the chamber through which the gases are removed. S_p is the rate at which the pressure decreases with time with no gas load, expressed as volume per unit time, for example, $1\,s^{-1} \cdot C_p$ depends on the cross-sectional area and length of the pipe, analogous to electric current, but also on geometry since tube bends can impede gas flow. Thus pipe curvature can affect C_p, depending on the geometric dimensions compared with the mean free path of the gas molecules. Formulae for C_p are available for tubes and orifices that depend on cross-sectional area, tube length, bend geometries, and atomic mass for molecular flow [3]. As with Kirchoff's electrical circuit laws, C_p of pipes in parallel equals the sum of the individual conductances, that is, $C_p = \Sigma_i C_i$, and pipes in series have conductance $1/C_p = \Sigma_i 1/C_i$. A pump in series with pipes has an effective pumping speed S_{eff} defined as $1/S_{eff} = 1/S_p + 1/C_p$.

Using the ideal gas law, $PV = nk_BT$ and differentiating with respect to time, one can show that the rate of gas molecules being pumped out of the chamber is

$$dn/dt = v_d A = d/dt(PV/k_BT) = (1/k_BT)(VdP/dt + PS_p) \tag{5.1}$$

At steady state, there is no further pressure change so that $dP/dt = 0$ and thus $P = k_BT\, v_d A/S_p$, which expresses the intuitive result that base pressure decreases with decreasing temperature, decreasing desorption rate, decreasing internal area, and increasing pumping speed.

Case Study: Effective Pumping Speed

What is the effective pumping speed of a system with pump speed $S_p = 500\,1\,s^{-1}$ and conductance $C_p = 125\,1\,s^{-1}$? How much smaller can S_p be and still maintain the same S_{eff} by doubling C_p?

Answer: $S_{eff} = (S_p \cdot C_p)/(S_p + C_p) = (500\,1\,s^{-1}) \cdot (125\,1\,s^{-1})/(625\,1\,s^{-1}) = 100\,1\,s^{-1}$.

$$100\,1\,s^{-1} = (x) \cdot (250\,1\,s^{-1})/(250 + x\,1\,s^{-1}) \rightarrow x = 166.7\,1\,s^{-1}$$

Hence a pump with one third the pumping speed can be used to achieve the same effective pumping speed, emphasizing the value of maximum conductance.

5.3   Manipulators

Manipulators are required to position specimens inside the UHV chamber for various processes and measurements. This hardware must have numerous capabilities including: (i) optical access, (ii) one or more sample holders to receive samples transferred from air into UHV and (iii) 3-dimensional freedom of movement to position these samples for various surface science and other measurements. Additional manipulator capabilities include heating, cooling, mechanical cleaving, line-of-sight gas dosing, and overlayer deposition.

5.4   Gauges

Different gauges are used to monitor pressure inside the vacuum chamber. Gauges to measure high pressures ($P > 10^{-4}$ Torr) include *Pirani gauges*, which operate by monitoring thermal conductivity changes of filament currents, and *diaphragm gauges*, which monitor volume changes of a flexible diaphragm chamber due to pressure differentials. *Ionization gauges* monitor pressure in the UHV range, that is, 10^{-9}–10^{-10} Torr and below. As shown in Figure 5.5a, the ionization gauge works by measuring a current of ionized molecules created by electrons from a cathode filament passing through a positively biased anode or *collector* grid. The collector current is proportional to the number of ionized molecules and hence the gas pressure. Figure 5.5b shows a "nude" ion gauge that is inserted into the chamber, whereas a glass-encased ion gauge shown in Figure 5.5a can be connected outside the chamber.

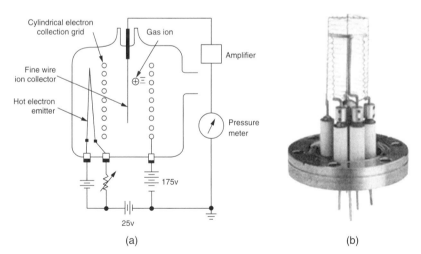

Figure 5.5 *Ionization gauge and electric circuit for measuring pressure from 10^{-4} to 10^{-10} torr and below. The electrodes can be positioned (a) with or (b) without a glass envelope. (Reproduced with permission of Agilent Technologies.)*

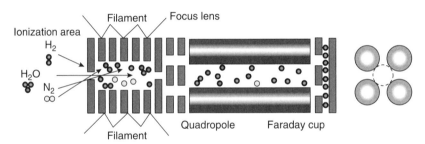

Figure 5.6   *Residual gas analyzer (RGA). Filaments ionize the molecules entering the analyzer with voltages applied to the quadrupole rods filter ions with different mass-to-charge ratio. Adapted from [4].*

5.5   Residual Gas Analysis

Besides the gas pressure inside the vacuum chamber, it is useful to know the composition of the residual gas molecules still present. For example, residual gas analyzer (RGA) detection of water vapor would signify the need for a bakeout in order to reach lower chamber pressures. The RGA can also be used as a leak detector. At chamber pressures below $\sim 10^{-7}$ Torr, a finely focused source of helium sprayed over the various seals or possibly damaged chamber components can reveal the presence of leaks by an increase in the RGA's otherwise low He molecule signal. In Figure 5.6, the RGA itself consists of filaments that ionize and focus incoming gas molecules into a quadrupole of four rods that electrostatically deflect the ions as they travel to a detector with an electron multiplier or channeltron that multiples the charge of the particles impinging on its plates. This process is analogous to that of a photomultiplier. Depending on the voltages applied to the quadrupole analyzer, only ions with a specific mass are deflected along a path that reaches the detector. This deflection process is analogous to that of electron energy analyzers to be discussed in Chapter 7 except ions rather than electrons are deflected.

Figure 5.7 shows a typical RGA spectrum for gases inside an unbaked vacuum chamber. The low-energy section of this spectrum shows clear peaks at 2, 17, 18, 28, 32, and 44 atomic mass units (AMU) that correspond to H_2, OH, H_2O, N_2, O_2, and CO_2. These molecules are characteristic constituents of air. Several lower intensity peaks correspond to various hydrocarbons. The high intensity of the AMU 18 H_2O peak indicates the presence of residual water vapor.

5.6   Deposition Sources

An important capability for UHV chambers is the deposition of new material on the specimen surface. Such materials may be categorized as: (i) surface adsorbates, (ii) metallic or insulating layers for electrical contacts and devices, or (iii) electrically- or magnetically-active materials in crystalline or amorphous form. A variety of sources and techniques are available for each type, including heated filament, crucible, electron beam

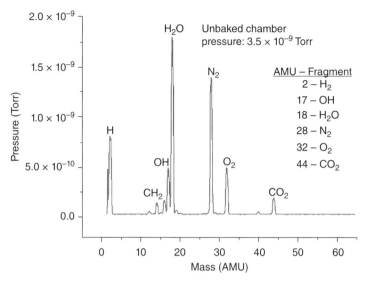

Figure 5.7   *Low mass section of an RGA spectrum showing common molecular constituents inside an unbaked vacuum chamber at a pressure of* 2×10^{-7} *Torr.*

heater, molecular beam epitaxy (MBE), chemical vapor deposition (CVD), RF sputtering and pulsed laser deposition (PLD) sources. Filaments and crucibles for the thermal evaporation of metals are easiest to install and generate minimal background gases, that is, *outgassing*, during operation. Furthermore, the stability of these sources permits control of even submonolayer film deposition. However, since the source material and heater are in direct contact, their metallurgical compatibility limits the source composition [5,6]. Thus Al alloys with W, the latter an otherwise common evaporation source material because of its high (3410 °C) melting point. Likewise, Ni and Pd alloy with refractory metal sources. For such reactive materials, ceramic containers are often used. In contrast, Au evaporates easily from most metallic sources.

Deposition of thick metallic, insulating, and magnetic layers typically involves electron beam, RF sputtering, atomic layer deposition (ALD) and PLD systems that are either integral or connected to the UHV chamber. Deposition of crystalline epitaxial films requires great deposition control of composition, which all but electron beam deposition typically can provide. RF sputtering, ALD, PLD and MBE can control thickness on the scale of single atomic layers. MBE can control individual layer composition and geometry on an atomic scale as well, assisted by *in situ* reflection high energy electron diffraction (RHEED) of the growing surface. Source material for RF sputtering and PLD can be in the form of target plates, whereas ALD, CVD, PLD, and MBE sources may be in elemental solid or organometallic form. Common MBE sources are *Knudsen cells*, composed of ceramic pockets wrapped with wires for resistive heating, and pneumatic shutters to control the sequence of elemental deposition. Accompanying Knudsen cells are *plasma generators* that direct beams of ionized gas to the sample. The ion extraction optics and nozzle maximize reactive molecules while minimizing charged species. *Leak valves* control gas flow.

Gas source MBE involves the flow of precursor gases across the substrate, where they dissociate and incorporate in the growing film.

5.7   Deposition Monitors

The control of atomic layer and film thicknesses requires *deposition monitors*. One of the most common monitors is the *quartz crystal microbalance* (QCM). The QCM consists of a thin wafer of piezoelectric material, typically quartz, sectioned along specific crystallographic planes that can generate an alternating voltage at a characteristic resonant frequency. Such resonant vibrational frequencies are analogous to those of a drumhead and are typically in the MHz range. Frequency counters that measure the AC voltage from the QCM have 1 Hz accuracy so that their precision can be equivalent to small fractions of a single atomic layer. The relationship between thickness and *vibration frequency f* depends on the specific piezoelectric crystal properties and the *density ρ* of the deposited material:

$$\Delta f = -f^2 \rho \Delta T / (N \rho_Q) \tag{5.2}$$

where ρ and ρ_Q are the densities of the deposited material and the crystal oscillator, respectively, ΔT is the thickness of deposited material, and N is the *crystal frequency constant*. For an "AT-cut" quartz crystal wafer, $\Delta f = \rho \Delta T / 1.77$, where ΔT is in Å units. Therefore for a frequency precision of 1 Hz, thickness precision can be much less than a single monolayer. Thus frequency changes per monolayer for Au, Al, and Ga are 44.4, 6.20, and 15.02 Hz, respectively.

Case Study: Quartz Crystal Oscillator

Calculate the frequency change corresponding to a 100 Å deposit of Pb on a quartz crystal oscillator. The density of Pb is $\rho = 11.35$ g cm^{-3} and the Pb lattice constant $a_0 = 4.950$ Å. How many Hz per Pb monolayer does that correspond to?

Answer: $\Delta f = \rho \, \Delta T / 1.77 = 11.35 \, \Delta T / 1.77 = 11.35 \times 100 / 1.77 = 641.2$ Hz
Number of monolayers $= 100$ Å$/4.95$ Å $= 20.2$

641.2 Hz$/20.2$ monolayers $= 31.7$ Hz/monolayer

Alternative methods of monitoring thickness include: (i) counting RHEED oscillations as epitaxial growth proceeds, monolayer by monolayer (see later chapters), (ii) monitoring the vapor stream by a residual gas analyzer (RGA) or (iii) measuring the intensity of laser-induced optical luminescence from the vapor stream. The RHEED beam can also excite X-ray photoemission, a novel method of monitoring composition during growth. The RHEED beam may also generate cathodoluminescence, which is sensitive to crystal stoichiometry and lattice imperfections on a parts per billion scale. The measurement of crystal stoichiometry and thickness is critical in order to produce high quality compounds, multilayer films, and quantum-scale superstructures with atomic abruptness and desired electronic properties.

5.8   Summary

A wide range of equipment is now available with which to prepare and study electronic material surfaces and interfaces in an atomically clean environment over extended periods of time. These tools include chambers, pumps, manipulators, gauges, and sources to alter surfaces and deposit new materials, free of unintentional adsorption or other impurities. This equipment makes it possible to design, create, and study material structures that can extend our understanding of surfaces and interfaces on an atomic scale.

5.9   Problems

1. At atmospheric pressure, room temperature and unity sticking coefficient, at what rate $(cm\ s^{-1})$ would a layer of air molecules grow on your body? Does this happen? Why or why not?
2. Calculate the room temperature pressure at which one monolayer of pure oxygen forms on the GaAs(100) surface in one second.
3. (a) Calculate the effective pumping speed of a system with a 400 1 s^{-1} pump linked to a chamber through a pipe a pipe and orifice of conductance 100 1 s^{-1}.
4. Cr evaporation onto a specimen and a quartz crystal oscillator with crystal frequency constant 1670 kHz mm results in a frequency change of 250 Hz. Assuming uniform coverage, how many monolayers thick is the Cr layer?
5. Pd is evaporated onto a clean surface using a quartz crystal oscillator with AT-cut quartz crystal frequency constant $N = 1670$ kHz mm, quartz density $\rho_Q = 2.65$ g cm^{-3}, and a frequency $f = 4.99 \times 10^6$ Hz, what is the change in frequency required to deposit 20.0 Å of Pd?
6. In molecular beam epitaxy, the flux density J (molecules cm^{-2} s^{-1}) from a Knudsen cell is given as $J = Ap \cdot \cos\theta / \pi L^2 (2\pi m k_B T)^{1/2}$, where A is the cell aperture, p the equilibrium vapor pressure in the cell, L the distance between the cell and the substrate, m the mass of the effusing species, and T the cell temperature. For Al atom deposition on a GaAs(100) substrate with unity sticking coefficient located 40 cm from a Knudsen cell, what is the maximum deposition rate for an aperture of 1 cm^2 and a cell temperature of 1000 °C? The equilibrium vapor pressure at this temperature is 10^{-4} Torr.

References

1. Lüth, H. (2001) *Solid Surfaces, Interfaces, and Thin Films*, 4th edn, Springer, New York, Panel I.
2. http://en.wikipedia.org/wiki/Vacuum_pump.

3. Holkeboer, D.H., Jones, D.W., Pagano, F., and Santeler, D.J. (1993) *Vacuum Technology and Space Simulation*, American Society Classics, American Insitute of Physics, New York.
4. http://www.stechoriba.com/rga%20principle.jpg.
5. Kern, W. (1991) *Thin Film Processes II (Vol. 2)*, Academic, New York.
6. https://www.rdmathis.com/PDF/Thin-Film-Evaporation-Source-Ref-Guide-watermark.pdf

Further Reading

Lafferty, J.M. (ed.) (1997) *Foundations of Vacuum Science and Technology*, Wiley Interscience, John Wiley & Sons, New York.

6

Surface and Interface Analysis

6.1 Surface and Interface Techniques

A wide range of experimental techniques are available to study the physical properties of electronic materials' surfaces and interfaces. The past half century has witnessed an explosion in the creation of new materials and material structures that have electronic applications. Thus it is now possible to create materials structures with atomic layer precision that have unique properties not found in nature. Researchers have recognized the impact of atomic-scale electronic structure on the electrical properties of these materials measured on a lab scale. Accordingly, they have developed techniques to measure local electronic structure, to monitor and control the effects of interfacial species, and to assess the properties of material structures, particularly at the nanoscale, designed theoretically and fabricated by various techniques. The continuous, rapid advances in the manufacture of optoelectronics and microelectronics have been a prime motivator of this research.

Table 6.1 lists the vast majority of techniques that provide information about surfaces and interfaces. Many of these techniques are widely used, while others that measure more refined properties and involve highly specialized equipment are less so. More than one technique can measure a particular physical property, yet each technique may have particular limitations. The availability of multiple techniques permits researchers to compare results between more than one technique, thereby avoiding possible artifacts of a particular technique in order to reach self-consistent conclusions.

Most of the techniques listed require UHV conditions to avoid ambient contamination during measurement as well as to allow excitation of the specimen under study. A few of the techniques require access to major instrumentation facilities, such as synchrotron storage rings, in order to excite their specimens with radiation not available in typical research labs. Several techniques listed can be adapted for both surface and interface measurements. For example, *static secondary ion mass spectrometry* (SSIMS), *Auger electron spectroscopy* (AES), and *cathodoluminescence spectroscopy* (CLS) can all measure features of the top

An Essential Guide to Electronic Material Surfaces and Interfaces, First Edition. Leonard J. Brillson.
© 2016 John Wiley & Sons, Ltd. Published 2016 by John Wiley & Sons, Ltd.
Companion Website: www.wiley.com/go/Brillson/

Table 6.1 *Surface/interface techniques and physical properties measured*

Technique	Surface Information
Auger electron spectroscopy (AES)	Chemical composition, depth distribution
X-ray photoemission spectroscopy (XPS)	Chemical composition and bonding states
UV photoemission spectroscopy (UPS)	E_F vs. E_C and E_V, Φ_M, valence band
Soft X-ray photoemission spectroscopy (SXPS)	Chemical composition, bonding, E_F vs. E_C and E_V, Φ_M, valence band states
Constant initial (CIS) & final (CFS) state spectroscopies	Empty states above E_F
Angle-resolved photoemission spectroscopy (ARPES)	Atomic bonding symmetry, Brillouin zone dispersion
Surface extended X-ray absorption Fine Structure (SEXAFS)	Local surface bonding coordination
Inverse photoemission spectroscopy (IPS)	Unoccupied surface state and conduction band states
Laser-excited photoemission spectroscopy (LAPS)	Band gap state energies, symmetries
Low-energy electron (LEED) & positron (LEPD) diffraction	Surface atomic geometry
X-Ray diffraction (XRD)	Near-surface and bulk atomic geometry
Surface photovoltage spectroscopy (SPV or SPS)	Gap states energies vs. E_C and E_V, Φ_M, qV_B
Infrared absorption spectroscopy (IR)	Gap state energies, atomic bonding and coordination
Cathodoluminescence spectroscopy (CLS)	Gap state energies, compound/phase band gap
Photoluminescence spectroscopy (PL)	Near-surface compounds, bandgap
Surface reflectance spectroscopy (SRS)	Surface dielectric response, absorption edges
Spectroscopic Ellipsometry (SE)	Composition, complex dielectric response
Surface photoconductivity spectroscopy (SPC)	Surface/near-surface gap state energies
Static secondary ion mass spectrometry (SSIMS)	Elemental composition, lateral mapping
Ion beam scattering spectrometry (ISS)	Energy transfer dynamics, charge density
Scanning tunneling microscopy (STM)	Atomic geometry, step morphology, filled and empty state geometry
Atomic force microscopy (AFM)	Morphology, electrostatic forces
Kelvin probe force microscopy (KPFM)	Φ_M, qV_B vs. morphology
Magnetic force microscopy (MFM)	Magnetic moment and polarization
Scanning tunneling spectroscopy (STS)	Bandgap states, heterojunction band offsets
Field ion microscopy (FIM)	Atomic motion, atomic geometry
Low energy electron microscopy (LEEM)	Morphology, atomic diffusion, phase Transformation, grain boundary motion
Total external X-ray diffraction (TEXRD)	Lattice structure, strain

Table 6.1   *(continued)*

Technique	Surface Information
Low-energy electron loss Spectroscopy (LELS)	Chemical reactions, plasmon, interband excitations
Depth-resolved cathodoluminescence spectroscopy (DRCLS)	Sub-surface gap states, interface compound/phase band gap
Cross-sectional cathodoluminescence spectroscopy (XCLS)	Heterojunction band offsets, interface states, Φ_M, qV_B
Ellipsometry	Sub-surface or interface dielectric response
Raman scattering spectroscopy (RS)	Interface bonding, strain, band bending
Confocal Raman spectroscopy (CRS)	Sub-surface bonding, strain
Rutherford backscattering spectroscopy (RBS)	Near-surface, bulk atomic symmetry, composition, depth distribution
Secondary ion mass spectrometry (SIMS)	Subsurface/interface elemental composition, depth distribution, lateral mapping
Ballistic electron energy microscopy (BEEM)	Barrier heights, heterojunction band offsets, barrier height lateral mapping
Scanning capacitance microscopy	Sub-surface doping, capacitance
Cross-sectional Kelvin probe force microscopy (XKPFM)	Φ_M, qV_B across interfaces (requires UHV)
High resolution transmission electron microscopy (HRTEM)	Interface lattice structure
Electron Energy Loss Spectroscopy (EELS)	Gap state energies, atomic bonding, lattice symmetry

few monolayers, yet each can measure similar features as a function of depth below the surface and through interfaces located below.

The vast majority of techniques listed can be organized in terms of an excitation versus measurement matrix. Most of these techniques involve electron, photon, or ion excitation. Similarly, electrons, photons, and ions carry the spectral information that these excitations generate. Table 6.2 represents this matrix of excitations and spectroscopies.

Several of these techniques are available with excitation beams focused on a nanometer scale. These include confocal *photoluminescence* (PL) and *Raman scattering* (RS), electron microscope-based CLS, electron energy loss spectroscopy (EELS), and AES, and focused ion beam secondary ion mass spectrometry (SIMS). The remaining techniques listed are scanned probe techniques, most of which employ atomically sharp tips as measurement devices. They detect either tunneling electric current (scanning tunneling microscopy (STM, BEEM), force-induced cantilever displacements (AFM, MFM), or capacitance displacement currents (KPFM, SPS). These techniques are described in chapters devoted to photon, electron, and ion excitation, scanned probe, and related techniques with particular emphasis on those contributing significantly to our understanding of electronic surfaces and interfaces.

Table 6.2   *Representative surface and interface spectroscopies*

	Photons Out	Electrons Out	Ions Out
Photons In	– Photoluminescence – Electroreflectance – Ellipsometry – Raman Scattering	– UV photoemission – X-ray photoemission – Soft X-ray photoemission – CIS, CFS, SEXAFS	– Photodesorption – Desorption induced electron transition
Electrons In	– Cathodoluminescence – Energy dispersive X-ray – Inverse photoemission	– Auger electron – Low energy electron loss – Low, Reflection high energy electron diffraction	– Field ion microscopy
Ions In	– Ion-induced luminescence	– Ion neutralization	– Rutherford backscattering – Secondary ion mass – Helium backscattering – Ion scattering

6.2   Excited Electron Spectroscopies

Many of the techniques listed in Tables 6.1 and 6.2 are excited electron spectroscopies that have formed the core of surface science research. Figure 6.1 schematically illustrates the excitation processes involved in four of these leading surface science techniques. The horizontal lines represent energy levels corresponding to E_{VAC}, E_F, and atomic core levels 1, 2, and 3. Hatched regions labelled V in these diagrams represent a range of closely spaced energy levels of the highest lying states filled with electrons.

UV photoelectron spectroscopy (UPS) involves photons with energy $h\nu$ incident on a solid surface that excite a high lying electron in V to a state above E_{VAC} (solid circle), leaving behind a hole (empty circle). An electrostatic analyzer collects the excited electron and measures its kinetic energy. In the UPS panel, the UV photon has only enough energy to excite electrons from the highest lying energy levels. In the *X-ray photoelectron spectroscopy* (XPS) panel, photons provide enough energy to excite electrons from deeper core levels of an atom to escape the solid, here exciting photoelectrons from levels 1, 2, and 3, leaving behind core level holes. The *kinetic energy E_K* of an excited electron leaving the atom and the solid is characteristic of that particular atom. This E_K is related to the *core level binding energy E_B* according to:

$$E_K = h\nu - E_B - (E_{VAC} - E_F) \tag{6.1}$$

Hence, the more tightly bound an electron is to the atom's nucleus, the higher is E_B and the lower is E_K for a given $h\nu$. A spectrum of photoelectron intensity versus E_K yields a

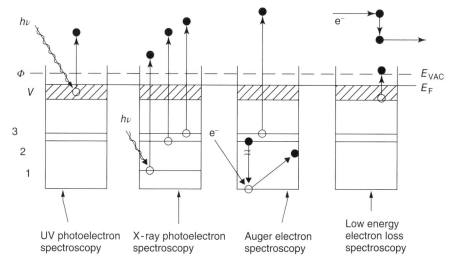

Figure 6.1   *Schematic energy band level diagram illustrating the four major excitation processes involving free electrons. (Brillson 2010. Reproduced with permission of Wiley.)*

set of peaks at energies equal to the difference between the incident photon energy $h\nu$ and the core level energy E_B, minus the additional energy to leave the solid, $E_{VAC} - E_F$, where binding energies E_B are defined with respect to E_F.

The *Auger electron spectroscopy* (AES) panel shows the 3-electron Auger process. An incident electron collides with and scatters a core electron, which leaves behind a core hole. A second, higher lying electron drops into the core hole, releasing energy to a third electron, which exits the solid. This Auger electron is also characteristic of the particular atom. For the energy levels 1, 2, and 3, kinetic energy E_{123}^{Auger} depends directly on the energy levels involved according to:

$$E_{123}^{Auger} = E_1(Z) - E_2(Z) - E_3(Z + \Delta) - E_{VAC} \qquad (6.2)$$

where $E_1(Z)$, $E_2(Z)$, and $E_3(Z)$ are the energies of the three core levels for elements with atomic number Z and Δ is the change in core level energy E_3 due to the screening of the atom's nucleus by one less electron. While AES involves a more complex process than XPS, it has several notable advantages: (i) Signal strengths are high because electron beam fluxes are generally higher than photon fluxes, permitting more rapid data acquisition and analysis; (ii) AES signals are characteristic of particular elements. The combination of high signal strength and rapid elemental identification enables rapid depth profiling and lateral mapping; (iii) electron beams can be focused down to the nanometer scale or below, permitting chemical analysis of grain boundaries, nanostructures, and even quantum-scale structures.

The last panel illustrates the *low-energy electron loss spectroscopy* technique (LEELS). Here, an electron incident on a solid surface loses some of its energy to electrons in the valence band or to energy levels just below. These excited electrons are excited to empty states in either the conduction band, the band gap, or the continuum of levels above E_{VAC}. The energies lost by exciting these transitions reduce the energy of the primary electron beam. All four of these excitation processes involve exiting electrons with energies that can be related to the energy levels within the atom and the solid. Some of these excited electrons lose energy by collisions within the solid and any other *inelastic* (non-energy conserving) processes that occur as the electrons travel from the excited atom to the solid's surface. The energies of such scattered or *secondary electrons* form a smoothly varying continuum with energies ranging from the incident beam energy down to zero. Because of their continuous nature, these can be readily distinguished from the peak features of the unscattered electrons.

6.3   Principles of Surface Sensitivity

All four of the surface science spectroscopies represented in Figure 6.1 involve excited electrons within a solid. In order to be detected, these electrons must first reach the free surface and escape into vacuum. As these electrons move through the solid, they scatter by collisions with the lattice, losing energy to plasmons, lattice vibrations, and electron–hole creation. The distance over which electrons can travel without scattering is highly dependent on their kinetic energy – a result of Coulomb scattering with the lattice and these energy loss mechanisms [1].

For kinetic energies in the range below a few thousand electron volts, experimental measurements of electron scattering lengths λ_e yield characteristic values shown in Figure 6.2. Here each data point represents a different element. While λ_e varies between different

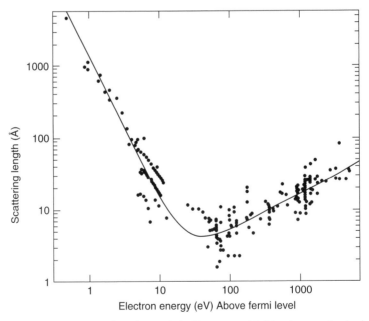

Figure 6.2 *Electron scattering length in solids. Electrons with kinetic energies in the range of 50–100 eV escape elastically only from the outer few Å of the surface. (Seah, M.P. and Dench, W.A. 1979 [2]. Reproduced with permission of Wiley.)*

elements, Figure 6.2 exhibits two key features: (i) there is a pronounced minimum in escape depth in the range 50–100 eV and (ii) this minimum corresponds to escape depths of only a few Å. What this means is that only excited electrons from the outer few Å of a solid can escape the solid without scattering. It is these electrons and the energy and momentum information they carry that the spectroscopies pictured in Figure 6.1 can measure. Conversely, those electrons that do scatter merely form the continuous background of secondary electrons. The result is that measurements of only electrons that escape without scattering are sensitive to the properties of only the outer few Å. This surface sensitivity is the basis for the most common surface science techniques.

6.4   Multi-technique UHV Chambers

Surface science experiments typically employ multi-technique UHV chambers. These chambers are designed not only to maintain a UHV environment above the sample surface but also to prepare and analyze the surface under different conditions. Typical chambers have at least one or more analytic techniques, such as XPS or AES, to measure surface composition. Besides excitation sources such as X-ray or electron guns, respectively, and an electron energy analyzer, these chambers often include the means to clean the surface *in situ*, either with an ion bombardment gun to ablate away surface contaminants, a cleaver to expose new surfaces for analysis, or gas or plasma treatments and/or thermal annealing. For studies of interfaces, these chambers may include facilities to evaporate metals on a submonolayer to monolayer scale, to chemically deposit new materials, and to measure the thicknesses of these deposits.

Additional equipment often added to these chambers include: (i) *low energy electron diffraction* (LEED) to measure patterns of surface atomic order, (ii) lasers to produce rapid, small area annealing, and (iii) electrical probes to measure the basic properties described in Chapter 3. Figure 6.3 provides a diagram and photo of a surface science system to illustrate how these various tools can be arranged inside the UHV chamber. Figure 6.3a provides a top view of the chamber, which has a *manipulator* extending down from the chamber top as shown in Figure 6.3b. This manipulator consists of a rotary flange outside the chamber held by a stage that calipers can move both by vertical and horizontal translation as well as by tilting. This rotary portion, termed a *rotary feedthrough*, links to a vertical shaft that passes from air into UHV and terminates in a horizontal platform or *carousel* to which several samples can be attached. Thus the manipulator is capable of positioning a sample in three dimensions inside the UHV environment. Figure 6.3a shows a variety of equipment arrayed around the manipulator including: X-ray, electron and ion guns, a UV monochromator, a LEED screen, crystal cleaver, a vibrating Kelvin probe driven by a speaker coil (top right of the manipulator in Figure 6.3b), and corresponding IR-near UV monochromator for *surface photovoltage spectroscopy* (SPS). The chamber also includes glass viewports to observe and position specimens for treatment and analysis.

Not shown are the pumps, electronics, and pressure gauges that maintain UHV conditions and power the various preparation and analysis techniques. Because of their weight, pumps are often placed beneath the analysis chamber – for example, below the table top pictured in Figure 6.3b. Finally, samples are attached to the manipulator, the chamber is sealed and pumped down, then baked to remove water vapor and achieve UHV pressures.

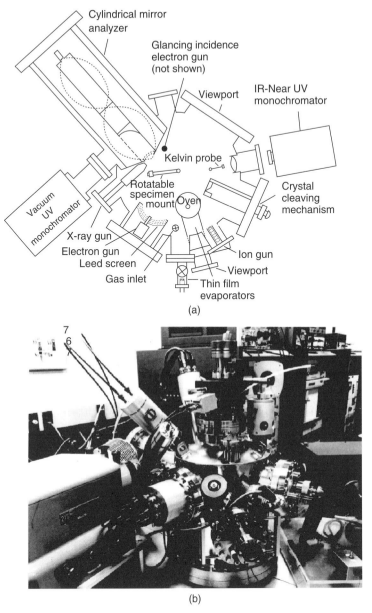

(a)

(b)

Figure 6.3   *Multi-technique UHV chamber (a) schematic top view slice through the main set of horizontal ports and (b) photo of the same chamber with a vacuum UV monochromator (left) and vibrating Kelvin probe (right). (Brillson 2010. Reproduced with permission of Wiley.)*

Interlocks that pass samples from air into UHV without exposing the UHV part of the chamber to air are often employed to avoid the need to bake out the chamber when specimens are introduced.

6.5   Summary

A wide range of techniques and tools are available to prepare specimens in a UHV environmentand to analyze the physical properties of their surfaces and interfaces, free of contamination due to air exposure. UHV chambers typically employ multiple methods that can prepare surfaces under well-controlled chemical and structural conditions. Ideally they also contain multiple techniques to characterize physical properties to provide as complete a picture of the surface or interface properties while also avoiding potential artifacts associated with a particular technique. UHV techniques that employ electron spectroscopies can achieve atomic scale surface sensitivity due to the pronounced minimum in electron escape depth at relatively low kinetic energies. Overall, UHV tools allow researchers to manipulate, process, and characterize the properties of electronic material surfaces on an atomic scale and their interfaces in three dimensions on a nanometer scale.

6.6   Problems

1. Calculate the kinetic energy of Ti 3d photoelectrons excited by a He I (21.2 eV) lamp. Calculate the kinetic energy of a Mg 1 s photoelectron excited by an Al Kα (1486.6 eV) X-ray source. Which photoelectrons escape from the top few monolayers? Assume a work function of 5 eV. (Hint: Use Track II.T7.1.)
2. Using Al Kα excitation of MgO, describe how to collect Mg 1 s electrons from just the top monolayer.
3. The O Auger KLL transition energy from the binding energies of the core level involved. (Hint: Use Track II.T7.1.) Compare with the energy measured experimentally.

References

1. Brillson, L.J. (2010) *Surfaces and Interfaces of Electronic Materials*, Wiley-VCH, Weinheim, Chapter 9.
2. Seah, M.P. and Dench, W.A. (1979) Quantitative electron spectroscopy of surfaces: A standard data base for electron inelastic mean free paths in solids. *Surf. Int. Anal.*, **1**, 2.

Further Reading

Briggs, D. and Seah, M.P. (1990) *Practical Surface Analysis*, 2nd edn, John Wiley & Sons, Inc. New York.

7

Surface and Interface Spectroscopies

This chapter introduces the major UHV techniques for studying surfaces and interfaces of electronic materials. These techniques encompass three types of excitations – photons, electrons, and ions, and use principles common to many related spectroscopies that employ these particles.

7.1 Photoemission Spectroscopy

7.1.1 The Photoelectric Effect

Photoelectron spectroscopy is one of the leading experimental techniques in surface science and in solid state research in general. As Table 6.1 indicates, it can provide a wide range of information on the physical properties of the surfaces and interfaces of electronic materials. The technique is based on the photoelectric effect, first explained by Albert Einstein in 1905 (for which he received the Nobel Prize) [1]. As shown in Figure 7.1a, light incident on a metal is capable of ejecting electrons from its surface. Figure 7.1b shows that the kinetic energy of these electrons increases linearly with increasing frequency/decreasing wavelength of the light. A biased electrode collects the electrons emitted from the metal surface in vacuum, gauging their maximum kinetic energy E_{max} by the voltage required to repel them.

No electrons are ejected for light below a minimum or *cutoff* frequency. These effects are independent of the light intensity, indicating the discrete nature of the energy transfer. Einstein recognized that the light could deliver its energy in quantized amounts, analogous to the quantized energies of lattice vibrations in a solid that Planck had postulated in 1901 to account for *black body radiation* (for which he received the Nobel Prize in 1918) [2]. These quantized packets of energy, termed *photons*, are defined by

$$E = h\nu \tag{7.1}$$

An Essential Guide to Electronic Material Surfaces and Interfaces, First Edition. Leonard J. Brillson.
© 2016 John Wiley & Sons, Ltd. Published 2016 by John Wiley & Sons, Ltd.
Companion Website: www.wiley.com/go/Brillson/

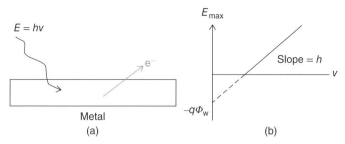

Figure 7.1   *The photoelectric effect. (a) Incident electrons with energy hν are absorbed in the metal and some are ejected with finite kinetic energy. (b) Maximum kinetic energy E_max increases linearly with slope h as frequency ν increases.*

where h = Planck's constant, equal to 6.626×10^{-34} J s and v = the light frequency, equal to the speed of light c divided by the light's wavelength λ. Defining $q\Phi_w$ as the minimum work required for an electron to escape the solid, termed the *work function*, the linear relationship shown in Figure 7.1b can be described by the relation

$$E_{max} = hv - q\Phi_w \tag{7.2}$$

7.1.2   Energy Distribution Curves

In terms of an energy band diagram such as Figure 2.3, the photoelectric effect consists of exciting an electron from an occupied state with *binding energy* E_i below the Fermi level E_F. [See Track II Table T7.1] for free atom subshell binding energies.

Figure 7.2 shows that, for incident photon energy hv, the electron exits the solid, in this case a metal, with kinetic energy E_K above the vacuum level E_{VAC}. Thus

$$E_K = hv - E_i - q\Phi_w \tag{7.3}$$

Since E_F corresponds to the highest occupied state in a metal, $E_i = 0$ so that $E_K = hv - q\Phi_w$ and measurable quantities hv and E_K yield $q\Phi_w = E_{VAC} - E_F$, the metal work function.

For a solid, electrons occupy a *density of states* (DOS) that varies with energy depending on the atomic orbitals that form states at those energies. For a metal, these states are filled up to the Fermi level. For a semiconductor, electrons fill states up to the top of the valence band E_V, as in Figure 2.3. In this case, E_F lies above E_V, that is, within the band gap. Figure 7.3 illustrates an *energy distribution curve* (EDC) generated by an incident photon of energy hv for an initial DOS. Without any experimental distortions, the initial DOS is shifted up with the same shape by energy hv. The resultant EDC is a direct indication of the DOS, modified by an additional contribution from secondary electrons that have scattered into a continuum of energies. The vacuum level E_{VAC} determines the low energy cutoff of the EDC, corresponding to kinetic energies $E_K < 0$ in Equation (7.3). For a metal, the highest electron energies appear at $hv - E_F$. For a semiconductor, they appear at $hv - E_V$, where E_V is the top of the filled valence band.

For a semiconductor with band gap E_G and electron affinity χ, E_V is defined relative to E_{VAC} and equal to $E_G + \chi$ as in Figure 2.3. Hence the width of the *EDC* (ΔEDC) shown in

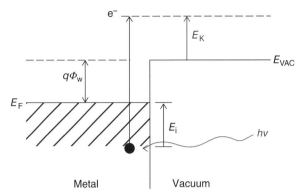

Figure 7.2   *The photoexcitation process for an electron from an energy level E_i below E_F in a metal into vacuum with kinetic energy E_K by photon hv.*

Figure 7.3 is $hv - (E_G + \chi)$ and for a semiconductor with a known band gap, measurement of ΔEDC provides a direct measurement of electron affinity. Likewise, measurement of the experimental E_F from a metal valence band and E_V from the highest energy of the semiconductor valence band yields $E_F - E_V$. UV photoemission spectroscopy (UPS) with helium lamp 21.2 eV (He I) or 40.8 eV (He II) excitation is well suited for these measurements, given their high cross sections for exciting valence electrons (see below), the easily measured ΔEDC ranges produced, and the narrow photon line widths that yield well-defined valence band edge and cutoff energies.

7.1.3   Atomic Orbital Binding Energies

X-ray photoemission spectroscopy (XPS) provides the higher energies needed to excite core level electrons out of the solid. Figure 7.4 illustrates how excitation of both core levels and valence band electrons yields the corresponding EDC features. The energies of features in an XPS spectrum are characteristic of specific elements, affording chemical characterization of the solid's near-surface region. Differences in the chemical bonding environment of these elements can produce small changes in their energies. These *chemical shifts* are useful in determining the nature of compound formation, adsorbate charge transfer, or metallization of deposited species on a surface. The XPS energy separation between sharp core level features and the valence band edge can be used for heterojunction band offsets measurements, to be discussed in Chapter 14. Again, scattered electrons form a background that increases with decreasing kinetic energy, atop which the core level features sit.

7.1.4   Photoionization Cross Sections

A major advantage of photoelectron spectroscopy is its quantitative nature, since one can calculate the strength of photoexcitation and the density of photoexcited electron states directly from physical principles. Such calculations use quantum mechanics to evaluate the probability that a transition takes place from an initial state to a final state of the solid. In quantum mechanical terms, the optical excitation process can be described by initial

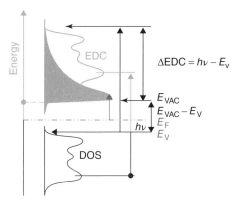

Figure 7.3   *Electron density versus energy for an initial DOS and the EDC due to excitation by a photon of energy hv. (shaded area). (Margaritondo, G. 1988 [3]. Reproduced with permission of Oxford University Press.)*

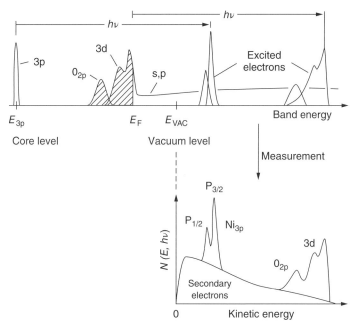

Figure 7.4   *Composite energy distribution curve of photoemitted electrons from a transition metal with adsorbed oxygen. Photons excite electrons from filled valence and core level states into empty states plus a secondary electron continuum. (Lüth, H. 2001 [4]. Reproduced with permission of Springer.)*

and final state wave functions $|\psi_i\rangle$ and $|\psi_f^v\rangle$, respectively, such that a photon of energy hv changes $|\psi_i\rangle$ into $|\psi_f^v\rangle$ or alternatively that $hv + X(i) \rightarrow X^+(f) + e^-$ for a solid with initial state $X(i)$ and final state $X^+(f)$. In this notation, $X^+(f) = |f\rangle_{solid} = |i\rangle_{solid} +$ a hole h^+ so that charge neutrality of the overall photoexcitation is preserved. The photoelectron part of $|\psi_f^v\rangle$ is a free electron state that can be described as $A_0 e^{i\mathbf{k} \cdot \mathbf{r}}$, that is, as a traveling wave with position $\mathbf{r}$ and wave vector $\mathbf{k}$ such that momentum $\mathbf{p} = \hbar\mathbf{k}$ and $\hbar = h/2\pi$. In photoemission spectroscopy, the aim is to obtain as much information as possible about $|\psi_f^v\rangle$, then process this data to extract information about $|\psi_i\rangle$, which contains the solid's physical properties before photoexcitation.

A key parameter to extract this information is the *photoionization cross section* $\sigma_{ij}(E)$, defined as the cross section for a photon of energy hv to produce a photoelectron with energy E from a solid with initial state i and final ionized state j. $\sigma_{ij}(E)$ can be expressed as

$$\sigma_{ij}(E) = (4\pi\alpha a_0^2/3g_i)\,(E + I_{ij})\,|M_{if}|^2 \tag{7.4}$$

where $\alpha =$ fine structure constant $= e^2/\hbar\, c (\simeq 1/137)$, $a_0 =$ the Bohr radius $= \hbar^2/me^2 (= 5.29 \times 10^{-9}\text{cm})$, $g_i =$ statistical weight (number of degenerate sublevels of the initial discrete state) and $I_{ij} =$ ionization energy $(E_i + q\Phi_w)$ expressed in Rydbergs, $Ry = 13.6\,\text{eV}$. The prefactor in Equation (7.4) is a constant, and for a given $hv = E$ and ionization energy, the key variable is $|M_{if}|^2$, which includes a matrix element $T = \sum_{i,f}|\langle f| \exp(i\mathbf{k}_v \cdot \mathbf{r}_u)\nabla_u |i\rangle|^2$ that represents the probability of the incident photon changing initial state i into final ionized state j. Here

$$|M_{if}|^2 = 4/(E + I_{ij})^2 \cdot \sum_{i,f}|\langle f| \exp(i\mathbf{k}_v \cdot \mathbf{r}_u)\nabla_u |i\rangle|^2 \tag{7.5}$$

where $\mathbf{r}_u =$ position coordinate of the uth electron, $\mathbf{k}_v =$ the photon's wave propagation vector, and $|\mathbf{k}_v| = 2\pi v/c$. Since $\mathbf{k}_v$ is known, the cross section calculation amounts to finding wavefunctions $|i\rangle$ and $|f\rangle$. Calculated atomic subshell photoionization cross sections are available for almost all elements and the most commonly used photon energies, both line source and soft X-ray, based on wave functions obtained theoretically, and these cross sections have proved useful experimentally [5]. A subset of these cross sections for many elements used in electronic materials appears in [Track II Table T7.2] for atomic subshell photoionization cross sections.

Photoionization cross sections can vary by orders of magnitude, depending on the incident energy hv and the particular atomic subshell. Cross sectional maxima and minima versus hv (available with continuously variable soft X-ray energies at a synchrotron) can be useful in identifying particular features with specific elements. Cross sections for valence band electrons are particularly high for low hv, as with UPS measurements.

Besides $\sigma_{ij}(E)$, the most important parameter measured by photoemission spectroscopy is the *density of states* (DOS) $\rho(E)$, the number of electron states at an energy E inside the solid, which depends on T according to *Fermi's Golden Rule*, see for example [6]. $\rho(E)$ is related to $\sigma(E)$ and T by

$$\rho(E) = [1/|T|^2]\sigma(E)/(\alpha/nN) = R(E)/|T|^2 \tag{7.6}$$

Here n is the solid's refractive index, N is the target particle density, and $R(E)$ is the rate of photoexcitation. Equation (7.6) shows that $R(E)$ is directly proportional to the density of electron states in the solid at a given energy. In turn, the number of photoelectrons actually measured is proportional to $R(E_K)$ for kinetic energy E_K of emitted electrons, where the EDC measured experimentally depends on parameters such as the instrumental resolution of the photoelectron analyzer, the energy broadening $\Delta h\nu$ of the incident excitation beam, whether or not the photoelectron has sufficient energy to escape the solid, and any distortions due to the excited electron's transport to the surface. Typical experimental resolutions are ~1 eV for un-monochromatized X-ray sources, <0.5 eV for monochromatized X-rays, ~0.1 eV for ultraviolet (UV) sources, and ≤0.1 eV for monochromatized soft X-rays.

7.1.5   Principles of X-Ray Photoelectron Spectroscopy

7.1.5.1   *Chemical Species Identification and Chemical Shifts*

XPS provides quantitative analysis of electronic materials in four ways: (i) using known binding energies and subshell photoionization cross sections, one can identify chemical species in survey spectra across a wide range of binding energies; (ii) focusing on a particular core level feature, one can use energy shifts in binding energy to identify chemical bonding environment; (iii) using the dependence of XPS features on the angle of photoemitted electrons, that is, the *take-off angle*, one can identify the elemental nature and binding geometry of near-surface species, and (iv) using rigid shifts of all XPS features, one can identify whether band bending or charging takes place. With the availability of soft X-rays and the ability to vary $h\nu$, one can also identify species from their known cross section variations as $h\nu$ varies.

Figure 7.5 illustrates two examples of atomic identification and chemical shifts. Figure 7.5a shows an XPS spectrum of Cr metal before and after oxidation. Since core level binding energy is relative to E_F, one adds the 4.5 eV known work function of Cr from Track II Table T7.2 to the 74.2 eV measured binding energy shown below to obtain a 78.7 eV binding energy relative to E_{VAC} – close to the 80 eV binding energy listed in Track II 1 Table T7.1 for the Cr 3s level. Similarly, Figure 7.5b shows an XPS spectrum of bare Si before and after oxidation. Again one adds the work function of Si, nominally 4.1 eV for n-type Si, to the 99.15 eV binding energy shown to obtain a 103.25 eV binding energy relative to E_{VAC} – close to the 104 eV binding energy listed in Track II Table T7.1 for the Si 2p level. Most work functions are in the 4–5 eV range so that binding energies relative to E_{VAC} can be approximated to within <1 eV without knowing the element in advance.

The oxidized spectrum of Cr in Figure 7.5a shows the appearance of a second peak shifted more than 4 eV to higher binding energy. This new feature at 78.2 eV corresponds to the Cr 3s level in the form of Cr_2O_3. This new bonding environment causes a charge transfer from Cr to O, increasing the Cr 3s binding energy since there are now fewer electrons shielding or "*screening*" the positively charged Cr nucleus. The lower screening increases the electron–nucleus electrostatic attraction and thereby the electron's binding energy. The presence of both peaks indicates that both metallic and oxidized species are present within the photoelectron's escape depth. Other core levels can exhibit similarly large chemical shifts, depending on the specific bonding environment. Likewise, the oxidized spectrum of Si in Figure 7.5b produces a binding energy shift of more than 4 eV as well. This new feature at 103.4 eV corresponds to the Si2p level in the form of fully oxidized SiO_2. Again the new bonding environment reflects a charge transfer from Si to O, increasing the Si

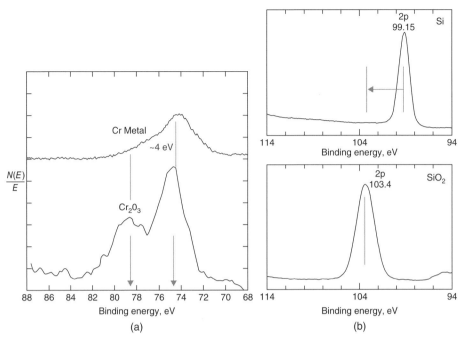

Figure 7.5 *(a) Chemical shift of Cr 3s core level with oxidation. The core level shifts by approximately 4 eV to higher binding energy with formation of Cr$_2$O$_3$. (b) Chemical shift of Si 2p core level between Si in bulk Si versus in SiO$_2$. (Wagner 1979 [7]. Reproduced with permission of Physical Electronics Industries.)*

2p binding energy. The presence of only one peak signifies that only fully oxidized Si is present within the photoelectron escape depth.

7.1.5.2   *Depth-Dependent Measurements*

Photoemission spectroscopy can provide depth-dependent information on chemical bonding within the near-surface region on a scale of a few monolayers by varying the take-off angle. Figure 7.6a provides an example of enhanced surface sensitivity with low electron take-off angle θ relative to the surface, which reduces the escape depth based on electron scattering from length λ at $\theta = 90°$ to $\lambda \sin \theta$. For the case of 1 monolayer SiO$_2$ on Si, the higher surface sensitivity at $\theta = 15°$ results in a higher XPS intensity of the SiO$_2$ layer, consistent with its surface film nature.

The attenuation of sub-surface photoemission intensity can also provide depth-dependent information. Figure 7.6b shows Al 2p core level spectra for a clean CdS surface, the same surface after deposition of first 1.5 Å Al, an additional 4.5 Å Al, and finally an additional 60 Å Al. For $hv = 140$ eV, the Al 2p binding energy of 75.5 eV, and $\chi = 4.79$ eV for CdS, the kinetic energy of Al 2p photoelectrons is then ~60 eV, corresponding to the scattering length minimum of $\lambda \simeq 4$ Å. The appearance of a single Al 2p peak at thickness $x = 1.5$ Å signifies a single Al chemical state. With a total of 6 Å Al, a second peak appears at 73 eV, corresponding to the Al 2p peak chemical shifted from Al-CdS to Al-Al bonding.

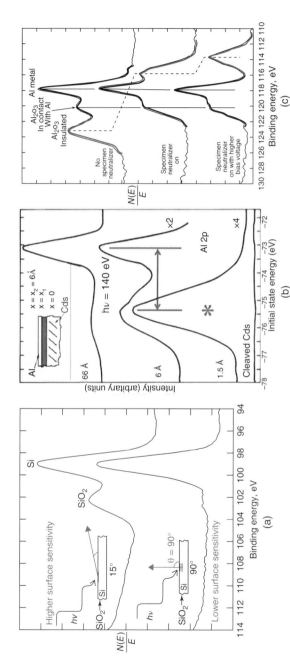

Figure 7.6 (a) The difference in XPS spectra for normal versus low take-off angle of approximately 1 SiO₂ monolayer on Si. Low XPS electron take-off angles yield significantly higher surface sensitivity. (Wagner 1979 [7]. Reproduced with permission of Physical Electronics Industries.) (b) Al 2p core level spectra with hν = 140 eV for cleaved CdS in UHV with increasing Al deposition. (c) XPS spectra of Al 2s metal core levels from insulating and conducting domains of Al₂O₃. Al₂O₃ peaks from insulating patches (red) shift with surface charging, whereas Al₂O₃ peaks from Al and conducting patches of Al₂O₃ do not shift. (Brillson 2010. Reproduced with permission of Wiley.)

With a total of 66 Å, this metallic Al peak dominates the spectrum, which indicates that the Al-CdS bonding remains localized at the metal–semiconductor interface. Quantitative analysis of the peak intensities in the 6 Å layer spectrum yields a thickness for the reacted interface layer. Assuming an unknown reacted layer thickness x_1, a total reacted plus metallic Al thickness $x_2 = 6$ Å and uniform thickness in the Figure 7.6b inset, the ratio of reacted-to-unreacted Al can be expressed as the ratio of two integrals over the corresponding depths of the two Al layers:

$$I_{\text{reacted}}/I_{\text{metal}} = \int_0^{x_1} e^{-x/\lambda(E)} \, dx / \int_{x_1}^{x_2} e^{-x/\lambda(E)} \, dx \qquad (7.7)$$

Fitting $I_{\text{reacted}}/I_{\text{metal}}$ with x_1 yields a reacted layer thickness of $\simeq 4$ Å. These results demonstrate that Al deposited on clean CdS at room temperature forms an ultrathin reacted interface layer. The presence of Al-CdS bonding at the interface indicates that charge transfer takes place across the interface. Such charge transfer can form a local dipole that can alter the measured Schottky barrier height as in, for example, Figure 3.4.

Besides chemical shifts, electrostatic charging can affect the energies of photoemission core levels for insulating or low conductivity samples. Net positive charge left by photoelectrons leaving the surface can increase the apparent binding energy if not conducted away. A neutralizer beam and/or electric bias can offset this charging and identify such insulating regions. The top spectrum in Figure 7.6c shows that a sample with both insulating and conducting patches of Al_2O_3 produces three Al 2s peaks corresponding to metallic Al, Al_2O_3 in contact with Al metal, and insulating Al_2O_3. A specimen neutralizer (an electron beam) shifts the peak corresponding to insulating Al_2O_3 to lower binding energy. If one increases the bias voltage of the neutralizer, the insulating Al_2O_3 peak energy decreases even further while the features from the conducting Al regions remain unchanged. The spectral shifts in Figure 7.6c are an indication that sample charging can be a major challenge to photoemission measurements of insulating samples.

7.1.5.3   Band Bending

Photoemission spectroscopy provides a direct method to measure band bending and Schottky barrier formation at semiconductor interfaces. For the metal–semiconductor interface, this technique involves monitoring the changes in the valence band edge and/or well-defined core level energies as metal layers are deposited on the semiconductor, layer by atomic layer. Figure 7.7 illustrates schematically how band bending affects these energy bands. Figure 7.7a shows how excitation of valence band electrons by a photon energy hv produces an electron with energy $E - E_F$ as measured by an energy analyzer grounded to the sample and therefore with the same E_F. For illustration purposes, the semiconductor is assumed to be n-type with no band bending initially so that E_F is near the conduction band E_C. Deposited metal or adsorbates at a clean semiconductor surface can induce charge transfer, thereby causing the energy bands within the surface space charge region to bend as in Figure 7.7b. In this case, the n-type band bending results in E_F moving toward to the valence band, decreasing $E_F - E_V$. Figure 7.7a shows that

$$hv = E - E_V = (E - E_F) + (E_F - E_V) \qquad (7.8)$$

so that

$$E_F - E_V = hv - (E - E_F) \tag{7.9}$$

As a result of n-type band bending, a photon hv excites a photoelectron from the valence band with higher kinetic energy $E - E_F$, decreasing $E_F - E_V$. Additional metal or other adsorbate deposition causes additional band bending and E_F movement as in Figure 7.7c. For a given $(E_F - E_v)$, n- and p-type Schottky barriers $q\Phi_B{}^n$ and $q\Phi_B{}^p$ respectively are then:

$$q\Phi_B{}^n = E_C - E_F = E_G - (E_F - E_V) \tag{7.10(a)}$$

$$q\Phi_B{}^p = E_F - E_V \tag{7.10(b)}$$

Figure 7.7d shows that, by measuring $(E - E_F)$ for a constant hv with increasing metal coverage, one can monitor the movement of E_F within the semiconductor band gap during the initial stages of band bending and Schottky barrier formation. Note that energy shifts of well-defined core levels can provide this information as well if $(E_F - E_V)$ is known initially. Indeed all core levels will exhibit a *rigid shift* in energy as E_F changes relative to E_V since the energy between each core level and E_V is a constant.

Figure 7.7e shows Si 2p core level spectra for a clean Si surface with increasing deposits of Pd. The Si $2p^{3/2}$ and Si $2p^{1/2}$ core levels both shift to higher kinetic energy with increasing Pd overlayer thickness, corresponding to a decreasing $E_F - E_V$. Significantly, the core level shifts are nearly complete with only a few monolayers of metal. In this case, band bending is complete within the first 10 Å of metal coverage [8]. Adding this $E_F - E_V$ shift to the starting value yields a final $E_F - E_V$ position that corresponds to the Schottky barrier height measured macroscopically for Pd_2Si on Si. The rapid change in band bending to its macroscopic value within the first few monolayers of metal coverage is actually quite common for metal–semiconductor interfaces, underscoring the value of surface science techniques to monitor such effects on an atomic scale.

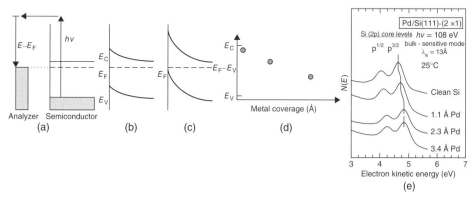

Figure 7.7 *Photoemission measurement of band bending. (a) Photoexcitation from filled valence states with analyzer-measured energy $E - E_F$. (b) E_F position in gap with band bending. (c) E_F position in gap with increased band bending. (d) $E_F - E_V$ versus metal coverage. (e) Rigid core level shifts and band bending at Pd/Si(111) interfaces. (Rubloff, G.W. 1983 [8]. Reproduced with permission of Elsevier.)*

7.1.6   Advanced Surface and Interface Techniques

Photoemission spectroscopy lends itself to a number of advanced surface and interface techniques that provide additional electronic, chemical, and structural information. Conventional XPS and UPS sources are capable of performing some of these techniques, while other techniques require soft X-rays available at synchrotron facilities. Advanced techniques such as angle-resolved photoemission spectroscopy and absorption spectroscopy have proven to be particularly useful for studying electronic materials.

7.1.6.1   Angle-Resolved Photoemission Spectroscopy

Angle-resolved photoemission spectroscopy (ARPES) can provide density of states information along different crystal orientations that can be converted into energy versus momentum (E versus k) diagrams. Such band diagrams are central to understanding the transport, optical, and magnetic properties of electronic materials. Figure 7.8a depicts a photoemitted electron collected at a take-off angle θ_p relative to the surface normal and azimuthal angle ϕ_p relative to a crystal direction. While momentum is not conserved normal to the surface plane, it is conserved in directions parallel to the plane. The kinetic energy E_K of free electrons is given by $E_K = \hbar^2 k^2 / 2m$ so that the momentum $k_{||}$ along a particular direction parallel to the plane of the crystal can be expressed as:

$$\overrightarrow{k_{||}} = (2mE_K/\hbar^2)^{1/2} \sin \phi_p \tag{7.11}$$

Hence one can obtain E versus k information by collecting photoelectrons (i) at constant $h\nu$ and varying angles relative to specific crystal directions or (ii) at constant angle and varying photon energy. For (i), a He I light source with $h\nu = 21.2\,\text{eV}$ is useful for valence band ARPES since E_K is ~16 eV for typical ionization energies, allowing $k_{||}$ measurements across the Brillouin zone with a practical range of ϕ_p angles. For (ii), a synchrotron source of soft X-rays can provide a tunable range of $h\nu$ at high enough energies to excite electrons from many deep core levels.

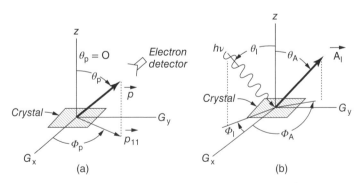

Figure 7.8   *(a) Angle-resolved photoemission geometry showing the position of electron detection and corresponding electron momentum. (b) Geometry showing angles of light propagation direction and of the polarization vector relative to the crystal plane. (Lapeyre 1979. Reproduced with permission of Elsevier.)*

Figure 7.8b illustrates the angles of incoming light θ_I and ϕ_I relative to the crystal plane and its polarization vector A if the light is polarized. Such polarization-specific orientations are useful in identifying the chemical bonding and dipole orientations of surface atoms. Polarized light is available with UV light using polarizers or with naturally polarized synchrotron radiation.

Several other advanced photoemission techniques listed in Table 6.1 are described in [9].

7.1.6.2   X-Ray Absorption Spectroscopy

The ability to tune excitation energies in the soft X-ray regime has enabled useful techniques that are specific to particular atoms. These include X-ray absorption spectroscopy (XAS) and total electron yield (TEY) spectroscopy. XAS measurements consist of monitoring the absorption of light intensity passing through the sample as hv varyies from just below to just above the absorption threshold energy of a particular atom in the solid. The absorption versus hv dependence can provide information on the atom's charge state and bonding environment as well as the energies of empty electron states above E_F. TEY measurements are based on the current generated by photoexcited electrons leaving the solid. When hv increases above the threshold energy for a core level excitation to occur, photoexcited electrons are emitted and a corresponding sample current to maintain charge neutrality is measured with a picoammeter. Here the hv dependence of core level excitation is monitored electrically rather than optically.

7.1.7   Excitation Sources: X-Ray, Ultraviolet, and Synchrotron

A variety of light sources are available for photoemission spectroscopy. UV gas discharge lamps can provide photon energies sufficient to excite electrons from the valence band and low binding energy core levels. Chief among these sources is a He discharge lamp, whose most intense emissions are at energies of 21.2 eV (He I line) and 40.8 eV (He II line). Figure 7.9a illustrates a cutaway view of a He discharge lamp, which consists of a high pressure He region in which an applied high voltage creates ionized gas molecules that emit light, a pumping line to control the gas pressure, second and third pumping lines that reduce the pressure, and a capillary through which light passes into the chamber. No window separates the lamp from the chamber since windows would strongly absorb the UV light. Differential pumping and the capillary are needed to minimize gas flow into the UHV chamber. Electrically biased wire around the insulating (glass) capillary can further reduce pressure by attracting ionized He atoms that would otherwise reach the sample.

Conventional X-ray sources operate by accelerating electrons produced by a filament so that they strike a target material at high voltage, generating X-rays. Figure 7.9b shows the inside of an X-ray gun. Typical targets are Al, which produces $K_{\alpha1,2}$ X-rays with 1486.6 eV energy and Mg, which produces $K_{\alpha1,2}$ X-rays with 1253.6 eV energy. Both lines consist of primary emissions and satellites with intensities 1–2 orders of magnitude lower. X-ray monochromators can decrease the spectral width of these lines by (i) reflecting the X-rays off bent crystals that disperse the emission lines and (ii) passing the light through a filter that strongly absorbs the higher energy satellites.

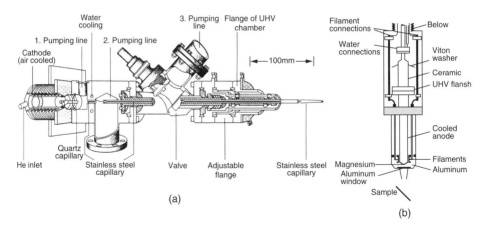

Figure 7.9   *(a) Cutaway view of a He UV discharge lamp for UV photoelectron spectroscopy. Milli-Torr pressures in the He discharge cavity decrease with differential pumping into the 10^{-9} Torr or lower range at the capillary's nozzle. (b) Basic design of an X-ray source for XPS. Filament connections to supply current and voltage pass through insulated feedthroughs to the anode region. Both Mg and Al anodes can be present in the same source. (Lüth, H. 2001 [4]. Reproduced with permission of Springer.)*

Synchrotron radiation sources available at national facilities provide a wide range of tunable photon energies. Depending on the energy of the storage ring used to generate the synchrotron radiation, photon energies can range from the infrared to the hard X-ray region. Furthermore, photon intensities can be many orders of magnitude higher than conventional light sources, enabling ultrahigh detection efficiencies, ultrafast measurement of dynamic processes, and X-ray imaging of electronic, chemical, and structural processes.

7.1.8   Electron Energy Analyzers

Several methods are available to measure the kinetic energies of photoemitted electrons. The simplest is the *retarding plate method*, which determines E_K from the bias voltage needed to repel the free electrons. More sophisticated methods provide rapid detection of photoelectrons across a range of kinetic energies and with high energy resolution. Figure 7.10 illustrates two common electron energy analyzers. The cylindrical mirror analyzer in Figure 7.10a passes electrons with energies within a narrow energy window ΔE by applying a voltage between concentric cylinders that prevents all electrons with energies outside this window from reaching an electron multiplier. Dashed lines illustrate the electron trajectory. This instrument is compact and relatively inexpensive.

Figure 7.9b illustrates a more advanced design of electron energy analyzer, consisting of electrically-biased concentric hemispheres that pass only electrons within a narrow band of energies. Dashed lines show the trajectory that the collected electrons follow. A microchannel plate and phosphor screen enable electron detection at multiple energies simultaneously, thereby increasing the speed and energy resolution of data collection.

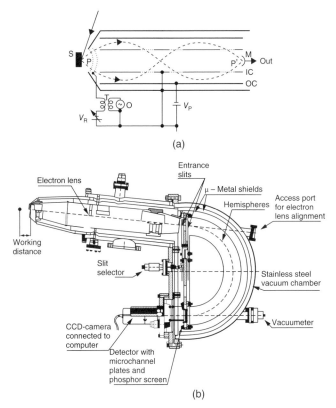

Figure 7.10   *(a) Cylindrical mirror analyzer in cross-section with electron trajectory; (b) hemispherical analyzer with electron trajectory.*

7.1.9   Photoemission Spectroscopy Summary

Photoemission spectroscopy is one of surface science's core techniques and is of considerable utility for studying electronic material surfaces and interfaces. Photoemission spectroscopy has contributed greatly to our understanding in this field because of its capability to measure electronic densities of states, chemical composition, chemical binding, and geometric structure quantitatively. Photoemission spectroscopy also provides a powerful complement to other surface science techniques in characterizing electronic materials under atomically controlled conditions.

7.2   Auger Electron Spectroscopy

Auger electron spectroscopy (AES) is the leading surface science technique based on electron excitation. The ability to focus the electron beam with high current in small areas

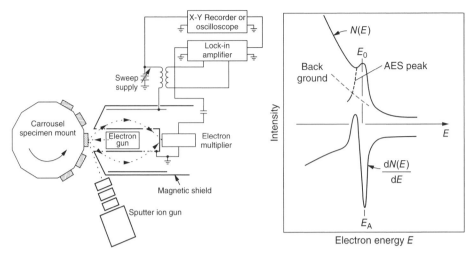

Figure 7.11 (a) Schematic arrangement of electron gun, sputter ion gun and CMA, all focused on a single spot of a sample for AES experiments [10]. (b) Top: AES peak sits on a smoothly varying background of secondary electrons. Bottom: differentiating this feature provides a dN/dE signal whose peak-to-peak magnitude defines the AES elemental intensity. (Lüth, H. 2001 [4]. Reproduced with permission of Springer.)

enables high signal strength for rapid chemical analysis both laterally and depth-wise. The ability to focus electron beams down to the nanometer scale also enables quantum-scale analysis and chemical mapping on a scale beyond the capability of photoemission. AES is used as a standard surface analytic technique for monitoring: (i) surface cleanliness in UHV, (ii) surface chemical composition, (iii) thin film growth uniformity, and (iv) depth profiles of elemental composition.

The basic experimental system for AES consists of an electron beam source, such as a glancing or normal incidence electron gun, and an electron analyzer such as one of those in Figure 7.10. In addition, an ion or "sputter" gun enables surface cleaning and depth profiling. Figure 7.11 illustrates an AES system configured to perform automated analysis of samples mounted on a carousel in UHV. Here a coaxial electron gun is mounted inside a cylindrical mirror analyzer (CMA), and an ion gun is focused on the same spot as the electron gun. The electrical circuit shown creates an oscillating voltage superimposed on a voltage ramp. This modulated voltage creates a differential component to the AES signal current $I(V + V_0 \sin \omega t) = I_0 + dI/dV V_0 \sin \omega t + \dots$ received by the electron multiplier, which enhances analog signal-to-noise since the differential signal discriminates against the otherwise large but slowly varying background due to secondary electrons (as in Figure 7.3). With computer data acquisition, differentiation of AES data can be performed digitally without the need for voltage modulation.

The electron beam enters the solid and generates a cascade of secondary electrons by impact ionization. As these electrons multiply, they lose energy by creating Auger

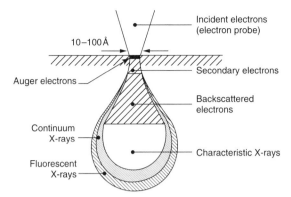

Figure 7.12 *Characteristic pear-shaped volume of secondary electrons and the various excitations they produce. (Lüth, H. 2001 [4]. Reproduced with permission of Springer.)*

electrons, secondary electrons, X-rays, plasmons, and electron–hole pairs. Once their energy decreases below the ionization energy of a free electron–hole pair, the excitation of additional electrons ends. Figure 7.12 illustrates these various processes and their spatial extent for an incident beam focused in the 1–10 nm range. In accordance with the scattering length dependence shown in Figure 6.2, only Auger electrons within the outer few nanometers of the free surface can exit the solid without scattering and losing energy. Scattered electrons become part of the broad energy continuum of secondary electrons that can create a variety of X-rays. As the electrons penetrate deeper into the solid, they spread out, forming a pear shape of excited volume that can extend for hundreds of nm or more. Only the near-surface excitations are confined to the initial focused beam dimensions. These processes form the basis for a number of electron beam spectroscopies besides AES, including *energy dispersive X-ray spectroscopy* (EDXS), *energy loss spectroscopy* (ELS), and *cathodoluminescence spectroscopy* (CLS).

For the AES excitation process shown schematically in Figure 6.1, an incident electron (or photon) collides with and knocks out a core level electron, a higher lying electron fills the hole created, causing emission of a third electron with the excess energy given up by the higher lying electron as it fills the core hole. Auger transitions can be categorized in terms of the three particular energy levels involved. Thus Figure 7.13 illustrates a variety of AES processes that can occur. These examples are labeled in terms of both their KLM nomenclature, which refers to atomic shells as well as s and p notation that indicates shell splitting into subshells by angular momentum and spin. Figure 7.13a shows a KL_1L_2 transition involving an L_1-subshell electron filling aK-shell hole with the energy released to an L_2-subshell electron that exits the solid. Figure 7.13b shows an $L_1M_1M_2$ transition involving an M_1-subshell electron filling an L_1 hole with the energy released to another M_1-subshell electron. Figure 7.13c shows an L_2-subshell electron filling an L_1-subshell hole with the energy released to an M_1 electron that escapes the solid. Finally Figure 7.13(d) shows an L_3VV transition involving a valence electron filling an L_3-subshell hole, exciting

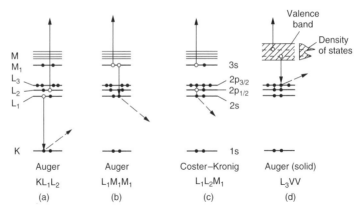

Figure 7.13   *Atomic-level representation of the Auger electron processes: (a)* KL_1L_2 *Auger, (b)* $L_1M_1M_2$ *Auger, (c)* $L_1L_2M_1$ *Coster Kronig, and (d)* L_3VV *Auger. (Lüth, H. 2001 [4]. Reproduced with permission of Springer.)*

another valence electron to leave the solid. In this case, the intensity versus energy spectrum can provide information about the valence band density of states.

   In addition to the AES processes in Figure 7.13, creation of a core hole and filling by a higher energy electron can result in emission of an X-ray, as indicated in Figure 7.12. Such X-ray emission provides chemical analysis as well, forming the basis for EDX with electron microscopes (to be discussed below).

7.2.1   Auger versus X-Ray Transition Probabilities

For the AES process, Fermi's Golden Rule again yields cross sections and rates of excitation based on the wave functions of the four electrons involved. In Figure 7.13a, they are the initial L_1 (2p), the final state K (1s), the initial state L_2 (2p), and final free electron. Instead of the photon's wave propagation vector in Equation (7.5), the rate of Auger excitation W_A involves the Coulomb potential $e^2/(r_1 - r_2)$ as the interaction potential, which in turn involves the positions r_1 and r_2 of the initial state electrons as well as the density of available electron states to which they can be excited. Calculation of W_A for a particular type of AES transition indicates that the Auger probability does not depend strongly on atomic number Z. On the other hand, the emission of an X-ray rather than excitation of a final state electron is strongly dependent on Z and competes with AES excitations. The lifetime τ of a core hole, that is, the rate $1/\tau$, depends on the rates of three types of transitions such that

$$1/\tau = W_X + W_A + W_K \tag{7.13}$$

where W_X, W_A, and W_K are the transition rates for X-rays, Auger electrons and Coster–Kronig processes. The ratio $W_X/(W_X + W_A) = w_x$ is termed the *X-ray yield,*

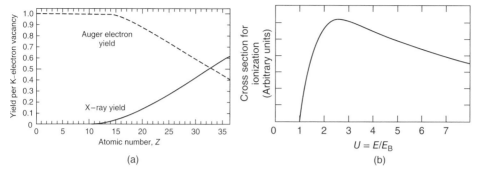

Figure 7.14   *(a) X-ray and Auger yields for de-excitation involving a K-shell core hole versus atomic number Z. Auger transitions dominate at low Z, while X-ray emission dominates at high Z [11]. (b) Ionization cross sections versus reduced energy U = E/E_B that exhibits a maximum at intermediate values. (Bishop and Riviere 2003. Reproduced with permission of American Institute of Physics.)*

which is found to be proportional to Z^4. Since W_A is relatively independent of Z, w_x increases and the Auger yield w_a decreases with Z. Figure 7.14a illustrates these relative yields versus Z, showing that AES processes are constant and dominant for low Z, while the X-ray proportion increases with Z, crossing over the Auger electron yield at $Z \sim 33$, and dominating at high Z. Figure 7.14a provides a useful guide to the most efficient method of chemical analysis using electron excitation, either AES or EDX.

From calculations of electron–electron scattering [13] the electron impact cross section σ_e for electrons with energy E more than a few hundred eV is

$$\sigma_e = \pi e^4 / E \cdot E_B = \pi(e^2)^2 / U E_B^{\,2} \tag{7.14}$$

where $U = E/E_B$ and E_B is the binding energy of an orbital electron. By definition, $\sigma_e \equiv 0$ for $U < 1$ since there is not enough energy to excite the core hole. Figure 7.14b illustrates how ionization cross sections vary with incident electron beam energy. Cross sections for ionization are non-zero for $U > 1$, reaching a maximum near $U \sim 3$–4. Typical AES chemical analysis involves detection of electrons with core levels in the energy range of <1 keV so that AES studies are often performed using incident beam energies of 2–3 kV.

7.2.2   Auger Electron Energies

The Auger transition energies can be calculated based on the energy differences of the three electrons, corrected for a change in screening due to the ionized atom. From Equation (6.2), repeated below,

$$E_{123\text{Auger}} = E_1(Z) - E_2(Z) - E_3(Z + \Delta) - E_{\text{VAC}} \tag{6.2}$$

where $E_1(Z)$, $E_2(Z)$, and $E_3(Z)$ are the energies of the three core levels for elements with atomic number Z and Δ is the change in core level E_3 due to the screening of the atom's nucleus by one less electron. For example, in the KL_1L_2 transition process pictured in Figure 7.13a

$$E^Z_{KL_1L_2} = E_K{}^Z - E_{L_1}{}^Z - E_{L_2}{}^Z - \Delta E(L_1L_2) \tag{7.15}$$

The correction term ΔE is due to many body effects, in particular the increase in binding energy of an L_2 electron when an L_1 electron is removed or that of an L_1 electron when an L_2 electron is removed. Thus

$$\Delta E(L_1L_2) = \tfrac{1}{2}[E_{L_2}{}^{Z+1} - E_{L_2}{}^Z + E_{L_1}{}^{Z+1} - E_{L_1}{}^Z] \tag{7.16}$$

$\Delta E(L_1L_2)$ represents the average increase in binding energy due to missing electrons. For a given Auger energy E, $\Delta E / E$ is a small correction, $< 1\%$.

Case Study: Auger Electron Energy for $E^Z_{KL_1L_2}$ of Iron Atoms

Using Equations (7.15) and (7.16) with core level energies relative to E_F for iron $(Z = 26)$ of $E_K{}^{Fe} = 7114$ eV, $E_{L_1}{}^{Fe} = 846$ eV, and $E_{L_2}{}^{Fe} = 723$ eV and cobalt $(Z = 27)$ of $E_{L_1}{}^{Co} = 926$ eV and $E_{L_2}{}^{Co} = 794$ eV, one obtains $\Delta E(L_1L_2) = \tfrac{1}{2}[(794 - 723) + (926 - 846)] = \tfrac{1}{2}[71 + 80] = 75.5$ eV   so that $E^Z_{KL_1L_2} = 7114 - 846 - 723 - 75.5 = 5469.5$ eV. The measured $E^Z_{KL_1L_2}$ is actually 5480 eV – within 0.2% of the calculated value.

Because there is a strong dependence of binding energies on Z, it is relatively straightforward to analyze AES spectra to determine elemental composition. Figure 7.15 shows a chart of principal Auger electron energies up to 2400 eV. Here the dots mark the energies of transitions for a given element while the dot size signifies the relative intensity of that AES feature. For each set of subshell transitions, these energies vary nearly linearly with Z, allowing the energies of measured features to be compared with those of these various AES transition energies. Figure 7.15 includes multiple dots to represent features with multiple peaks. When more than one element has AES energies that match the experimental spectrum, one can eliminate candidates using: (i) the dominant peak energy, (ii) the number and intensities of multiple features that appear close in energy to the dominant peak, and (iii) the appearance of any other subshell feature at the expected energy. Most often, a chart such as Figure 7.15 is sufficient to identify elements even with only the approximate energies shown. In this way, AES can provide a "fingerprint" associated with a given element. More precise energies are available when needed in, for example, Reference [10]. Also note that $Z = 1$ (H) and $Z = 2$ (He) are not included in Figure 7.15 since at least 3 electron states are required for AES processes.

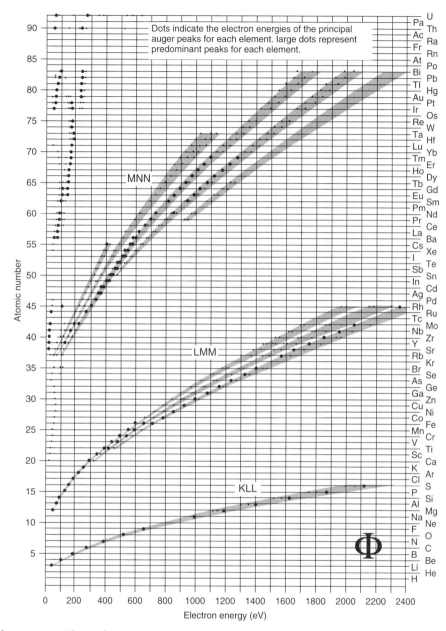

Figure 7.15 *Chart of principal Auger electron energies up to 2400 eV. (Davis 1976 [10].
Reproduced with permission of Physical Electronics Industries.)*

Case Study: Elemental Identification from Peak Energies

Consider an AES spectrum that shows dominant peaks at 68 and 1396 eV plus only two minor features at 1345 and 1380 eV. Identify the element.

From Figure 7.15, the peak at 68 eV could be associated with many elements, but the 1396 eV peak is near to those of only three: Al, Br, or Er. Br has a strong feature near 1400 eV but no strong feature near 68 eV, plus it has 2 minor features each centered near 1200 and 1300 eV. Therefore it is not consistent with the observed spectrum. Likewise, Er has a major peak near 160 eV and three minor peaks near 1060, 1220, and 1400 eV. The extra features are again not consistent with the observed AES spectrum. On the other hand, Al has peak features that agree with all four energies in Figure 7.15 and no additional features. On this basis, only Al is consistent with the observed peak features.

7.2.3   Quantitative AES Analysis

As shown in Figure 7.11b, AES spectra provide $dN(E)/dE$ or $N(E)$ features whose amplitudes provide a measure of surface composition. In order to determine surface composition

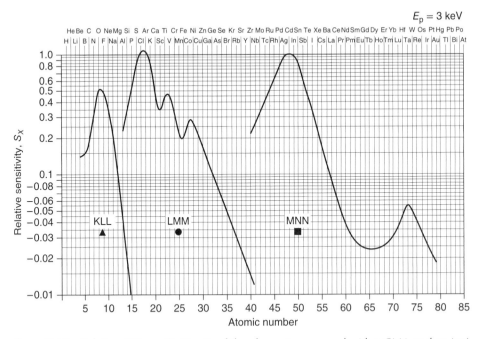

Figure 7.16   *Relative AES sensitivities S_x of the elements measured with a CMA and an incident beam energy of 3 KeV (Davis 1976 [10]. Reproduced with permission of Physical Electronics Industries.)*

quantitatively, one requires *sensitivity factors* for each major AES feature. These sensitivity factors vary by nearly 2 orders of magnitude with Z and exhibit characteristic variations as each atomic subshell is filled. Figure 7.16 displays these variations for AES spectra obtained with a cylindrical mirror analyzer and a primary beam energy of 3 keV [10]. This follows directly from the dependence of electron scattering cross section on incident electron energy displayed in Figure 7.14b. Note the high KLL sensitivities of AES for carbon, oxygen, nitrogen, a useful feature for surface science, where these constituents of air are usually to be avoided. Si, a basic microelectronic constituent, has a high LMM (but not KLL) sensitivity.

Figure 7.16 provides a general guide to these AES sensitivity factors, which permit surface chemical analysis to within ~1% accuracy. However, their precise values may vary somewhat between electron analyzers, given differences in transmission functions and energy resolution. Precision measurements of composition require sensitivity factors for the specific analyzer as well as the mode $-N(E)$ or $dN(E)/dE$, of data collection.

Case Study: Chemical Composition Analysis of Silicon Surface

Consider the 3 keV AES spectrum of a silicon surface cleaned by argon ion sputtering. The main feature at 92 eV corresponds to the LMM feature of Si according to Figure 7.15. In addition, another pronounced feature appears at 1619 eV with two less intense features at slightly lower energies, that match the $Z = 14$ intersection with the KLL curve in Figure 7.15, confirm this identification with Si. Weak features also appear at intermediate energies that can be identified with C_{KLL}, O_{KLL}, and even implanted Ar_{LMM}. The peak-to-peak intensities in Figure 7.17 provide $dN(E)/dE$ values that must be normalized by their Figure 7.16 sensitivity factors. Then all the normalized elemental values are then normalized to 100% total composition.

$$C_{X1} = [dN(E)/dE]_1/S_{X1}/\{[dN(E)/dE]_1/S_{X1} + [dN(E)/dE]_2/S_{X2}$$
$$+ [dN(E)/dE]_3/S_{X3} + \cdots [dN(E)/dE]_n/S_{Xn}\} \qquad (7.17)$$

where n = total number of different atomic constituents on the surface. The table below shows the results.

Auger Feature	Intensity	Sensitivity S_x	I/S_x	%
Si_{LMM}(92 eV)	140	0.37	378	97.6
Ar_{LMM}(220 eV)	5	1.05	4.8	1.2
C_{KLL}(273 eV)	0.5	0.2	2.5	0.6
O_{KLL}(510 eV)	1	0.5	2	0.5

Thus the silicon surface has 1.1% residual carbon and oxygen and 1.2% imbedded Ar from the ion sputter cleaning.

(continued)

(continued)

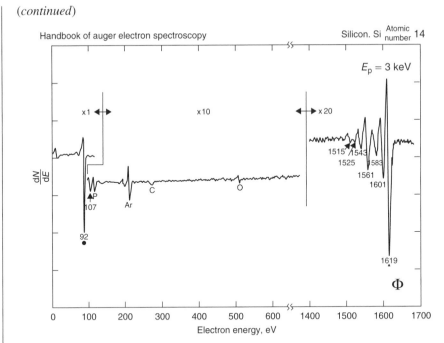

Figure 7.17   *AES spectrum of a sputtered "clean" silicon surface with 3 KeV primary beam. (Davis 1976 [10]. Reproduced with permission of Physical Electronics Industries.)*

7.2.4   Auger Electron Spectroscopy Summary

AES is a widely used technique that is often employed with other surface science techniques. It can use the same energy analyzer used for photoemission spectroscopy and requires in addition only an electron gun to generate spectra. The large differences in the energies and multiple features in AES spectra for different elements provide "fingerprints" for rapid elemental identification. Sensitivity factors enable quantitative analysis of surface elemental composition. AES's high signal strength enables rapid chemical analysis, and its ability to focus and scan surfaces permits the generation of elemental maps and depth profiles.

7.3   Electron Energy Loss Spectroscopy

Electron energy loss spectroscopy encompasses a broad range of inelastic scattering mechanisms. The technique has various acronyms depending on the loss mechanism and its corresponding energy range, as listed in Table 7.1. Here the incident electron beam loses energy in materials by exciting surface phonons, plasmons, electronic band transitions involving interface states, and core level transitions to unoccupied states above the Fermi level. Figure 7.18 displays electron energy loss features corresponding to the

Table 7.1   *Electron energy loss spectroscopies, the physical excitations they measure, and their relevant energy ranges*

Measure	Technique	Energy Range (eV)
Surface phonons (lattice, adsorbate)	High resolution electron energy loss spectroscopy (HREELS)	< 0.01 – 0.1
Interface state transitions	Low energy electron energy loss spectroscopy (LEELS)	0.1 – 10
Near-surface interband transitions	Electron energy loss spectroscopy (EELS)	0.1 – 10
Interband transitions	Electron energy loss spectroscopy (EELS)	1 – 50
Plasmons (surface, bulk)	Electron energy loss spectroscopy (EELS)	10 – 50
Core level transitions	Electron energy loss spectroscopy (EELS)	1– > 1000
Cross sectional, atomically-resolved interface transitions	High resolution transmission electron microscopy electron ELS (HRTEM-EELS)	>> 1000

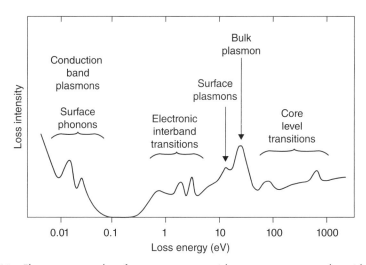

Figure 7.18   *Electron energy loss features across a wide energy range on a logarithmic scale. (Lüth, H. 2001 [4]. Reproduced with permission of Springer.)*

major excitation mechanisms over a wide dynamic range [4]. Different electron beam sources and monochromator resolving power are needed, depending on the energy range and spectral line widths of the excitations involved. Thus surface phonon measurements require energy resolutions ΔE of both analyzer and source comparable to phonon linewidths – typically <1 cm^{-1} (1 eV = 8065 cm^{-1}) but incident energies E_B of only a few eV. Interband transitions require typically $\Delta E = 0.1$ eV spectral resolution but E_B in the range of 100 eV. Core level transitions usually involve $\Delta E = 0.1 - 0.3$ eV but $E_B > 10^3$ eV for core level transitions and E_B much higher for atomically-resolved interface transitions (to be discussed later). The low energies employed in measuring surface excitations take advantage of the short scattering length of electrons in the 50–100 eV kinetic energy range. Likewise, the electron guns and energy analyzers typically used in conventional surface science for AES and XPS are also useful for ELS. Excitation of deep core level and atomically-resolved energy loss features requires much higher energies plus energy filtering of the scattered electron beam.

7.3.1   Dielectric Response Theory

Depending on E_B, the inelastic scattering mechanisms that give rise to these features can involve both short- and long-range potentials. Short-range interactions are termed *virtual excitations* since they involve an excited state of an atom in an intermediate stage of the scattering event. Long-range interactions involve scattering potentials that include the oscillating dipole fields from collective excitations such as surface and bulk lattice vibrations, charge oscillations termed *plasmons*, and the dynamic dipole moments of vibrating adsorbed atoms or molecules. These long-range interactions can be distinguished by their angular distributions, that is, by their near-specular (equal incoming and exiting trajectory angles) scattering.

The inelastic electron scattering in long-range dipole fields can be treated by a theory in which the dielectric properties of the solid are described by a complex dielectric function. The energy loss process is then described in terms of an energy transfer rate

$$dW/dt = \text{Re}\left\{ \int dr\mathscr{E} \cdot dD/dt \right\} \tag{7.18}$$

where $\boldsymbol{D}$ is a displacement field vector of the moving electron, $\varepsilon(\omega) = \varepsilon_1 + i\varepsilon_2$ is the corresponding complex dielectric permittivity of the dielectric medium, $\mathscr{E}$ is an electric field vector, and $\boldsymbol{r}$ is a displacement vector of the excitation. Thus dW/dt represents the change in density of the Coulomb field inside the solid, shielded by the dielectric medium.

Both the electric field $\mathscr{E}$ and the dielectric permittivity ε depend on both frequency ω and wave vector $\boldsymbol{q}$. Hence $\boldsymbol{D}(\omega, \boldsymbol{q}) = \varepsilon_0 \varepsilon(\omega, \boldsymbol{q}) \, \mathscr{E}(\omega, \boldsymbol{q})$. Substituting into the expression for dW/dt, one can show [4] that the energy transfer rate is proportional to the imaginary part of the inverse dielectric permittivity:

$$dW/dt \propto \text{Im}\{-1/\varepsilon(\omega, \ q)\} = \varepsilon_2(\omega)/[\varepsilon_1{}^2(\omega) + \varepsilon_2{}^2(\omega)] \tag{7.19}$$

The energy loss rate exhibits maxima as in Figure 7.18 when $\varepsilon_2(\omega)$, the imaginary component of ε, increases due to optical excitations and absorption such as interband transitions, excitons, and phonons. It also exhibits maxima when ε_1, the real part of ε, is $\simeq 0$ and $\varepsilon_2(\omega)$

is small and smoothly varying, which occurs for longitudinal collective oscillations such as plasmons. In the case of surface scattering, the semi-infinite half plane of the solid with $\varepsilon(\omega)$ is bounded by a half plane with $\varepsilon = \varepsilon(0) = 1$. The boundary condition imposed by the continuity of D results in

$$\mathrm{d}W/\mathrm{d}t \propto \mathrm{Im}\{-1/[\varepsilon(\omega) + 1]\} = \varepsilon_2(\omega)/\{[\varepsilon_1(\omega) + 1]^2 + \varepsilon_2(\omega)^2\} \qquad (7.20)$$

and the surface loss function again exhibits maxima with ε_2 maxima but also when $\varepsilon_1(\omega) = -1$ in frequency regions of small $\varepsilon_2(\omega)$. This condition determines the frequencies of surface collective excitations such as surface plasmons.

7.3.2 Surface Phonon Scattering

High resolution electron energy loss spectroscopy (HREELS) is useful for identifying adsorbed species, their binding sites and geometries at the surfaces of electronic materials. Vibrational frequencies for gas molecules bonded at surfaces typically range from a few hundred to a few thousand cm^{-1} (a few tenths of an eV) and correlate well with gas phase frequencies. For comparison, phonon frequencies of semiconductors are typically in the few hundred cm^{-1} range. Cryogenic sample cooling not only condenses gas molecules on the surface but also reduces thermal broadening of vibrational features, reducing their line width and increasing their peak intensities. Binding sites can be deduced from the frequency deviations introduced by the motion of the substrate atoms. As an example, for H bonded to Ga at the GaN surface, the motion is entirely in the adsorbate, whereas the vibration of H bonded to the much lighter N atoms deviates due to the center of mass motion of both H and N atoms [14].

HREELS can also provide molecular orientation and binding information. Equation (7.18) shows that energy loss is directly proportional to the coupling between the displacement vector D of the moving electron with the electric dipole of the excitation. Figure 7.19 illustrates the dipole orientation of molecular vibrations parallel to or normal to the solid surface. Vibrating molecules introduce dynamic dipoles as their bonds stretch and charge redistributes within the molecule. Since the incident electron beam is primarily normal to the surface, coupling and thereby energy loss are strongest to molecules vibrating normal to the surface. Oblique incidence electrons can also couple to dipoles parallel to the surface.

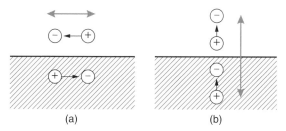

(a)　　　　　(b)

Figure 7.19 *Molecular vibrations parallel (a) or normal (b) to a metal surface. A normal incidence electron beam couples strongly with dipoles in (b) but not (a). Image forces in metals partially compensate dipoles in (a) but amplify them in (b). (Lüth, H. 2001 [4]. Reproduced with permission of Springer.)*

Figure 7.20 *Spherical shell of radius r and thickness δr within a dielectric medium (light shading). Fluctuations in r produce a restoring force at a characteristic plasmon frequency. (Brillson 2010. Reproduced with permission of Wiley.)*

These geometric coupling features introduce *dipole selection rules* for surface scattering (within the approximations of dielectric response theory) that enable orientation analysis of the molecular vibrations.

7.3.3   Plasmon Scattering

EELS is an effective tool for measuring plasmons in electronic materials. Plasmons represent the collective motion of electrons, both the charge density of valence band electrons and, for highly n-type semiconductors, conduction band electrons in the solid. The plasmon frequency of oscillation reflects the electrostatic restoring force on fluctuations in charge density within the dielectric medium. Figure 7.20 represents a shell of such charge within the uniform charge density. Here the number of electrons in a shell of radius $r = 4\pi r^2 \delta r \cdot n$ where n is the electron concentration. Fluctuations that expand or contract this sphere produce electric fields

$$\mathscr{E} = e \cdot \delta n/r^2 = e \cdot 4\pi r^2 \delta r \cdot n/r^2 = 4\pi \, ne \, \delta r \tag{7.21}$$

that act as a restoring force $F = -e\,E$ to this charge redistribution. In terms of mass m and acceleration, $F = m\,d^2\delta r/dt^2$ so that

$$d^2\delta r/dt^2 = -4\pi \, ne^2 \delta r/m = -\omega_p{}^2 \delta r \tag{7.22}$$

where the plasmon frequency $\omega_p = (4\pi \, ne^2/m)^{1/2}$ and $\varepsilon_0 = 1$. The solution to Equation (7.22) is $\delta r = r_0 e^{i\omega pt}$ so that the charge oscillates with frequency ω_p and amplitude r_0. Bulk plasmon frequencies for metals correspond to energies in the range 10–20 eV due to the high density of outer shell electrons in these solids.

Case Study: Metal Plasmon Energies

For a metal with $n \cong 10^{23}$ cm^{-3} electrons/cm^3, $\omega_p = [4\pi \, (10^{23}cm^{-3})(2.3 \times 10^{-19}$ erg cm)/(9.1 \times 10^{-28}$g)$]^{1/2} = 1.78 \times 10^{16}$ s^{-1} so that $\hbar\omega_p = (1.062 \times 10^{-27}$erg s$)$

$(1.78 \times 10^{16} \text{ s}^{-1}) = 1.89 \times 10^{-11} \text{erg} = 11.8 \text{ eV}$. A useful approximation used here is: $e^2 = 1.44 \times 10^{-7} \text{eV cm} = 1.44 \times 1.6 \times 10^{-12} \text{ erg eV}^{-1} \times 10^{-7} \text{ eV cm} = 2.3 \times 10^{-19}$ erg cm. For Mg, $\hbar\omega_p(\text{Mg}) = 10.6$ eV and for Al, $\hbar\omega_p(\text{Al}) = 15.3$ eV. [See Track II.T7.3 Metal Plasmon Energies.]

For surface plasmons, the restoring force is only half that of the bulk since only a semi-infinite half plane of dielectric is present, that is, $F_{\text{surface}} = {}^1\!/\!{}_2 F_{\text{bulk}}$ so that $d^2\delta r/dt^2 = -2\pi ne^2/m\delta r = -\omega_{\text{sp}}{}^2\delta r$ and surface plasmon frequency $\omega_{\text{sp}} = \omega_{\text{bp}}/\sqrt{2}$. In other words, $\omega_{\text{sp}} = \omega_p/\sqrt{2}$. This direct proportionality enables ready identification of surface plasmon features from their bulk counterparts.

Case Study: Semiconductor Plasmons

Plasmons are also present in semiconductors, associated with the valence electrons of the atoms comprising the semiconductor. For example, the Si lattice has 4 atoms/unit cell on each of two face-centered cubic (FCC) sublattices for a total of 8 atoms/unit cell. Given the cubic lattice constant for Si of $a = 5.43 \times 10^{-8}$ cm, there are $8/a^3 = 8/(5.43 \times 10^{-8}\text{cm})^3 = 5 \times 10^{22}$ atoms/cm^3. With 4 electrons/atom, the density of Si electrons $= 2 \times 10^{23}$cm^{-3} so that $\omega_p = \sqrt{2} \cdot 11.8 = 16.69$ eV compared with measured values of 16.4–16.9 eV.

Plasmon energies for conduction band electrons are considerably less since free carrier densities are typically in the range of $10^{15} - 10^{18}$cm^{-3}. Compared with 10^{23}cm^{-3}, plasmon frequencies and energies will be 100–1000 times smaller. Nevertheless, highly degenerate semiconductors with carrier densities of $\sim 10^{21}$cm^{-3} can yield $\omega_p \cong 1.2$ eV, a useful range for coupling light and charge for optoelectronics.

7.3.4   Interface Electronic Transitions

EELS can also provide information on electronic transitions between filled and empty electronic states. At $E_B \sim 100$ eV, EELS can probe interband and collective excitations weighted toward the Brillouin zone center (wave vector $k \cong 0$), whereas for E_B of hundreds of kV, electrons can transfer sufficient momentum during excitation to probe significant regions of a semiconductor's Brillouin zone [15]. The EELS sensitivity to both bulk and surface excitations enables one to probe interfaces as well. Here one acquires EELS of a semiconductor as a function of deposited overlayer thickness, layer by atomic layer. As Figure 7.6b showed, metal overlayers can produce local chemical and electronic changes that may be confined to only a few atomic layers. For higher thicknesses, the properties of the overlayer quickly resemble the bulk properties of the overlayer. Given the very short scattering length of electrons for $E_B < 100$ eV, low energy EELS is sensitive to only a few atomic layers. For deposited layer thicknesses above this, EELS is no longer sensitive to the interface region. Hence EELS spectra as a function of atomic layer thickness can provide information on the bulk substrate, the interface region, and the growing bulk overlayer. This sequence of measurements can then reveal unique electronic properties of the interface with suitable modeling of dielectric properties. [See Track IIT7.4Al on CdSe: A Reactive Interface.]

7.3.5   Transmission Electron Microscopy Energy Loss Spectroscopy

Electron energy monochromators are now available for the transmission electron microscope (TEM) that can provide EELS spectra on an atomically-resolved scale and with energy resolution of several hundred meV or less. Thus researchers can probe band gaps and localized states at interfaces in cross section on a monolayer by monolayer scale for comparison to theoretical models of specific localized charge redistribution. In addition, TEM-based EELS can measure the losses near core level threshold energies, which can provide characteristic "fingerprints" of that atom's charge state. The charge state of atoms near the interface versus the bulk then provides a guide to the charge transfer across the interface between specific atoms. Such information is useful for understanding many interface phenomena, including Schottky barrier formation, interface-induced magnetism, dielectric polarization, and super conductivity.

7.3.6   Electron Energy Loss Spectroscopy Summary

EELS encompasses a significant range of experimental techniques capable of measuring transitions between electronic states as well as collective excitations. At high resolution and low beam energies, HREELS can identify surface bonding and vibrational modes of surface atoms and adsorbates. At intermediate energy resolution and higher beam energies, EELS can measure bulk and surface plasmon modes, interfacial dielectric constants, and the presence of interfacial chemical reactions. At somewhat lower energy resolution and atomic-scale spatial resolution, HRTEM-EELS can measure the charge state of individual atoms and the charge redistribution to form localized electronic states.

EELS energies can be understood in terms of dielectric response theory where the energy transfer rate is proportional to the imaginary component of the solid's complex dielectric constant features. Loss features are modified near surfaces and interfaces in order to take into account electrostatic boundary conditions between materials (or vacuum) of different dielectric permittivity.

EELS has provided strong evidence for interface reactions between metals and semiconductors, detecting interfacial layers with new dielectric properties produced near room temperature with only a few monolayers of deposited metals. Coupled mode analyses of EELS features involving vacuum, metal, interfacial dielectric, and semiconductor can provide the dielectric constant of the reacted interface layer. TEM-based EELS measurements of threshold energies for core level transitions near interfaces can identify charge transfer between specific atoms on an atomic scale.

7.4   Rutherford Backscattering Spectrometry

Rutherford backscattering spectrometry (RBS) is the leading ion beam technique for measuring the absolute elemental composition versus depth of electronic materials and their interfaces on a micron scale. It has been used extensively in the development and process-ing of electronic material interfaces and devices. RBS involves charged ion scattering in a Coulomb field and provides chemical analysis of thick films nondestructively, that is, without ablating away the surface atoms. The RBS technique involves a classical scatter-ing process of high energy, for example, 2 MeV helium, ions with atoms in the solid. A simple kinematic (i.e., "billiard ball") analysis is possible since the wavelength of the inci-dent ions is much smaller than the range of the lattice's scattering potential. According to the *De Broglie relation*

$$\lambda = h/p = h/mv \tag{7.22}$$

where h is *Planck's constant*, m is the ion mass, and v it velocity. Incident ion beam energy yields velocity according to $E_{kinetic} = mv^2/2$. For light ions with $E_{kinetic} > 1$ MeV, wave-lengths λ are on the order of 10^{-4} Å, much smaller than the Å scale of atomic spacings and electrostatic scattering potentials.

Case Study: High Energy Ion Wavelength

Calculate the wavelength of a 2 MeV helium atom. Using $h = 6.63 \times 10^{-34}$ J s and $m = 4 \times 1.67 \times 10^{-27}$ kg:

$$v = [2\,E_{kinetic}/m]^{1/2} = [\,2\,(2 \times 10^6 \text{ eV})(1.6 \times 10^{-19}\text{J eV}^{-1})/(6.68 \times 10^{-27}\text{kg})]^{1/2}$$

$$= 9.79 \times 10^6 \text{m s}^{-1}$$

$$\lambda = h/mv = 6.63 \times 10^{-34}\text{J s}/(6.68 \times 10^{-27}\text{kg})(9.79 \times 10^6\text{m s}^{-1}) = 1.01 \times 10^{-4}\text{Å}$$

7.4.1   Theory of Rutherford Backscattering

In Rutherford backscattering, the positively charged ions accelerated into the specimen scatter by Coulomb repulsion from the positive target nuclei. Figure 7.21 illustrates this charged particle scattering for positively charged He^+ ions by an oxygen atom. Based on a Thomas–Fermi–Moliere (electron screening) scattering potential, He^+ ions approaching the target produce a shadow cone of He^+ ions past the oxygen atom. One can calculate the closest approach of the incoming He^+ projectile by setting its kinetic energy $E_{kinetic}$ equal to the Coulomb repulsion energy of the ion with the target oxygen atom.

$$V = (Z_1 Z_2 e^2)/(4\varepsilon_0 \pi r_{min}) = E_{kinetic} \tag{7.23}$$

where Z_1 and Z_2 are the ionic charges of the ion and target. The closest approach is then

$$r_{min} = (Z_1 Z_2 e^2)/(4\varepsilon_0 \pi E_{kinetic}) \tag{7.24}$$

Equation (7.24) shows that the higher the incident projectile energy, the closer the approach. For example, a 2 MeV helium atom scattering off a silicon atom, $r_{min} = 2 \times 10^{-4}$ Å. [See Track II T7.5 Rutherford Scattering of He^+ with Si.]

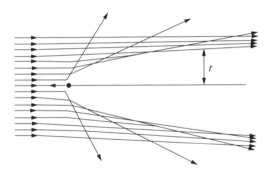

Figure 7.21   *Calculated scattering trajectories of He^+ ions moving toward an oxygen atom from the left and the shadow cone of scattering He^+ ions they produce to the right of the target. (Lüth, H. 1995. Reproduced with permission of Springer.)*

Because RBS is based on classical scattering in a central-force field, the motion of the projectile and target atom can be treated using conservation of energy and momentum, which in turn show that the energy of the scattered incoming particle is characteristic of the scattering atom. By conservation of energy, the initial and final energies must be equal. Thus

$$\tfrac{1}{2}M_1 v^2 = \tfrac{1}{2}M_1 v_1^2 + \tfrac{1}{2}M_2 v_2^2 \tag{7.25}$$

where M_1 and M_2 are the masses of the incident particle and target atoms, respectively, v and v_1 are the incident particle's velocity before and after scattering, respectively, and v_2 is the target atom's recoil velocity after the scattering event. By conservation of momentum,

$$M_1 v = M_1 v_1 \cos \theta + M_2 v_2 \cos \varphi \tag{7.26a}$$

$$0 = M_1 v_1 \sin \theta + M_2 v_2 \sin \varphi \tag{7.26b}$$

Combining Equations (7.25) and (7.26), one can solve for the ratio of incident to final particle velocities in terms of just the scattering angle θ and masses M_1 and M_2.

$$v_1/v = [\pm(M_2^2 - M_1^2 \sin^2\theta)^{1/2} + M_1 \cos \theta]/(M_1 + M_2) \tag{7.27}$$

Likewise, the ratio of the final to initial incident particle energies (for $M_1 < M_2$) is

$$E_1/E_0 = \{[(M_2^2 - M_1^2 \sin^2\theta)^{1/2} + M_1 \cos \theta]/(M_1 + M_2)\}^2 \tag{7.28}$$

$E_1/E_0 = K$ is termed the *kinematic factor*, which is determined only by the masses of the incoming particle, the target atom and the scattering angle.

Experimentally, energy E_0 is set by the incident particle acceleration while E_1 is measured by a nuclear particle detector set at various angles θ. See Figure 7.22. The detector measures energies in terms of the number of ionized electron–hole (e–h) pairs created by the incident particle (*i.e.*, by impact ionization) within the depletion region of a reverse-biased Schottky diode. The higher the particle energy, the larger the number of e–h pairs and the stronger the charge pulse generated by the incident particle. The energy resolution of the particle detector is determined by statistical fluctuation in the number of e–h pairs, typically 10–20 keV for ^{4}He ions in the MeV energy range. The number of pulses

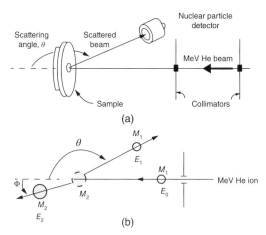

Figure 7.22 *(a) Schematic diagram of RBS scattering experiment. A collimated incident beam of He atoms scatters off a sample at scattering angle θ into a nuclear particle detector. An incident particle with mass M_1, energy E_0, and velocity v collides with a target atom of mass M_2 at rest. (b) Diagram of energy and momentum conservation in a two-particle scattering event. After the collision, the incident particle scatters at angle θ, energy E_1, and velocity v_1. The target atom recoils at angle φ, energy E_2, and velocity v_2 [16].*

generated per second determines the intensity of scattering. The pulse width must be short enough to discriminate between pulses at high scattering rates. The full width at half maximum (FWHM) is determined by the time required for an applied electric field to sweep the carriers across the particle detector's reversed biased diode.

Since M_1, v, and E_0 are known while v_1, E_1, and θ are measured, Equations (7.27) and (7.28) yield the mass M_2 of the target atom. For direct backscattering, $θ = 180°$ and Equation (7.28) becomes

$$E_1/E_0 = \{(M_2 - M_1)/(M_1 + M_2)\}^2 \tag{7.29}$$

which produces the largest change in the scattered particle energy E_1 and hence the highest possible changes between target atoms of different mass M_2.

Besides the energy ratio E_1/E_0, the intensity of the backscattered signals provides additional information. This *yield ratio* of backscattered intensities for different elements in the same target materials is proportional to the ratio of cross sections for particle–particle scattering given by [17]

$$σ(θ) = (Z_1 Z_2 e^2 / 4E)^2 / \sin^4(θ/2) \tag{7.30}$$

Thus for the same incident ion, ion energy, and scattering angle θ in Equation (7.30), the scattering cross sections for two elements with atomic numbers Z_{2A} and Z_{2B} will be proportional to $(Z_{2A}/Z_{2B})^2$. Together with the energy dependence of backscattered ions on target ion mass, this proportionality of scattering intensity on atomic number can be used to characterize both the elements present and their distribution in depth within a material.

7.4.2   Rutherford Backscattering Equipment

Figure 7.23 shows a typical geometry of an RBS accelerator. Here negative He$^-$ ions extracted from a source are accelerated through a foil stripper that converts them to He^{++} ions. These ions, focused on the sample, scatter and produce backscattered He ions that are collected by the particle detector. Light ions such as ionized He are used to minimize effects of electron screening of the target nuclei, maximize mass differences with target ions, and enable particle acceleration to high energies that increase the depth of penetration.

The combined energies and intensities of the backscattered ions provide concentration profiles of different elements as a function of depth from the free surface. The primary ions lose energy continuously as they pass through a solid, mainly by electronic excitations such as plasmon creation. Thus MeV He ions lose energy along their path at a rate $dE/dx \sim 30 - 60$ eV Å^{-1} [18]. For thin films, the total energy loss ΔE at a depth t is proportional to t.

$$\Delta E_{\text{incident}} = \int dE / dx \cong dE/dx|_{\text{incident}} \cdot t \tag{7.31}$$

so that $E(t) = E(t) = E_0 - t \cdot dE/dx \mid_{\text{incident}} t$ and the energy width of scattered incident ions of energy E_0 is $\Delta E_0 = \Delta t[S_0]$ where the *backscattering energy loss factor* S_0 depends on the energy loss rate of both the incident particle before and after scattering. [See Track II T7.6 Backscattering Energy Loss Factor].

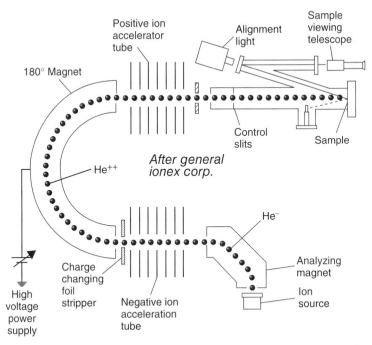

Figure 7.23   *RBS accelerator. He atoms are ionized, charge changed, accelerated to a specific high positive voltage, and focused on a sample. The backscattered He ions are energy analyzed with a nuclear particle detector.*

7.4.3   RBS Experimental Spectra

Figure 7.24 illustrates how RBS of a thin homogeneous film of equal atom concentration translates into an RBS spectrum with different scattered particle yields, energies, and energy widths. In Figure 7.24a, a binary compound film composed of a heavy and a light element masses M and m, respectively, scatters the incident particle beam with different kinetic energies and yields. In this simplified example, the film consists of equal concentrations M and m. (Figure 7.24b). However, because $M > m$, the yield ratio of M exceeds that of m by the ratio $(Z_M/Z_m)^2$. Furthermore, the highest kinetic energy K of M is larger than that of m according to Equation (7.29). Finally, the width of each element's yield distribution differs since the energy loss per unit path length dE/dx in the range of energies $K_M E_0$ and $K_m E_0$ is different. These energy loss rates can be calibrated for a given element

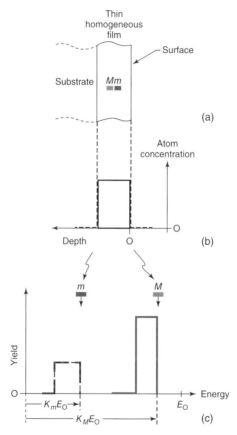

Figure 7.24   *Conversion of atomic composition versus depth for a homogeneous binary film to an RBS yield versus energy spectrum. (a) Schematic of homogeneous film composed of elements A and B with atomic masses M and m. (b) Atomic concentration for both masses m and M. (c) Yield versus energy showing energy separation and intensities for yields of masses m and M. $K_M E_0$ and $K_m E_0$ signify leading edges of scattering energies for masses M and m, respectively* [19]

of known thickness in its KE_0 energy range. Hence the continuous energy loss of a scattered particle along its path through a crystal provides a simple relation between scattering depth and energy loss.

7.4.4   RBS Interface Studies

RBS is useful for understanding the metallurgical reactions that can occur between semiconductors and metals used in microelectronics. Metals can interdiffuse or chemically react with semiconductors at their interfaces, producing undesirable electronic effects. These interfaces must be stable at the high temperatures typically used in the device fabrication process, a requirement that has become ever more important as electronic devices shrink into the nanometer scale.

Case Study: Metal–GaAs Interdiffusion at Elevated Temperature

Compare the interface chemical stability of GaAs with Au versus W overlayers. Figure 7.25 illustrates RBS spectra for the two interfaces before and after thermal annealing for 2 hours.

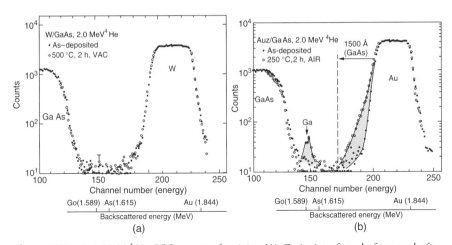

Figure 7.25   *2.0 MeV ^{4}He RBS spectra for (a) a W–GaAs interface before and after annealing at 500 °C for 2 hours versus (b) a 1500 Å Au–GaAs interface before and after annealing 250 °C for 2 hours. Even with lower annealing temperature, the Au–GaAs junction exhibits strong interdiffusion versus no interdiffusion for the W–GaAs interface. (Sinha and Poate 1973 [20]. Reproduced with permission of American Institute of Physics.)*

In Figure 7.25a, the Ga and As KE_0 are similar and appear at much lower energies than that of the much heavier W. A 500 °C anneal produces no apparent changes. In Figure 7.25b, these KE_0 are similarly below that of the much heavier Au. However,

after annealing, dramatic changes occur in both Ga and Au profiles. The spreading
of the Ga profile to higher energies signifies movement of Ga out of the GaAs into
the Au. Likewise, the expanded Au profile to lower energies indicates Au diffusion
into the GaAs over a range of 1500 Å. Together, these changes signify strong interdiffusion between Au–GaAs at 250 °C but much thermal stability between W and
GaAs at 500 °C.

Beside the extent of chemical interaction, the RBS depth profile reveals how the atoms
move at the interface. In the case study above, both Au and Ga mix in an intermediate
layer consistent with formation of a thermodynamically stable Au–Ga eutectic at elevated
temperatures. At some metal–Si interfaces, reaction products with well-defined proportions
can form. The RBS profiles can establish the stoichiometry of these compounds and how
the metal and Si atoms diffuse into each other – important information in forming abrupt
ohmic contacts.

7.4.5   Channeling and Blocking

RBS can help assess lattice crystallinity and dopant substitution based on the phenomena
of channeling and blocking. Figure 7.26 illustrates a crystal lattice viewed along a high
symmetry direction. Along such directions, the lattice appears to have "channels" between
columns of atoms through which incident particle scattering is greatly reduced. Scattering can increase by orders of magnitude along random, non-channeling directions as the
comparison in Figure 7.26b shows. Such *blocking* or non-channeling scattering increases
if the crystal lattice is either partially disordered or *amorphous*, which can both occur following ion implantation to dope the semiconductor. Subsequent annealing can improve the

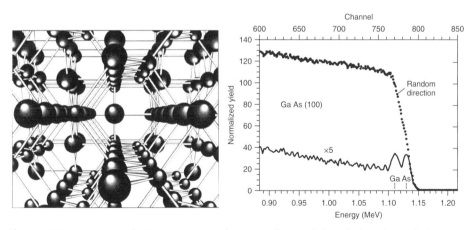

Figure 7.26   *(a) Open lattice structure of a crystalline solid and the channels between
columns of atoms. Scattering yields along these symmetry directions are relatively low. (b)
RBS spectrum of 1.4 MeV He⁺ for a GaAs (100) surface along a random non-channeling
direction versus along the surface <100> normal direction. The channeling spectrum is an
order of magnitude lower in amplitude [19].*

lattice order, and channeling measurements can then assess the recovery. Similarly, channeling along high symmetry directions can determine the efficiency of dopant substitution since the dopant blocking will be low if dopant atoms occupy substitutional sites within the columns rather than interstitial sites between the columns.

The increases Ga and As yield at their leading edges in Figure 7.26b also show that the scattering along symmetry directions occurs primarily from surface and near-surface atoms at the head of each column. Thus RBS is sensitive to the rearrangement of surface atoms, termed *surface reconstruction*. RBS studies of such reconstruction are typically performed at lower E_0 in order to minimize the ablation of atoms by the incoming particles. Such medium energy RBS studies can even be used to measure the "shadowing" of surface atoms by adsorbate atoms.

7.4.6   Rutherford Backscattering Spectroscopy Summary

RBS is a relatively non-destructive technique for measuring the composition of electronic materials and the distribution of elements versus depth from the free surface. The classical or "billiard ball" scattering follows classical kinematics so that the elements and their depth distributions can be derived from simple calculations. RBS is particularly useful for assessing lattice perfection and dopant ordering of crystalline semiconductors.

The RBS technique has been central to the development of materials and processing for microelectronics. It is particularly useful for monitoring the metal–semiconductor interfaces used for ohmic and rectifying contacts in electronic devices since it can help identify process conditions that either minimize unwanted chemical reactions or that promote reactions with desirable properties.

7.5   Surface and Interface Technique Summary

This chapter introduced the primary photon, electron, and ion beam spectroscopies that form the core of surface and interface studies of electronic materials. Photoelectron spectroscopy represents a powerful technique for measuring the surface chemical and electronic properties of surfaces and interfaces as well as the densities of electronic states of the bulk materials themselves. AES provides a high signal method to rapidly assess surface cleanliness, composition, and thin film growth uniformity in UHV, lending itself to high spatial resolution studies based on the ability to focus the electron beam to nanometer dimensions. EELS in low, medium, and high energy ranges can provide information respectively on near-surface lattice vibrations, electronic transitions unique to surfaces, thin films, and their interfaces, as well as charge transfer processes at interfaces important for ferroelectricity, magnetism, and superconductivity. The chapters to follow describe first the use of these techniques to probe physical properties of electronic materials below their free surfaces, then the scanned probe techniques that complement these tools at the nanometer and atomic level, and then an array of optical methods that extend these photon, electron, and ion techniques in new directions.

7.6 Problems

1. A clean metal film in ultrahigh vacuum is illuminated by monochromatic light. A parallel plate biased relative to the metal collects photoemitted electrons and monitors the collected current versus applied voltage. (a) For a light wavelength $\lambda = 1550$ Å incident on an Au film, what is the minimum voltage V_0 that suppresses all photoelectrons? (b) Sketch the photocurrent as a function of voltage difference between the Au and the collector plate. Sketch several curves with different light intensities. (c) For He I photons incident on this film, what is the highest energy and velocity of a photoemitted electron?

2. For a clean single crystal Si surface illuminated by He I line emission, what is the expected room temperature width of the measured energy distribution curve?

3. A semiconductor with a band gap of 2.4 eV, an electron affinity of 4.0 eV, and a bulk doping such that $E_C - E_F = 0.1$ eV. After surface treatments, valence band spectra show $E_F - E_V = 0.5$ eV. What is the band bending? Is it n-type or p-type band bending? What is the work function? What is the ionization potential? Suppose this surface shows a work function of 6.1 eV. What conclusion can you draw?

4. In order to monitor an Al–Si system by XPS using an Al K_α source, which core levels would provide the strongest photoemission signals and why?

5. You are given a GeSi alloy whose composition you are to measure using XPS with an Mg K_α source. Identify core levels with the same symmetry and the highest photoelectron cross sections in both constituents with this excitation. Calculate and compare the photoelectron cross sections for these core level electrons.

6. (a) From Track II T7.1, what core level would you use with Al K_α X-rays to measure Ni atoms on a surface? (b) For highest cross section *and* high surface sensitivity, which core level and photon energy would you use?

7. An unknown semiconductor yields Mg K_α XPS doublet peaks with kinetic energies E_{KE} relative to the vacuum level at 1103/1109, 1143/1147, 1177/1177, 1208/1209, and 1233/1234 eV. Identify this (conventional) semiconductor.

8. A ZnO surface exhibits XPS spectra with peaks at 531.15 and 533.2 eV binding energy relative to the Fermi level with deconvolved peak intensities of 4.8 and 9.2, respectively. In addition, peaks are present at 1025.35 and 283.5 eV binding energy. After oxygen plasma exposure, the low binding energy peak disappears, and peaks are now positioned at 530.4 and 1024.6 eV. Describe the surface composition, bonding, and electronic changes (a) before and (b) after surface treatment.

9. To study InP surfaces with photoemission spectroscopy, which photon source would be more effective – a 40.8 eV He II or a 151.4 Zr ζ source? Assume comparable photon intensities.

10. In an experiment to probe GaAs surfaces and interfaces with high surface sensitivity SXPS, which core levels and soft X-ray photon energies would you use?

11. In the SXPS data below, the bulk- and surface-sensitive attenuation of P 2p core levels at $h\nu = 140$ and 170 eV, respectively, is deconvolved to show the behavior of the substrate, a transition region, and a phosphide reaction product for Cr (top) and Fe

(bottom) overlayers. Calculate the electron scattering lengths from the substrate atten-
uations. How do these compare with the universal escape depth curve for $hv = 170$ eV
excitation of the P 2p core level?

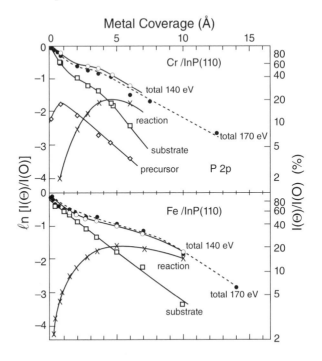

12. Pt, Au, Mo, and Al thin films ($\Phi_M = 5.65$, 5.1, 4.6, and 4.28 eV, respectively)
    are deposited in UHV on a clean, low defect semiconductor surface with electron
    affinity of 3.5 eV and a band gap of 4 eV. SXPS calibration of a metallic standard
    in electrical contact with the semiconductor places the initial Fermi level 0.1 eV
    below the conduction band edge. Rigid core level shifts with metal deposition are 1.6,
    1.5, 1.0, and 0.7 eV. Calculate and plot the Schottky barrier heights and compare to
    that expected classically. What do you conclude about any near-surface states in the
    band gap?
13. (a) Calculate the momentum of an electron collected at angle $\theta_p = 45°$ with photon
    energy $hv = 12$ eV and energy level $E_B = 5$ eV below the valence band edge E_{VBM},
    which is located 4 eV below the vacuum level. Assume $m^* = m_0$. $\mathbf{k} = \mathbf{k}_\perp + \mathbf{k}_{||}$.
14. In angle resolved photoemission experiments, momentum perpendicular to the surface
    is not conserved due to the barrier for electrons to escape the solid, that is, the work
    function Φ. Energy is still conserved so Schroedinger's equation for initial and final
    wavevectors $\mathbf{k}_i$ and $\mathbf{k}_f$ can be expressed as: $\hbar^2 k_i^2/2m_0 - q\Phi = \hbar^2 k_f^2/2m_0$. Calculate the
    change in k perpendicular to the surface for a surface with $q\Phi = 4$ eV and $\theta_p = 45°$.
15. Calculate the Bohr–Bethe range R_B of 3 keV incident electrons in GaN. Compare this
    result with the corresponding range in GaSb.

16. For Cu K_α radiation ($E = 8.04$ keV) incident on Al, calculate the photoelectric cross section σ_{ph} for the K shell of Al ($E_B = 1.56$ keV) and compare its value with the electron impact ionization cross section σ_e for 8.04 keV electrons.

17. You are given a 200 Å thick layer of $Ga_xAl_{1-x}As$ on an InP substrate about 1 mm thick and are asked to determine the Ga to Al ratio. You can carry out XPS, AES, or EDXS analysis using 20 keV electrons and an Al K_α X-ray source. In order to compare the different techniques, you carry out the following calculations or comparisons: (a) What is the cross section ratio σ_{Ga}/σ_{Al} for the K shell electron impact ionization and L shell photoeffect? (b) What is the fluorescence yield ratio $\omega_X(Ga)/\omega_X(Al)$ for the K-shell hole? (c) What is the ratio of 20 keV Ga/Al Auger yields? Photoelectron yields? X-ray yields? (d) Which technique will provide the clearest determination of the Ga to Al ratio?

18. Calculate the range of Auger energies for the LM_1M_2 transition of gallium. Calculate the KL_1L_2 transition energy of oxygen. How well do these calculations agree with experiment?

19. The three 3 keV Auger electron spectra show a UHV-cleaved semiconductor surface (top), after a further cleaning process (middle), and after a chemical reaction (bottom). Label all the elemental features in all three spectra. Identify the semiconductor, the change induced by the further cleaning process, and the type of final chemical reaction. Calculate the percent composition of each element on the chemically-reacted surface.

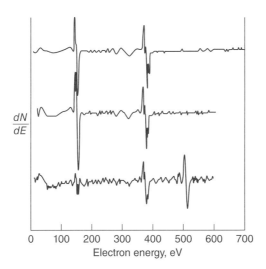

20. The three Auger electron spectra below show a semiconductor surface after etching and subsequent 60 s 200 °C and 250 °C anneals in UHV. (a) Label the elemental features in all three spectra. (b) Identify the semiconductor. (c) Is there any C or O contamination? (d) What is the effect of the UHV annealing? Calculate the percent composition of each element in each of the three spectra. (e) Which is the most stoichiometric surface?

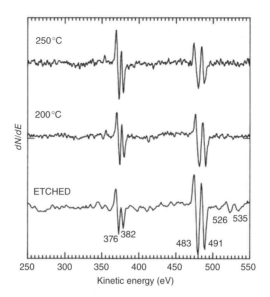

21. Calculate the velocity and wavelength of a 3 keV O atom.

22. Prove that the K-shell radius of an element with atomic number Z equals a_0/Z, where a_0 is the Bohr radius ($\hbar^2/me^2$). Hint: use the balance of electrostatic and centripetal forces and the condition that momentum $mvr = n\hbar$ is quantized. What is the Bohr radius for S?

23. Calculate the closest approach of a 1 MeV helium atom to a gold atom. How does it compare to the atom's radius?

24. A Rutherford backscattering experiment involves incident 2 MeV He$^+$ ions backscattered from a SiO$_2$ target. (a) Calculate the kinematic factors for the backscatting of each element. (b) Calculate the yield ratio of Si versus O.

25. Draw the backscattering spectrum for 2 MeV He$^+$ ions from a Ge target alloyed with 10% Au.

26. For an Au overlayer of thickness 1500 Å on GaAs and an energy loss per path length of 60 eV Å^{-1}, draw the backscattering spectrum for 2 MeV He$^+$ ions. How does this compare with Figure 7.26b?

References

1. Einstein, A. (1905) Concerning an heuristic point of view toward the emission and transformation of light. *Ann. Phys.*, **17**, 132.

2. Planck, M. (1901) On the law of distribution of energy in the normal spectrum. *Ann. Phys.*, **4**, 551.

3. Margaritondo, G. (1988) *Introduction to Synchrotron Radiation*, Oxford University Press, New York.

4. Lüth, H. (2001) *Solid Surfaces, Interfaces, and Thin Films*, 4th edn. Springer, Berlin, pp. 266–270.

5. Yeh, J.J. and Lindau, I. (1985) *Atomic Data and Nuclear Data Tables*, **32**, 1.

6. Brillson, L.J. (2010) *Surfaces and Interfaces of Electronic Materials*, Wiley-VCH Verlag GmbH, Weinheim, Appendix 6.

7. Wagner, C.D., Riggs, W.M., David, L.I., *et al*. (1979) *Handbook of X-Ray photoelectron Spectroscopy*, Physical Electronics Industries, Eden Prairie.

8. Rubloff, G.W. (1983) Microscopic properties and behavior of silicide interfaces. *Surf. Sci.*, **132**, 268.

9. Brillson, L.J. (2010) *Surfaces and Interfaces of Electronic Materials*, Wiley-VCH Verlag GmbH, Weinheim, Chapter 8.

10. Davis, L.E., MacDonald, N.C., Palmberg, P.W., *et al*. (1976) *Handbook of Auger Electron Spectroscopy*, 2nd edn, Physical Electronics Industries, Eden Prairie.

11. Siegbahn, K., Nordling, C.N., Falhlman, A., *et al*. (1967) *ESCA, Atomic, Molecular and Solid State Structure Studied by Mean of Electron Spectroscopy*, Almqvist and Wiksells, Uppsala.

12. Chang, C.C. (1974) Analytic Auger electron spectroscopy, in *Characterization of Solid Surfaces*, Chapter 20 (eds P.F. Kane and G.R. Larrabee), Plenum Press, New York.

13. Brillson, L.J. (2010) *Surfaces and Interfaces of Electronic Materials*, Wiley-VCH Verlag GmbH, Weinheim, Chapter 9.

14. Grabowski, S.P., Nienhaus, H., and Mönch, W. (2000) Vibrational and electronic excitations at GaN{0001} surfaces. *Surf. Sci.*, **454–456**, 498.

15. Schnatterly, S.E. (1979) *Solid State Physics*, Vol. 34, Academic Press, Inc., New York, p. 275.

16. Feldman, L.C. and Mayer, J.W. (1986) *Fundamentals of Surface and Thin Film Analysis*, Chapter 2, Prentice Hall. Upper Saddle River.

17. Brillson, L.J. (2010) *Surfaces and Interfaces of Electronic Materials*, Wiley-VCH Verlag GmbH, Weinheim, Chapter 9.

18. Feldman, L.C. and Mayer, J.W. (1986) *Fundamentals of Surface and Thin Film Analysis*, Prentice Hall, Upper Saddle River, NJ, Chapter 2.

19. Chu, W.K., Tu, K.N., and Mayer, J.W. (1978) *Backscattering Spectrometry*, Academic Press, New York.

20. Sinha, A.K. and Poate, J.M. (1973) Effect of alloying behavior on the electrical characteristics of n-GaAs Schottky diodes metallized with W, Au, and Pt. *Appl. Phys. Lett.*, **23**, 666.

Further Reading

Lüth, H. (2001) *Solid Surfaces, Interfaces, and Thin Films*, 4th edn, Springer, Berlin.

Feldman, L.C. and Mayer, J.C. (1986) *Fundamentals of Surface and Thin Film Analysis*, Prentice Hall, Englewood Cliffs, NJ.

Ibach, H. (2006) *Physics of Surfaces and Interfaces*, Springer, Berlin.

8

Dynamical Depth-Dependent Analysis and Imaging

8.1 Ion Beam-Induced Surface Ablation

While the techniques presented in Chapter 7 provide electronic, chemical, and structural information with high surface sensitivity, they can also yield similar information as a function of depth below the surface. This can be accomplished using *ion milling*, that is, by bombarding the sample surface with noble gas ions to remove any surface adsorbates as well as to etch away the outer atomic layers. Noble ions, such as Ar^+ or Xe^+, are typically used as bombarding particles in order to avoid any chemical bonding with the specimen or to other chamber components. Layer-by-layer ablation of surface layers is often used with Auger electron spectroscopy (AES) and secondary ion mass spectrometry (SIMS) but this dynamical depth-dependent analysis can also be performed using X-ray photoemission spectroscopy (XPS).

Care is required using the ion milling technique, since it can introduce several artifacts: (i) the ablation may not be uniform such that more than one layer is probed at a time; this could reduce the depth resolution of the analysis; (ii) the impact of ions with surface atoms may introduce mixing of several layers as well as drive surface atoms further below the free surface, essentially producing a "snow plow" effect; this would alter the actual distribution of atoms; (iii) the sputtering process can preferentially remove one or more of the constituents, distorting the overall composition measured; (iv) the disruption of atomic bonds and charging can change the charge state of atoms at or near the surface, complicating the measured chemical bonding information. These effects are most pronounced for ion bombardment at normal incidence and at high incident ion energies. To reduce or avoid such effects, ablation can be performed using glancing incidence ions, relatively low ion energies and low energy electron discharging if necessary. Under the low energy, glancing incidence conditions, however, the removal rate of atoms decreases, leading to a trade off between the speed of analysis and profiling accuracy. This chapter provides examples of the

An Essential Guide to Electronic Material Surfaces and Interfaces, First Edition. Leonard J. Brillson.
© 2016 John Wiley & Sons, Ltd. Published 2016 by John Wiley & Sons, Ltd.
Companion Website: www.wiley.com/go/Brillson/

information available using dynamical depth profiling with each of the photon, electron, and ion analysis techniques.

8.2 Auger Electron Spectroscopy

Auger electron spectroscopy (AES) depth profiling is widely used in electronic materials research to: (i) determine elemental composition within a particular layer, (ii) measure the extent of diffusion between adjoining layers, and (iii) identify the presence of chemical reaction near these interfaces. AES can also detect changes in chemical bonding since the AES transitions involve core levels whose energies can "*chemically shift*" with changes in bonding environment.

A common application of AES depth profiling is to monitor the presence of chemical reactions at metal–semiconductor interfaces. Whereas RBS studies of such interface reactions are non-destructive and effective for micron-scale thin film overlayers and multilayers, AES requires sputtering to remove overlayers on a scale that is typically on a sub-micron scale. On the other hand, AES depth profiling has several comparative advantages: (i) depth resolution at surfaces and near interfaces can be on an Å scale, (ii) AES is very sensitive to many elements common to conventional microelectronics, and (iii) the electron beam can be focused and scanned on a nanometer scale, permitting analysis of extremely confined test structures and scanning to produce maps of chemical composition and bonding.

Figure 8.1 illustrates the use of AES depth profiling to detect an interfacial reaction at a microelectronic interface – Pd metal deposited on Si [1]. Here the known thickness of the deposited Pd film provides a calibration of the sputter time with depth from the free Pd surface to its interface with Si. AES spectra acquired at various time intervals during the sputtering process yield features associated with Pd MNN, Si LVV, and a chemically shifted Si LVV transition attributed to a Pd silicide. The 2:1 ratio of Pd versus Si amplitude at the depth corresponding to the peak of this Pd silicide signal indicates the formation of Pd_2Si. This depth profile shows that Pd deposited in ultrahigh vacuum (UHV) on clean Si produces a non-abrupt interface that includes an interfacial reacted layer that is more than a hundred Å thick – even though the junction was formed near room temperature. For such thicknesses, detection by RBS or X-ray diffraction techniques would be difficult. Also, comparison of the silicide peak with the crossover of Pd and Si signals indicates that the reaction extends into the Pd rather than the Si. Such interfacial reactions between metals and Si near room temperature are quite common, reflecting low activation energies for the reactions to proceed [2].

AES depth profiling can provide useful measurements of chemical activity even for ultra-thin metal overlayers on semiconductors. In order to avoid disrupting the metal layer / semiconductor structure by the sputtering process, one can use low incident energy ions at glancing incidence. Figure 8.2a illustrates AES spectra at three different sputter times for 30 Å Au deposited at room temperature on a clean, UHV-cleaved (110) surface of InP [3]. Before Ar^+ bombardment, the 0 minute spectrum displays a strong Au signal but also significant features due to P and In segregated to the surface. After 8 minutes, the Au signal increases, the LMM P signal has disappeared, but the MNN In signal remains. Finally, after

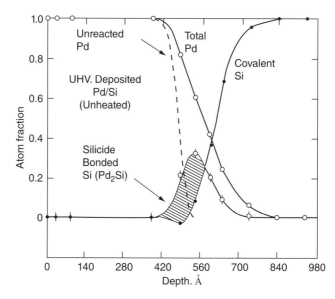

Figure 8.1 *AES Depth profile for room temperature deposited Pd on Si(100) obtained with Ar⁺ sputter etching. The proportion of silicide-bonded Si at the interface indicates the formation of Pd_2Si. (Roth and Crowell 1978 [1]. Reproduced with permission of American Institute of Physics.)*

92 minutes, the spectrum shows only In and P, but no Au as expected after sputtering away the metal overlayer.

Figure 8.2b shows the complete depth profiles of Au, In, and P AES peak intensities for different thicknesses of Au on InP. For the 10 Å Au layer, there is strong surface segregation of both In and P. With increasing Au thickness, both In and P continue to diffuse into the Au with In concentration higher than that of P. In and P segregated to the free Au surface are evident for all Au thicknesses, particularly for P, but their concentrations decrease with increasing Au thickness. Thus Figure 8.2 shows that there is a thickness-dependent diffusion of semiconductor atoms into the metal overlayer with In diffusing out of InP into Au over tens of Å, while P diffuses into and out of the Au film to the free surface. These effects become more apparent with increasing Au overlayer thickness. The asymmetry between the In and P profiles shows that the Au films are uniform overlayers. If bare InP were exposed through uneven thicknesses or pinholes in the metal overlayer, equal and increasing concentrations of In and P would be observed rather than the segregated species shown.

The outdiffusion of semiconductor anion and cation atoms into metal overlayers – even near room temperature – is actually a general phenomenon. It has significant implications for electronic material interfaces since such outdiffusion from the semiconductor lattice must leave behind vacancies of these atoms. Such native point defects and their complexes with other defects at neighboring lattice sites are often electrically active and can therefore play a role in the charge transfer involved in Schottky barrier or heterojunction band offset formation.

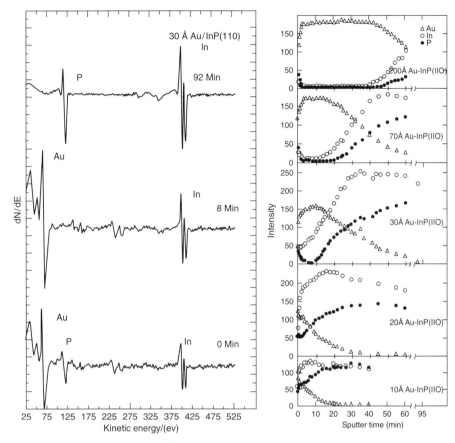

Figure 8.2 *(a)AES spectra of 30 Å Au–InP(110) versus sputtering time. Segregated In and P is present at the free Au surface. (b) AES depth profiles for Au–InP(11) junctions with different Au overlayer thicknesses. Segregated In and P vary with Au thickness on a nm scale. (Shapira and Brillson 1983 [3]. Reproduced with permission of American Institute of Physics.)*

8.3 X-Ray Photoemission Spectroscopy

Sputter profiling with XPS can also provide important depth profile information, particularly as an alternative to Rutherford backscattering spectrometry (RBS), the most common method of establishing thin film stoichiometry. Based on Equations (7.29) and (7.30), RBS is relatively insensitive to light elements such as oxygen, which is a primary constituent of oxides and especially complex oxides. For such complex oxides, the ability to measure stoichiometry precisely is important in order to avoid parasitic phases, which can form in the absence of line compounds with exact composition ratios [4]. Furthermore, XPS and AES are more sensitive than RBS to surface composition and atomic bonding. In addition to the known XPS or AES sensitivity factors, one can use line compounds with well-defined compositions to calibrate the non-line compound spectra.

XPS is particularly sensitive to lattice damage. The appearance of peak features due to charge states not normally present in the compounds signifies the onset of lattice damage. For example, one can measure the compound stoichiometry of Sr_2FeMoO_6, an important magnetic oxide, with a sufficiently gentle sputter energy to (i) remove surface contamination but (ii) avoid the onset of lattice damage. [See Track IIT8.1 XPS Depth Profiling of Sr_2FeMoO_6.]

8.4   Secondary Ion Mass Spectrometry

Secondary ion mass spectrometry (SIMS) is the most sensitive of the dynamical depth analysis techniques. As with AES and XPS depth profiling, SIMS involves ablating away surface layers during the analysis. However, SIMS is capable of orders of magnitude higher sensitivity than AES or XPS. Depending on the element and the host materials, SIMS is capable of detecting elemental densities of parts per billion. For semiconductors with nominally 10^{22} atoms/cm^{-3} atom densities, the SIMS detection limit for some elements may lie in the range of 10^{13}–10^{15} cm^{-3}, below the limit of typical doping densities. SIMS is therefore an important tool for determining doping densities in semiconductors as well as the presence of unintentional impurities that could alter the intended doping concentrations. On the other hand, while the detection limit of SIMS is several orders of magnitude better than AES or XPS, the quantitative precision of the latter techniques is significantly higher, particularly at surfaces, and relatively independent of the host matrix.

8.4.1   SIMS Principles

As with AES and XPS depth profiling, the SIMS technique involves a beam of ions incident on a surface. Typical primary ions are Ar^+, Cs^+, and O^- with energies in the range of 1–10 keV. Upon impact, these ions transfer energy to surface atoms that become "secondary" ions, which are ejected and analyzed. Unlike AES and XPS, the analysis is based on these secondary ions rather than Auger or photoexcited electrons. Figure 8.3 illustrates the SIMS sputtering process. A primary ion impinges on a surface, creating a cascade of single-particle collisions. The cascade removes atoms on and just below the surface.

The atoms leaving the target emerge as either neutral or charged secondary ions depending on the primary ion. Adsorbed atoms may leave either as individual atoms or in combination with target atoms at the surface. The collisions also produce lattice defects and implant ions in and below the surface.

In *static SIMS* (SSIMS), the primary ion sputters atoms only from the topmost target layer. The ion current density must be low enough to remove only a fraction of surface atoms over periods of minutes. A rule of thumb is that only 0.1% of surface atomic sites should be affected by the measurement. Based on surface atomic area densities of 10^{14}–10^{15} cm^{-2}, primary ion current densities must be orders-of-magnitude smaller, typically 10^{10} ions/cm^2 s, which for singly charged ions translates into ion currents of 10^{-9} A cm^{-2} or less. Detection limits for static SIMS can be ~10^9 cm^{-2}. Figure 8.4 illustrates a scanning tunneling microscope (STM) image (see Chapter 9) of a Si surface before and after a SSIMS measurement. The bombarded surface in (b) shows evidence of atom removal but still exhibits much of the ordered atomic pattern in (a).

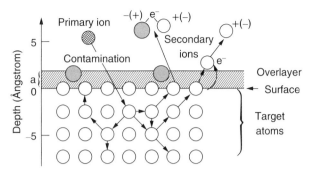

Figure 8.3   *The SIMS sputtering process represented schematically. (Lüth, H. 2001, figure 8.3. Reproduced with permission of Springer.)*

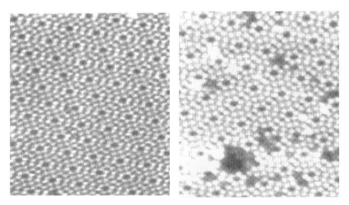

Figure 8.4   *Scanning tunneling microscopy image of Si(7x7) surface before and after exposure to 2×10^{12} ions/cm^2 [5]. (Zandvliet et al 1992 [5]. Reproduced with permission of Wiley.)*

In *dynamic SIMS* (DSIMS), orders-of-magnitude higher ion currents are used that sputter through multiple atomic layers into the target. Typical ion currents are 10^{-4}–10^{-5} A cm^{-2}. Here the primary beam produces a crater and a second beam of either positive or negative ions is focused on a relatively small spot in the flat center of the crater to produce the secondary ions. These secondary ions arise from surface areas that are proportional to their size. Thus surfaces areas for atoms are ~1 nm, small molecule areas are ~5 nm, and large molecule areas are ~10 nm or more.

8.4.2   SIMS Equipment

The SIMS technique requires equipment to: (i) produce a collimated beam of primary ions that can be focused on a target, (ii) extract, steer, filter, and focus the resultant secondary ions, and (iii) detect and analyze these secondary ions. Besides the quadrupole mass analyzer described in Chapter 5, there are two major types of ion mass separator – the magnetic sector and the time-of-flight mass spectrometer.

Figure 8.5 illustrates the SIMS components in the magnetic sector spectrometer. The ionization chamber contains filaments to ionize the initially neutral atoms, then accelerate

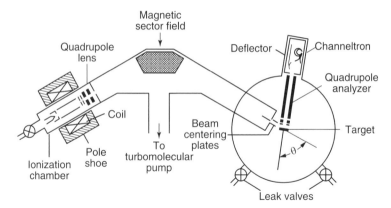

Figure 8.5   *Secondary ion mass spectrometer layout for controlling trajectory of secondary ions and elemental analysis. (Lüth, H. 2001. Reproduced with permission of Springer.)*

them with high voltage electrodes through a quadrupole lens. This lens focuses the ions, now with much higher kinetic energy, into the field of a magnetic sector, which deflects those ions with desired mass-to-charge ratio, toward the target. Any additional impurity atoms accelerated into the magnetic sector field are deflected in different directions, thereby filtering the primary beam. The now filtered and focused primary beam strikes the target, producing secondary ions that are separated in mass either by another magnetic sector or by a quadrupole analyser (*e.g.,* Figure 5.6).

A second type of SIMS spectrometer involves time-of-flight separation of different secondary ions. In time-of-flight SIMS (TOF-SIMS), the primary ions arrive at the target in pulses. See Figure 8.6. A high bias voltage V_A applied to the sample accelerates the secondary ions pulses into a channel that functions as a racetrack. A detector records the arrival of the secondary ions after traveling a distance d and the elapsed time t from the initial pulse. The elapsed time is a direct function of the secondary ion's mass. The ion has a kinetic energy $E = \frac{1}{2}mv^2$ equal to qV_A. Therefore the elapsed time t is

$$t = d/v = d/(2E/m)^{1/2} = d/(2q\,V_A/m)^{1/2} \tag{8.1}$$

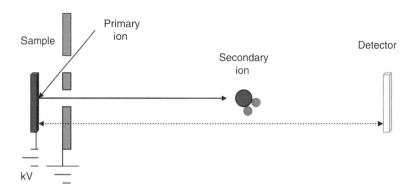

Figure 8.6   *Time-of-flight method of determining particle mass in a TOF SIMS instrument*

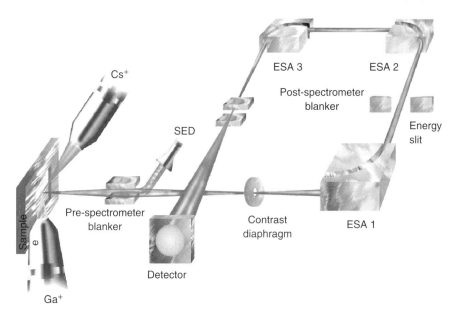

Figure 8.7   *Schematic diagram of ion trajectory inside a PHI TRIFT III TOF-SIMS. (Reproduced with permission of Physical Electronics Industries.)*

so that $m = 2qV_A t^2/d^2$ plus any time delay introduced by the detector circuit. Then for a given q, V_A and d, the longer the time delay, the higher the mass. TOF-SIMS measurements became feasible with the introduction of electronics and interconnects capable of timing pulses in the nanosecond regime. The timing can be calibrated by fitting the time delays in a spectrum with several peaks of known mass in the $m = at^2 + b$ relationship to obtain constants a and b. Figure 8.7 illustrates the TOF-SIMS apparatus schematically for a primary beam of Cs^+ ions and an ablating beam of Ga^+ ions. At each depth of the ablated surface, the secondary ions produced by the Cs^+ beam pass through an electrostatic blanker that stops all ions except those produced by the initial pulse. The ions then pass through a series of apertures and electrostatic bends that force the ions around the race track and on to the detector. A secondary electron detector records the image of secondary electrons generated at the sample surface by an electron gun.

The magnetic sector and TOF-SIMS instruments each have unique strengths. See Track II Table T8.2 for a comparison of mass spectrometers. The mass range of the magnetic sector mass spectrometer exceeds that of both TOF-SIMS and quadrupole analyzers, permitting analysis of many organic molecules. TOF-SIMS has higher mass resolution $\Delta m/m$ than the others, exceeding 10^4 and permitting clear distinction of different isotopes of the same element. It also has higher transmission than either magnetic sector or quadrupole due to its parallel rather than sequential signal detection.

Track II T8.3 Interface Broadening at a Double Heterostructure illustrates the value of $\Delta m/m > 10^4$ at InP / InGaAs heterojunctions. GaAs and InP have nominally the same atomic mass, 145.8 amu, but their 0.03 amu separation is easily distinguished, permitting detailed analysis of interface broadening at their interface.

Table 8.1   *Comparison of XPS, AES, DSIMS, and SSIMS*

	XPS	AES	DSIMS	SSIMS
Probe Beam	Photons	Electrons	Ions	Ions
Analyzed Beam	Electrons	Electrons	Ions	Ions
Sampling Depth	0.5–5 nm	0.5–5 nm	0.1–1 nm	0.1–1 nm
Detection Limits	1×10^{-4}	1×10^{-4}	1×10^{-9}	1×10^{-6}
Information	Elemental	Elemental	Elemental	Elemental
	Chemical	Chemical	Structural	Molecular
Spatial Resolution	1–30 μm	10 nm	50 nm	120 nm
Materials	All Solids	Inorganics	Inorganics	All Solids

Reproduced with permission of Physical Electronics Industries.

Track II T8.4. Interdiffusion at an $AlGaN/Al_2O_3$ Heterojunction illustrates the value of SIMS' high sensitivity to trace elements. The growth of AlGaN on sapphire (Al_2O_3) can result in the outdiffusion of O from Al_2O_3. Since O is a relatively shallow donor in GaN alloys, this outdiffusion causes degenerate doping in the AlGaN, far exceeding the intended doping level yet still below the detection limit of other surface and interface techniques.

Table 8.1 provides a comparison between the two SIMS techniques, DSIMS and SSIMS, versus XPS and AES. Each technique has particular advantages, depending on the information and sample under test. All four provide chemical information on the elements present. XPS and AES provide chemical bonding information via chemical shifts, whereas SIMS fragments provide indications of molecular species. The spatial resolution of AES can be <10 nm with a focused electron microscope beam, whereas the focused SIMS beam is in the 50–100 nm range, and conventional XPS is in the low micron range. In terms of sensitivity, SIMS detection is orders of magnitude greater than either XPS or AES. SIMS detection can reach 10 ppb, particularly for DSIMS, whereas AES and XPS can detect at best fractions of a percent surface coverage. SSIMS and XPS can probe all solids, both organic and inorganic, whereas AES and DSIMS are subject to charging and beam damage.

8.4.3   Secondary Ion Yields

The ablated atoms must be ionized for the applied voltage V_A to extract them into the racetrack. This ionization requires charge transfer between the secondary ion and the lattice from which it leaves. In turn, the efficiency of this ionization depends sensitively on the electronegativity of the primary ion. This SIMS ionization efficiency is termed *ion yield* and equals the fraction of sputtered atoms that become ionized. Depending on the ablated ion, the ion yield can vary by orders of magnitude. Figure 8.8 displays the variation in ion yield for a wide range of elements. For positive ions produced by O^- primary ions, Figure 8.8a shows that as the energy required to ionize the sputtered atom, the *ionization potential*, increases, the positive ion yield decreases by orders of magnitude. For negative ions produced by Cs^+ primary ions, Figure 8.8b shows that as the ease with which the sputtered ion attracts an electron, the *electron affinity*, increases, the ion yield increases exponentially. Thus O^- bombardment increases the ion yield of positive ions while Cs^+ bombardment increases the ion yield of negative ions. The partial Periodic Table in Track II. Table T8.5

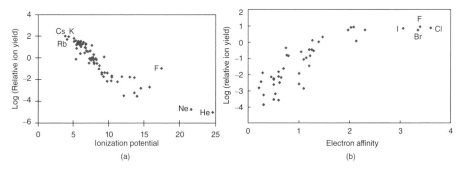

Figure 8.8 *Positive (a) and negative (b) ion yields with O ion or Cs ion sputtering, respectively relative to those for Si in Si [6]. (Reproduced with permission of Physical Electronics Industries.)*

SIMS Primary Ion Table displays the elements most likely to produce the higher ion yield with an O^- versus Cs^+ primary beam. Physically, the O-induced increase in positive ion yield can be understood in terms of oxygen–metal bond formation with the increase of O concentration in the surface layer. When these bonds break in the emission process, the high electron affinity and low ionization potential of O results in charge transfer to the O, so that the metal ions are positively charged. Likewise, Cs^+ bombardment and increased Cs in the surface layer results in a work function decrease that increases the excitation of secondary electrons over the surface barrier, leading to more negative ion formation [6].

Depending on the primary ion selected, the sensitivity of SIMS to a particular element is given by a relative sensitivity factor (RSF). The RSF for a particular element is defined in terms of secondary ion intensity I_E for an element E at a concentration C_E of that element compared with a reference secondary ion intensity I_R of an element R at a concentration C_R. Thus $I_R/C_R = RSF_E \cdot I_E/C_E$. The lower the RSF_E, the higher the sensitivity of SIMS to that element. If one takes the reference element R to be the host matrix element M, then this relation can be re-expressed as

$$C_E = RSF \cdot I_E/I_M \tag{8.2}$$

so that the concentration C_E of element E is the ratio of the element versus the matrix element intensities, all multiplied by the RSF for those elements. RSFs for most elements across the Periodic Table appear in Track II Table T8.5, SIMS Periodic Table of Elements with RSFs for O_2^+ primary ions and Track II Table T8.6, SIMS Periodic Table of Elements with RSFs for Cs^+ primary ions. These sensitivities can vary by orders of magnitude. For O_2^+ primary ions, Table T8.6 shows that Column III elements such as Al, Ga, and In as well as the Column I elements have very low RSF, indicating that SIMS using O_2^+ primary ions will have high sensitivity. For Cs^+ primary ions, Track II Table T8.7 shows that Column VI elements such as S, Se, and Te as well as the Column VII elements have relatively low RSF, indicating that SIMS using Cs^+ primary ions will have high sensitivity. Furthermore, ion yields depend sensitively on the matrix in which the particular atom is imbedded. Therefore, quantitative SIMS analysis requires standards of the same element with a known concentration in the same matrix. Such standards can be prepared by an ion accelerator that implants known concentrations of the element to be calibrated into the same material as the host matrix.

8.4.4   Organic and Biological Species

Inorganic electronic materials such as semiconductors are relatively straightforward to analyze by SIMS compared with organic and biological molecules. The latter are typically composed of complex molecular groupings with atomic weights that can be interpreted in more than one way. Track II.T8.8, SIMS of Organic Species: Polystyrene illustrates the various fragments and their atomic weights from SIMS analysis of polystyrene, a relatively simple polymer structure. Identification of organic molecules is especially challenging since the constituents are primarily C, O, H, N, and little else. Libraries of spectra are available to catalog the various fragments generated during secondary ion emission and their relative intensities for a particular organic or biological molecule, the latter consisting of C, O, H, N, and possibly P and S [7].

8.4.5   SIMS Summary

Dynamic SIMS is the leading technique to measure dopant and impurity concentrations in electronic materials quantitatively. This is based on its parts-per-billion sensitivity, which far exceeds the chemical sensitivities of other surface science techniques. However, SIMS requires standards – known concentrations of the same impurity in the same matrix as that of the host material under test – in order to be quantitative. SIMS' ability to detect impurities at concentrations comparable to doping levels is particularly useful at metal–semiconductor and semiconductor heterojunction interfaces where even trace amounts of impurities can affect transport and optical properties significantly.

8.5   Spectroscopic Imaging

Mapping of chemical distributions in two and three dimensions is possible with depth profiling combined with AES, XPS, or SIMS. Figure 8.9 illustrates schematically how full m/z scans obtained pixel by pixel can be assembled into two-dimensional maps of specific elements at a given surface. Here the primary ion beam is rastered across an area of the surface, revealing regions with different total ion intensity and atomic composition. In addition, individual peak intensities I within each I versus m/z spectrum provide maps of composition for a specific mass across the surface. Chemical maps 1 and 2 illustrate such maps for different species on the same surface.

Such imaging near real-space electronic materials interfaces can provide critical information about any chemical interactions taking place. For example, Track II T8.9 SIMS Imaging of Stainless Steel illustrates how one can assess the effect of precipitates on the surrounding stainless steel matrix. Analogous mapping is possible using AES and XPS. For example, Track II T8.10 AES Imaging of GaN/ Al_2O_3 Junction illustrates both chemical diffusion and reaction at a cross section at the interface between GaN grown at very high temperature and its sapphire substrate. The ability to obtain full SIMS, AES, or XPS spectra, pixel by pixel, at each stage of depth profiling, then assemble them into two-dimensional intensity maps at a selected energy or m/z ratio, is termed *hyperspectral imaging* (HSI) and has been extended to optical spectra as well.

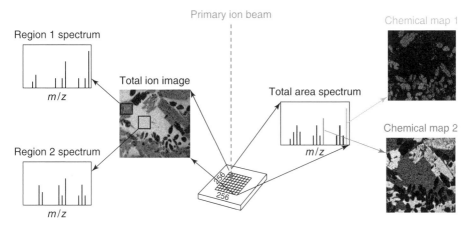

Figure 8.9 *TOF-SIMS imaging for primary ion beam rastered across a specimen surface, yielding I versus m/z spectra at specific locations and maps of total ion as well as individual element versus position chemical maps [8]. (Reproduced with permission of Physical Electronics Industries.)*

8.6   Depth-Resolved and Imaging Summary

Dynamical depth-resolved analysis and imaging have been important techniques in the development and processing of electronic materials. Advanced electronic devices involve structures that must be atomically abrupt and thermally stable under growth, processing and operating conditions. Sputter depth profiling using AES, XPS, and SIMS provides measurements of the elemental structure of thin films and their interfaces on an atomic scale and the changes that take place with interdiffusion and chemical reaction. This information has guided technologists in developing materials designs and device construction over many decades.

8.7   Problems

1. Assuming a primary O^- current of 10^{-9} A cm^{-2} impinging on a GaAs(100) surface and one atom per incident ion removed by a primary ion, what is the maximum time available to acquire a static SIMS spectrum without significantly disrupting the surface?
2. The time-of-flight mass spectrometer is capable of $>10^4$ mass resolution $\Delta m/m$. For mass 100 amu, 2 keV applied voltage, and a flight path equal to 2 m, what clock speed of the detection electronics is required in order to achieve this resolution?
3. Which ionization source(s) would you use to probe for C, O, Mg, Cl, and Co using SIMS with host semiconductor of GaAs?
4. Which ionization source(s) would you use to probe for interdiffusion across (a) a CdS/Si interface? (b) a GaAs/InP heterojunction?
5. A silicon wafer is implanted with Ni atoms. Between AES, XPS, SIMS, RBS, and TEM, evaluate the best technique to determine the presence of this layer for the following layer

thickness and depth below the Si surface: (a) a 20 nm thick layer 50 nm deep or (b) a 20 nm thick layer 3 μm deep.

6. A mm-thick Si wafer containing oxygen forms SiO_2 precipitates that are amorphous and spherical with characteristic sizes ~10 nm. Describe the best technique to determine their presence, amorphous nature, and composition.

References

1. Roth, J.A. and Crowell, C.R. (1978) Application of Auger electron spectroscopy to studies of the silicon/silicide interface. *J. Vac. Sci. Technol.*, **15**, 1317.
2. Tu, K.N. and Mayer, J.W. (1978) Effect of thin film interactions on silicon device technology in *Thin Film Reactions and Diffusion*, (eds J.M. Poate, K.N. Tu, and J.W. Mayer), Wiley, New York, Ch. 2, pp. 13–55.
3. Shapira, Y. and Brillson, L.J. (1983) Auger depth profiling studies of interdiffusion and chemical trapping at metal-InP interfaces. *J. Vac. Sci. Technol.*, *B*, **1**, 618.
4. Rutkowski, M., Hauser, A.J., Yang, F.Y. *et al.* (2010) X-ray photoemission spectroscopy of Sr2FeMoO6 film stoichiometry and valence state. *J. Vac. Sci. Technol. A*, **28**, 1240.
5. Zandvliet, H.J.W., Elswijk, H.B., van Loenen, E.J., and Tsong, I.S.T. (1992) *Proceeding of the 8th International Conference on SIMS*, (eds A. Benninghoven, J. Tjuer and H.W. Werner), John Wiley & Sons, Ltd. Chichester, pp. 3–9.
6. http://www.eag.com/mc/sims-secondary-ion-yields-primary-beam.html.
7. Belu, A., Graham, D.J., and Castner, D.G. (2003) Time-of-flight secondary ion mass spectrometry: techniques and applications for the characterization of biomaterial surfaces. *Biomaterials*, **24**, 3635.
8. Physical Electronics Industries, Chanhassen, MN, http://www.phi.com.

Further Reading

1. Feldman, L. and Mayer, J.M. (1986) *Fundamentals of Surface and Thin Film Analysis*, Prentice Hall, Englewood Cliffs, New Jersey.
2. Wilson, R.G., Stevie, F.A., and Magee, C.W. (1989) *Secondary Ion Mass Spectrometry: A Practical Handbook for Depth Profiling and Bulk Impurity Analysis*, Wiley, New York.

9

Electron Beam Diffraction and Microscopy of Atomic-Scale Geometrical Structure

Electron beams enable several important probes of geometrical structure at electronic material surfaces and interfaces. These structural probes are based on electron diffraction, secondary electron imaging, transmission electron atomic imaging and atom-specific electron energy loss. The geometrical arrangement of atoms at electronic material surfaces is important because they can possess very different electronic properties – properties that determine the nature of charge transport normal to and along the surface.

9.1 Low Energy Electron Diffraction – Principles

Low energy electron diffraction (LEED) is a standard technique for obtaining structural information of surfaces. LEED has provided the vast majority of surface structural data, which is now supplemented by scanning tunneling microscopy (STM) and other techniques. The LEED technique is based on the coherent scattering of electrons at surfaces of crystalline solids in the energy range $10 < E < 1000$ eV. One can view the crystal as a set of geometrically-equivalent planes parallel to the surface. Within each plane, atoms comprise regular two-dimensional arrays that act as diffraction gratings. In light diffraction by a grating, the relationship between the spacing d of grating lines and the diffracted wavelength is

$$n\lambda = d \sin \theta \tag{9.1}$$

where θ is the angle between the incident and diffracted light and $n\lambda$ is a multiple of the incident wavelength. When the path length of light reflected from adjacent grating lines equals a multiple of this wavelength, the reflected light interferes constructively, increasing the intensity of *diffracted* light at those angles.

An Essential Guide to Electronic Material Surfaces and Interfaces, First Edition. Leonard J. Brillson.
© 2016 John Wiley & Sons, Ltd. Published 2016 by John Wiley & Sons, Ltd.
Companion Website: www.wiley.com/go/Brillson/

The two-dimensional array of surface atoms also acts as a diffraction grating where d is the separation between atoms. Davisson and Germer received a Nobel Prize for their discovery in 1927 that electrons at surfaces also diffracted. Their result was of fundamental importance since it demonstrated the wave nature of electrons and the particle-wave duality of matter, a fundamental tenet of quantum mechanics. According to the de Broglie relation,

$$\lambda = h/p \tag{9.2}$$

where h is *Planck's constant* $= 6.626 \times 10^{-34}$ J s $= 4.136 \times 10^{-15}$ eV s and $p = mv$ is the particle momentum. For electron mass $m = 9.1 \times 10^{-31}$ kg and $E = \frac{1}{2} mv^2$,

$$\lambda = \sqrt{(150.4/E)} \tag{9.3}$$

where λ is in Å and E in eV. According to Equation (9.3), an electron with energy 150.4 eV has a wavelength λ of 1 Å. Thus according to Equation (9.1), electrons must have low energies in this range in order for θ to be large enough to produce diffraction patterns with spots that are visibly separated. Likewise, the diffracted angles of individual spots change as the incident beam energy changes.

9.1.1   Low-Energy Electron Diffraction Techniques

Figure 9.1 illustrates the basic principles of LEED. Here an incident beam of electrons with a single kinetic energy impinges on a crystal surface, resulting in beams of electrons diffracted by the rows and columns of the surface atoms. Unlike a grooved diffraction grating, the surface atoms are regularly spaced in two dimensions rather than just one, resulting in a two-dimensional array of spots rather than a one-dimensional array of lines. In Figure 9.1a, diffracted electrons project backwards towards a screen with a fluorescent coating. A high voltage accelerates the electrons between this screen and a grounded wire mesh to create a visible diffraction pattern. Furthermore, at energies of a few hundred electron volts or less, the incident electrons scatter mainly from the outer few Å of the surface (see Figure 6.2), so that the diffracted intensity contains information about the atomic arrangement within just that near-surface layer. Hence the intensity of each visible spot changes with incident beam energy as Figure 9.1b represents.

To illustrate the spot pattern at a given energy, Figure 9.1c shows a LEED pattern for 92 eV electrons diffracted from a Si surface with a surface geometry termed (2×1) according to the pattern's vertical versus horizontal periodicity. From the diffraction pattern, one can measure (i) the two-dimensional surface geometry, that is, the *periodicity* of the surface – the *surface unit mesh*, from the spot symmetry and separation, (ii) the lattice spacing of surface atoms from their angular dependence on energy, and (iii) the periodic rearrangement of groups of atoms that have reorganized within the outer few surface layers from detailed analysis of the spot intensity versus energy dependence.

9.1.2   LEED Equipment

Figure 9.2 illustrates schematically the equipment used to generate the LEED patterns. Here a set of retarding and accelerating grids act as filters to pass and display the diffracted electron beams as a function of energy. A high voltage between a grounded

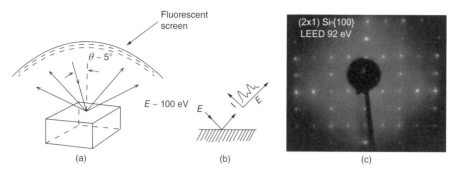

Figure 9.1 *Schematic illustration of LEED measurement. (a) A 100 eV incident electron beam generates diffracted beams toward a fluorescent screen. (b) Dynamical LEED spectrum of diffracted beam intensity versus incident beam energy. (c) Photograph of reverse-view LEED pattern for 92 eV electrons bombarding a (2 × 1) Si(100) surface. The center region is shadowed by the electron gun producing the pattern. (Brillson 2010. Reproduced with permission of Wiley.)*

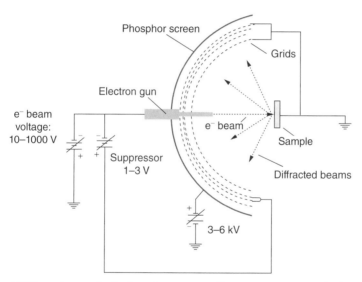

Figure 9.2 *LEED equipment including electron gun, fluorescent screen at high voltage, plus variable voltage suppressor, acceleration, and grounded screens to control kinetic energy of diffracted electrons.*

semi-transparent wire grid and the fluorescent screen accelerates the electrons to produce visible spot patterns. Electron beam spot size is typically ~1 mm with ~1 µA currents and energy width $\Delta E \sim 0.5$ eV due to thermal broadening by the hot electron gun filament. At higher electron gun voltages coupled with an additional variable voltage grid and corresponding ground grid, this LEED "optics" is configured to retard incident electrons and filter their energies for AES analysis. However, this AES arrangement has higher signal-to-noise than the electron analyzers described in Chapter 7. With advances in

LEED equipment, channel plates have replaced fluorescent screens, amplifying intensity signals and providing digital images with computer-controlled beam tracking at much lower (10^{-10} A) beam currents.

9.1.3   LEED Kinematics

The angular displacement of electron beams relative to the incident beam represents a momentum transfer *parallel* to the surface by the crystal lattice. Equation (9.4) represents this conservation of momentum parallel to the surface

$$k_{i,||} = k_{f,||} + g \tag{9.4}$$

where $k_{i,||}$ = vector momentum of incident electron parallel to the surface, $k_{f,||}$ = vector momentum of the diffracted electron parallel to the surface, and g = the vector of the two-dimensional reciprocal lattice (termed the Bravais lattice) associated with the surface plane. Since $E = \hbar^2 k^2 / 2m$, the momentum parallel to the surface is

$$k_{i,||} = (2mE/\hbar^2)^{1/2} \sin \theta \tag{9.5}$$

and $g = hA + kB$, where A and B are *primitive vectors* of the reciprocal lattice and h and k are integers.

Figure 9.3a illustrates one such spot pattern for the GaAs (110) surface with each spot labelled according to h, k integers along the A, B vector directions. From Equation (9.5) therefore, the spatial distribution of diffracted beams relative to the incident beam depends on the translational symmetry and spacing of atoms in the outermost crystal layer. One can then use the LEED pattern to extract the dimensions of the unit mesh from the dependence of the angle θ for each spot along symmetry directions versus the incident beam's energy. These spots can be viewed according to a construction termed the *Ewald Sphere* in Figure 9.3b, which provides a geometric representation of the momentum conservation relationships described by Equations (9.4) and (9.5) in terms of k_i and k_f versus A and B. In this figure, lines parallel to $k_{i,||}$ represent momenta supplied by multiples of primitive vector B in the horizontal direction such that diffracted beams have momentum $k_{f,||}$ due to momentum transfer S. The intersection of these lines in two dimensions (*rods* in three dimensions) defines the direction and magnitude of $k_{f,||}$. The symmetry of the spot pattern can also provide: (i) adsorbate atom site symmetry and (ii) evidence for surface domains and steps.

9.1.4   LEED Reconstructions, Surface Lattices, and Superstructures

LEED can also provide information on the atomic rearrangements at semiconductor surfaces. These rearrangements can involve atom movements both parallel to and perpendicular to the surface, termed *reconstructions*, and can be modified further by adsorption of additional species. The conservation of momentum expressed by Equation (9.4) parallel to the surface does not apply for momenta perpendicular to the surface since the presence of the surface destroys the translational invariance along the surface normal. Furthermore, multiple scattering effects prevent extension of X-ray diffraction theory to low energies.

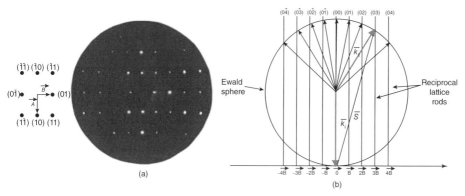

Figure 9.3   *(a) Photograph of normal incident LEED pattern of the cleaved GaAs(110) surface using 150 eV incident electron beam energy. To the left, each spot is indexed according to its primitive vectors.(b) Ewald sphere with rods of the two-dimensional reciprocal lattice, incident wave vector $\mathbf{k}_i$, diffracted wave vector $\mathbf{k}_f$, and a momentum transfer $\mathbf{S}$ along the $\mathbf{B}$ direction, shown here for the (03) beam. (Kahn, A. 1983 [1]. Reproduced with permission of North Holland.)*

In order to extract surface geometry normal to the surface using LEED, one can take advantage of the strong interaction of low energy electrons with the first few layers of the solid. This requires multiple scattering ("dynamical") models that can interpret the energy and temperature dependence of LEED spot intensities. Such models involve calculating bond angles and bond lengths and simulating beam profiles iteratively to find the best match to experimental profiles. One notable success of dynamical LEED theory has been to establish the three-dimensional rearrangement, termed *reconstruction,* of the GaAs (110) surface. Figure 9.4 illustrates the side view of the GaAs (110) surface both before (a) and after (b) reconstruction.

The side view of the reconstructed surface shows the bond tilting arrangement, in which anions rotate out of the plane and cations rotate inwards with minimal change in bond length. This bond rotation results in a redistribution of charge from cation to anion and a lowering (raising) of anion (cation) dangling bond energies. These changes have the net effect of reducing the surface energy due to electrostatic repulsion by moving the additional negative charge on the outermost anion further from the surface. This reconstruction is characteristic of III-V compound semiconductors in general [3] and represents an important success in surface science. It also implies that the surface bond energy minimization depends primarily on geometric structure rather than the detailed electronic structure of a particular reconstruction. The nature of this reconstruction for these compound semiconductors has major significance for electronic materials since it explains the absence of intrinsic surface states otherwise expected at an ideally-terminated III-V compound surface.

Surface atomic layers can have different point group symmetries, Bravais lattices, and periodicities from the bulk lattice. Figure 9.5 illustrates three representative types of surface reconstructions: (a) a relaxation of surface atoms toward such that bond length $c_1 < c_2$, (b) formation of dimers from pairs of surface atoms, and (c) a surface layer with every other atom missing.

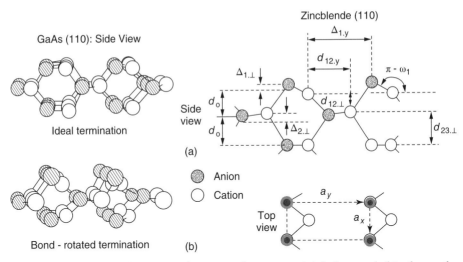

Figure 9.4 *Schematic illustration of GaAs surface atoms (a) before and (b) after surface reconstruction for (110) surfaces of compound semiconductors with the zincblende structure. The schematic of atomic angles and bond lengths (right) represents the independent structural variables that describe the reconstructed atomic geometries. (Duke, C.B. 1977 [2]. Reproduced with permission of American Institute of Physics.)*

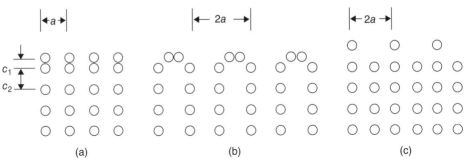

Figure 9.5 *Side views of representative surface reconstructions: (a) atomic relaxation, (b) dimer formation, and (c) missing rows for a cubic lattice with bulk lattice constant a.*

In Figure 9.5b and c as well as in Figure 9.4, the surface periodicity is twice as large as the underlying lattice and could introduce diffraction asymmetries such as the (2×1) LEED patterns in Figures 9.1c and 9.3a. Note that since more than one atomic arrangement can introduce such a periodicity, the LEED pattern alone is not sufficient to distinguish between these possibilities.

Figure 9.6 illustrates the additional complexity introduced by surface overlayers, showing side views of adsorbates on a crystal lattice with different ratios of adsorbate versus substrate crystal lattice constants. Such surface lattices with different periodicity from that of the substrate are termed *superlattices*. The integer superlattice with $b/a = 2$ has a periodicity twice as large as the underlying lattice as with cases (b) and (c) in Figure 9.5.

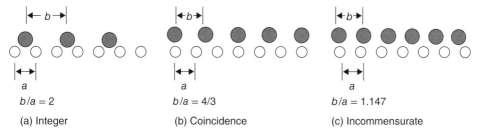

Figure 9.6 *Adsorbate superstructures with various ratios of adsorbate-to-substrate lattice constants.*

The non-integer superlattices produce more complex LEED patterns. A standard notation to describe superstructures by translational vector ratios and angles is the Wood notation. According to this notation, the surface superstructure of an element or compound M with surface Miller indices (*hkl*) can be described as

$$M(hkl) - \alpha[(a_s/a) \times (b_s/b)]\zeta - S \qquad (9.6)$$

where M(*hkl*) is the (*hkl*) crystal face of substrate M, α is either a *p* (primitive) or *c* (centered) unit mesh of the composite system, $[(a_s/a) \times (b_s/b)]$ are the ratios of overlayer to substrate vectors, *S* is an overlayer element if present, and ζ is an angle between the overlayer and substrate mesh vectors a_s and a. Track II.T9.1 Adsorbate-Substrate Superlattices provides a more complete description of the unit mesh vectors and representative ordered adsorbate–substrate superlattices.

9.1.5 Representative Semiconductor Reconstructions

LEED studies of electronic materials have focused primarily on Si and binary compound semiconductors. The Si surface is of high interest because of the pervasive use of Si in microelectronics and the importance of its interfaces with metals and dielectrics. Binary compound semiconductors, such as GaAs, InP, GaN, and now ZnO, are of similar interest due to their widespread optoelectronic, sensor, and high speed microelectronic applications. Considerable research has focused on the Si (100) surface since it is the wafer surface

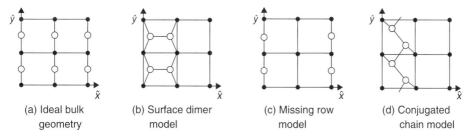

Figure 9.7 *Top view of Si (100) (2 × 1) (a) ideal bulk geometry, (b) surface dimer model, (c) missing row model, and (d) conjugated chain model.* ○ (●) *symbols signify Si atoms buckled outward (inward).*

used in device technology. This surface exhibits a (2×1) reconstruction for which several alternative structures have been proposed to account for its periodicity. Figure 9.7a illustrates a top view of the ideal bulk-terminated surface geometry with Si atoms buckled inward and outward. Figures 9.7b–d illustrate three alternative models that have the same (2×1) periodicity but different electronic band structures. Low temperature STM studies confirmed the presence of surface dimers on this (100) surface, supporting the surface dimer model shown in Figure 9.7b.

The Si (111) (7×7) surface presented a much more challenging surface reconstruction that was the subject of extensive investigation over several decades. Figure 9.8a presents an example of the fascinating LEED pattern for this reconstruction, whose seven-fold periodicity gives it its name. The nature of this reconstruction was finally established by Takayanagi *et al.* using a transmission electron microscopy technique with surface sensitivity [4]. Figure 9.8b shows a top view of the dimer-adatom-stacking fault (DAS) model used to describe the (7×7) reconstruction. STM measurements (Chapter 10) of this reconstructed surface are consistent with this model. The reconstruction involves three double layers of atoms. The circles in Figure 9.8b indicate atoms whose decreasing diameter signifies positions lower down, away from the top surface. Figure 9.9 illustrates the configurations of the various surface atoms in cross section.

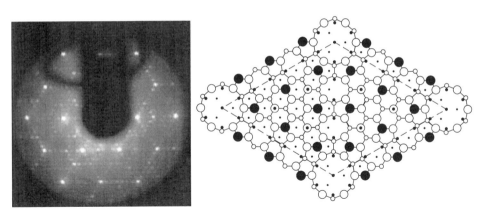

Figure 9.8   *(a) Si (7×7) LEED pattern. (Quinn, J.E http://www.matscieng.sunysb.edu/leed/7x7.html.) (b) Top view of DAS model for the (7×7) surface reconstruction. (Takayanagi et al 1985 [6]. Reproduced with permission of American Institute of Physics.)*

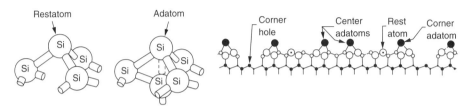

Figure 9.9   *(a) Three-dimensional representation of rest and adatoms in Si(111) (7×7) surface and (b) cross-sectional view of corner hole, center, rest, and corner adatoms. (Takayanagi et al. 1985 [6]. Reproduced with permission of American Institute of Physics.)*

The complexity of this DAS fault model is surprising, given that all atoms are the same and that the (111) is a relatively low index plane. Nevertheless, this complexity illustrates a significant point: Extended, coherent features of the surface topology can help minimize strain and overall energy for structural stability.

Compound semiconductors introduce a different type of complexity since their stoichiometry – the balance of anions and cations – at a surface can change with growth and processing conditions as well as with crystal orientation. Surface reconstructions of III-V compound semiconductors are of high interest due to the use of these compounds in many optoelectronic as well as high speed, high power microelectronic devices. Of these semiconductors and their orientations, the (100) surface of GaAs is used more widely than any other in compound semiconductor crystal growth and heteroepitaxy. Researchers have found many different surface reconstructions on this surface, depending strongly on how they are grown and processed. Thus the growth of GaAs by molecular beam epitaxy (MBE) produces a variety of reconstructions that depend sensitively on the vapor flux ratio, that is, beam equivalent pressure (BEP), of As_4 relative to Ga and substrate temperature. Track II.T9.2 GaAs Reconstructions versus MBE Growth Conditions illustrates a phase diagram of GaAs (100) reconstructions versus this flux ratio and temperature.

Surface reconstructions also display a strong dependence on the Ga-to-As surface coverage. Track II T9.3 Reconstruction Dependence on Surface Stoichiometry illustrates this correlation between atomic structure and chemical composition. One can obtain different reconstructions of the GaAs after growth by annealing in arsenic or vacuum, then cooling to room temperature. Significantly, a particular reconstruction is observable over a range of stoichiometries, suggesting either that the surface is composed of more than one reconstruction or that the unit cells are more complex than a simple extension of the primitive unit cell.

Case Study: – The GaAs(001) $(2 \times 4) - c(2 \times 8)$ Reconstruction

To show how the same reconstruction can occur for varying stoichiometries, Figure 9.10 provides an example of a geometric structure determined by an extended array of primitive unit cells. Here the GaAs (001) surface exhibits the same LEED pattern, termed a "$(2 \times 4) - c(2 \times 8)$" reconstruction, for a range of stoichiometries. Figure 9.10a shows the atomic arrangement of Ga and As atoms in an unreconstructed As-rich GaAs (001) surface measured by STM. In contrast, Figure 9.10b displays a reconstructed surface that includes both (2×4) and $c(2 \times 8)$ reconstructions. The (2×4) unit cell consists of three As dimers and one missing dimer in the top layer. The bottom two rows of these unit cells exhibit an "in-phase" arrangement. "Anti-phase" arrangements of these unit cells result in $c(2 \times 8)$ domain boundaries that are twice as large. Slight atomic variations at the boundaries of these two unit cell arrangements can also occur and are termed "kinks." Since the (2×4) and $c(2 \times 8)$ unit cells have different stoichiometries, such surfaces provide an explanation for the range of stoichiometries observed with the same LEED pattern.

(continued)

(continued)

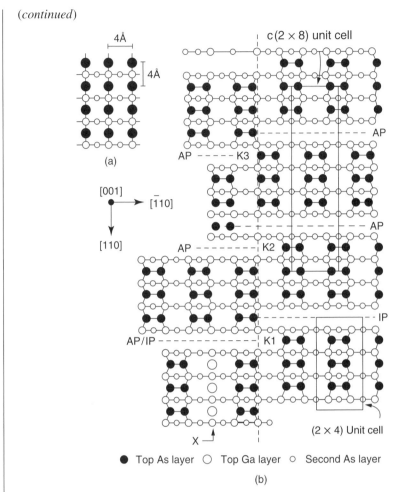

Figure 9.10 *Geometric structure for the GaAs(001) (2 × 4) – c(2 × 8) reconstruction indicated by STM: (a) structure of the unreconstructed GaAs (100) and (b) the missing dimer model for the GaAs (100) (2 × 4) surface producing (2 × 4) or (2 × 8) structures, depending on the in-phase (IP) or antiphase (AP) nature for the missing dimer boundary, respectively. Different boundary kinks – K1, K2, and K3 – can arise, depending on the the interaction of IP and AP domain boundaries. Disorder in the As pairing is denoted by X. (Pashley et al 1988 [7]. Reproduced with permission of American Physical Society.)*

Other features can contribute to LEED patterns, including adsorbed trimers and facets. In general, a wide range of reconstructions are observable for specific orientations of different semiconductors. See the Track II Table T9.2 Binary Compound Semiconductor Treatments and Surface Reconstructions. These surfaces are typically prepared either by epitaxial growth, Ar$^+$ bombardment and annealing, or crystal cleaving. Common reconstruction features are found for the same crystal orientation within the same family of III-V compounds,

Overall, a wide variety of features contribute to the geometric structure of a perfect, intrinsic semiconductor surface.

9.2   Reflection High Energy Electron Diffraction

Reflection high energy electron diffraction (RHEED) uses a grazing incidence electron beam at much higher energies than LEED, typically 20–30 kV, to monitor surface order during epitaxial growth. Because RHEED involves grazing incidence, momentum transfer with the lattice and surface sensitivity are similar for both techniques. RHEED has the additional advantage of providing direct feedback during the growth process so that composition, temperature, and growth rate can be adjusted in real time in order to maintain a desired reconstruction and crystal perfection.

9.2.1   Principles of RHEED

Figure 9.11a shows a schematic of the RHEED electron beam incidence and diffraction geometry. Crystalline islands or "droplets" in (b) produce transmission electron diffraction spots rather than (c) streaks from a flat surface. (d) These streaks correspond to the intersection of the Ewald sphere with the reciprocal lattice rods transverse to the incident beam direction. Streaks are due to the angular spread 2β and energy spread ΔE of the primary beam plus surface deviations such as phonons and lattice defects. Figure 9.12 a–c illustrate the different RHEED patterns for the same $SrTiO_3$ epilayer on Si(100) along three different azimuths. Figure 9.12d–f illustrate the evolution of RHEED patterns from streaks to spots

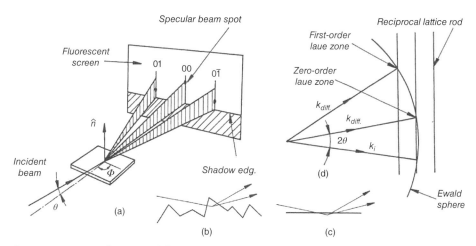

Figure 9.11   *(a) Schematic of the RHEED experimental set-up, (b) bulk scattering by a three-*
-dimensional crystalline island on top of the surface, and (c) surface scattering on a flat surface.
*(d) The Ewald sphere and intersection points of the reciprocal lattice rods (hk) at specific **k***
vectors. The sphere radius is much larger than the distance between the reciprocal lattice rods.
The elongation of spots corresponds to the intersection of the Ewald sphere with 01, 00, and
0 $\bar{1}$ rods. (Herman et al 1989 [8]. Reproduced with permission of Springer.)

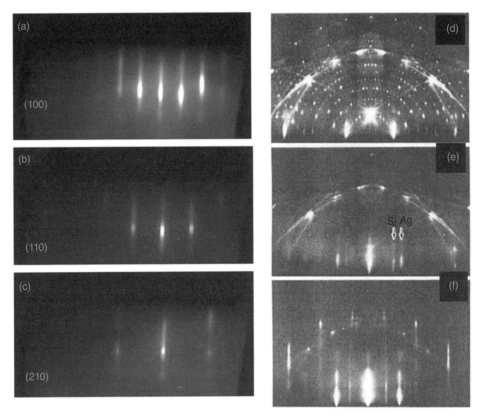

Figure 9.12 *SrTiO₃ on Si(100) RHEED patterns taken with 15 keV along (a) (100), (b) (110), and (c) (210). RHEED patterns taken with a primary energy of E = 15 keV incident along the [112] direction] on a Si(111) surface: (d) Clean Si(111) surface with a (7 × 7) superstructure. (e) After deposition of nominally 1.5 monolayers (ML) of Ag streaks due to the Ag layers are seen on the blurred (7 × 7) structure. (f) After deposition of 3ML of Ag [9]. The textured structure is due to the Ag layers that develop in place of the (7 × 7) structure. (Hasagawa et al 1987 [9]. Reproduced with permission of Elsevier.)*

as Ag is deposited on a clean Si (111) surface with a (7 × 7) reconstruction and becomes textured with an effective thickness of 3 ML.

9.2.2   Coherence Length

Both β and ΔE determine the coherence length Δr_c such that $\Delta r_c \cong \lambda / \{2\beta \sqrt{[1 + (\Delta E/2E)^2]}\}$, the radius over which surface atoms contribute in phase. For both LEED and RHEED, $\Delta E \sim 0.5$ eV, while 2β is 10^{-2} radians for LEED versus 10^{-4}–10^{-5} radians for RHEED, yielding $\Delta r_c \sim 100$ Å for LEED and $>\sim 200$ Å for RHEED. RHEED streak patterns differ with different azimuths φ, permitting one to distinguish between closely related reconstructions. RHEED spots can also be used to study surface corrugation and film growth modes.

9.2.3   RHEED Oscillations

Because RHEED can be performed during growth, it has the additional ability to monitor the layer-by-layer growth of the crystal. Figure 9.13 illustrates the principle behind this phenomenon. The diffraction intensity oscillates between high to low intensity as each atomic layer is completed. The vertical sequence on the left shows how the surface roughness

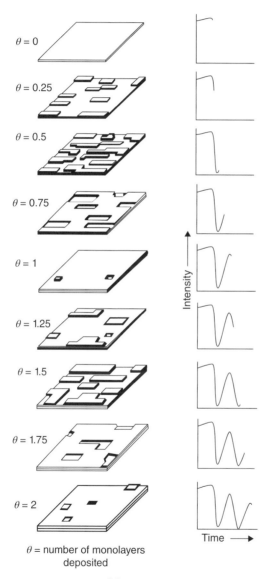

θ = number of monolayers deposited

Figure 9.13   *Schematic representation of the RHEED intensity oscillations associated with the laminar growth of epitaxial monolayers. Diffuse scattering is a maximum at half-layer coverage and a minimum at full monolayer coverage, leading to the oscillations shown. (King and Woodruff 1988. Reproduced with permission of Elsevier.)*

changes as atoms fill in each monolayer. Between completed layers, the RHEED streak intensities are more diffuse and have lower intensity. With increasing time, the start and completion of each monolayer then produces a characteristic oscillation, as shown by the corresponding RHEED intensity versus time plots on the right. These are pronounced with initial deposition and gradually decrease with further deposition as the contrast between completed and half-completed layers diminishes. These oscillations allow the growth operator to "count" the number of atomic layers and the time interval between layers.

9.3   Scanning Electron Microscopy

Scanning electron microscopy (SEM) is a standard technique to obtain micro- and nanoscale morphological information at surfaces and cross-sectional interfaces. The SEM also provides electron excitation for a number of spectroscopies such as AES, XPS, energy dispersive X-ray analysis (EDAX), and cathodoluminescence spectroscopy (CLS). Figure 7.12 illustrates the variety of interactions that the electron beam entering the solid can generate.

9.3.1   Scanning Auger Microscopy

With a UHV SEM, the electron beam can produce full AES spectra, pixel by pixel, across a scanned area. Within each spectrum, one can then produce maps of the intensities at kinetic energies corresponding to particular elements. At surfaces, such maps can probe the formation of varying chemical phases. At interfaces, these maps can identify the presence of atomic interdiffusion and chemical reaction. See, for example, the GaN/Al_2O_3 SEM AES maps in Track II T8.10.

The electron beam also produces X-rays with characteristic core shell energy transitions that are element-specific. As with AES, the spatial resolution is limited by electron beam spread illustrated in Figure 7.12 unless the specimen thickness is less that the cascade diameter. CLS provides a wide range of electronic information including band-to-band transitions, band-to-defect level transitions, and transitions unique to specific surfaces or interfaces, depending on the incident beam energy. See Chapter 11.

9.3.2   Photoelectron Microscopy

The incident beam of electrons can also excite photoelectrons from microscopic regions of a sample's outer few monolayers. This photoelectron microscopy (PEM) technique uses an electron analyzer similar to those used in UPS, XPS, or SXPS but with a high extraction voltage, which enables 50–100 nm spatial resolution. Within the analyzer, apertures enable one to select a subset of wavevectors for ARPES [10]. As an example, Figure 9.14 shows how the variation in film thickness of a few layers of graphene provides significant visual contrast because of differences in work function. Photoemission from different layer thicknesses denoted, from top to bottom, by circles located on 4, 3, and 1 ML thick layers in Figure 9.14a, yield the corresponding micron-scale PEM valence band spectra shown in Figure 9.14b.

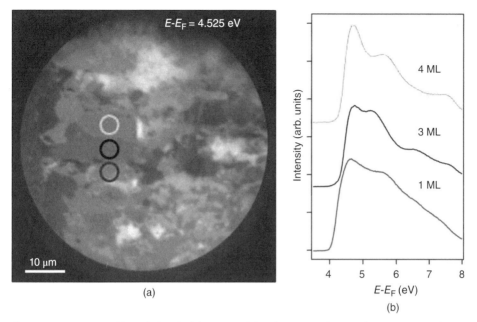

Figure 9.14 *Energy filtered threshold image of few layer graphene. Three regions of interest (ROIs) are defined corresponding to, from top to bottom, 4, 3 and 1 ML graphene; (b) local secondary electron spectra from the ROIs defined in (a) [11]. Barrett et al. 2013 [11]. Reproduced with permission of Elsevier.).*

Besides secondary electron images and localized work function measurements, PEM can provide band structure imaging and dispersion analogous to ARPES from micron-scale regions [12]. Multiphoton and time-resolved PEM is also possible [13].

9.4   Transmission Electron Microscopy

9.4.1   Atomic Imaging: Z-Contrast

TEM provides one of the most direct methods of imaging interfaces. The general expression for particle–particle scattering cross section used for RBS

$$\sigma(\theta) = (Z_1 Z_2 e^2/4E)^2/\sin^4(\theta/2) \tag{7.30}$$

can also be extended to electron scattering with atoms and shows that electron scattering is proportional to the square of the atomic charge, Z^2. Because of this atomic Z^2 dependence, cross-sectional TEM images of ultrathin films only a few nanometers thick can show strong contrast between columns of the same element. Such Z-contrast images provide direct measurement of abruptness on an atomic scale, the formation of new interfacial phases, and/or evidence of atomic interdiffusion.

Case Study: – The Sr_2FeMoO_6 (001) – $SrTiO_3$ (001) Interface

The Sr_2FeMoO_6 – $SrTiO_3$ interface provides an example of Z-contrast imaging for a junction involving five elements in a complex structure. Both constituents have the perovskite crystal structure shown in the schematic drawing of Figure 9.15 [14] and are termed *complex oxides*. Figure 9.15a shows this interface is atomically abrupt and that both perovskites are of single phase near the junction. Figure 9.15b illustrates the uniform atomic ordering of the Sr_2FeMoO_6 layer on an extended scale. Figure 9.15c demonstrates the Z-contrast imaging used to identify the individual elements comprising the Sr_2FeMoO_6 lattice near the $SrTiO_3$ interface.

9.4.2   Surface Atomic Geometry

TEM combined with lattice diffraction can provide structural information near surfaces with very high spatial resolution. This TEM approach has been successful in determining surface structures [10] either by *profile* or *plan-view imaging*. In *profile imaging*, atomic columns are imaged parallel to the surface to obtain information about the size of reconstruction perpendicular to the surface. In *plan-view imaging*, the surface is imaged along the surface normal. Each method has its drawbacks: In profile imaging, quantitative measurements of atomic displacements and relaxation require detailed simulations and the observed surface structure may not be at equilibrium or representative. In plan-view imaging, the transmission geometry has relatively low signal intensity since only a few atomic layers contribute to the overall signal. Radiation at the high incident beam energies and momenta can also damage the crystal lattice.

Besides image processing to obtain high resolution plan-view images of surfaces, beam intensity analysis of diffracted beams can be combined with structure-determining techniques, known as *crystallographic direct methods*, to extract surface structures [12]. These make use of constraints on the phases of measured reflections, such as atomicity (calculated charge densities should have regions of larger charge densities separated by charge-free regions), positivity (charge density in real crystals should always be positive), and localization (significant atomic displacements from bulk positions should occur only in the near-surface region). The maps of diffraction intensities based on these calculated charge densities and atomic displacements can then be compared with experimental images.

9.4.3   Electron Energy Loss Spectroscopy

In addition to Z-contrast imaging, it is also possible to use electron energy loss spectroscopy (EELS) to image the arrangement of atoms at interfaces. Figure 9.16 shows a spectroscopic image of an $La_{0.7}Sr_{0.3}MnO_3/SrTiO_3$ multilayer whose atoms can be identified from the EELS features at energies corresponding to core level thresholds [16]. Thus the maps in Figures 9.16a–c correspond to the EELS threshold intensities at the La M, the Ti L, and the Mn L edges, respectively, in Figure 9.17. An overlay of these images produces the map shown in Figure 9.16d. Lines of different shading indicate Mn–Ti intermixing

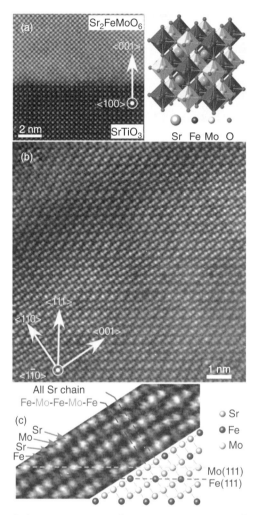

Figure 9.15 Unfiltered *aberration-corrected HAADF STEM images of (a) a* Sr_2FeMoO_6 *(001) film grown on a* $SrTiO_3$ *(001) substrate with an atomically sharp interface, (b) a* Sr_2FeMoO_6 *(111) epitaxial film on* $SrTiO_3$ *viewed along the* $(1\bar{1}0)$ *direction with bright "triplet" patterns indicative of atomic number contrast, and (c) an enlarged STEM image highlighting the triplets (dashed box), each of which is a bright Sr-Mo-Sr chain due to their high atomic numbers separated by a darker Fe (lighter atom) atomic column. Here the Mo-Fe ordering is separated by a Sr chain. The schematic in (c) is the projection of the DP lattice along the* $(1\bar{1}0)$ *direction, which matches the pattern seen in the STEM image of (c). The orientations in (c) are the same as in (b) indicated by the dashed lines. Hauser et al. 2011 [14]. Reproduced with permission of American Physical Society.) Color online: Track II TC 9.15.*

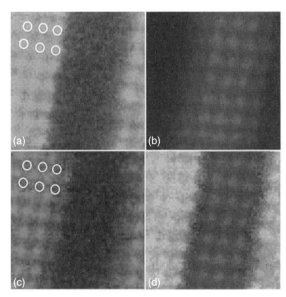

Figure 9.16   *Spectroscopic imaging of a* $La_{0.7}Sr_{0.3}MnO_3/SrTiO_3$ *multilayer, showing the different chemical sublattices as an image extracted from 650 eV-wide electron energy-loss spectra recorded at each pixel. (a) La M edge; (b) TiL edge; (c) Mn L edge; (d) false color image obtained by combining the rescaled Mn, La, and Ti images. Off-color lines at the interface in (d) indicate Mn–Ti intermixing on the B-site sublattice. The white circles indicate the position of the La columns,showing that the Mn lattice is offset. Field of view, 3.1 nm. (Muller et al. 2008 [16] Reproduced with permission of AAAS.) Color online: Track II TC 9.16.*

on the B sublattice of the ABO_3 perovskite multilayer interfaces. EELS spectra are also sensitive to energy losses at band gaps as well as collective excitations such as plasmons. Crystal distortions can also produce image aberrations that can be interpreted to yield strain measurements, notwithstanding effects of *focused ion beam* (FIB) thinning to prepare the samples for TEM.

9.5   Electron Beam Diffraction and Microscopy Summary

LEED and RHEED have provided a wealth of information on surface atomic reconstructions, which depend on surface orientation, growth temperature, surface stoichiometry, and thermal annealing. These reconstructions can involve bond rotations, missing atoms, dimerization, and composites of simpler domain structures. Several of these reconstructions involve large numbers of surface atoms in aggregate structures that help minimize strain and overall energy for structural stability. Within classes of binary III-V and II-VI compounds, the bond rotation model can be used to account for their similar reconstructions. The appearance of different reconstructions with the same surface atom periodicity highlights the importance of multiple techniques to distinguish between different reconstruction models.

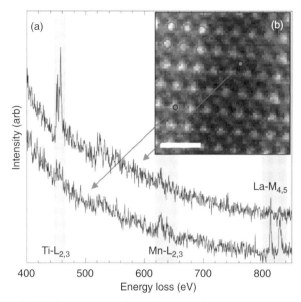

Figure 9.17   *(a) Individual EELS spectra from the series shown in Fig. 9.16. (b) The simultaneously recorded annular dark-field (ADF) image. The large circles show that the La signal from Figure 9.16 is correctly peaked where the annual dark field scattering is strongest. The small red dots indicate the spatial locations from which the EELS spectra were selected. Scale bar, 1 nm. Reproduced with permission of AAAS.) Color online: Track II TC 9.17.*

LEED and STM have revealed the complex and often beautiful organization of atoms on semiconducting and metallic surfaces, providing researchers with insight into their electronic, chemical, and structural properties. RHEED extends electron diffraction to surfaces while the crystals are growing, providing crystal growers with a valuable tool to optimize their growth processes in real time as well as to monitor the crystal growth, monolayer by monolayer. SEM enables imaging on nanometer scale using scanning AES of surfaces and interfaces.

The SEM can also provide electronic information via photoelectron spectroscopy on a sub-micron scale by the PEM technique. TEM together with electron diffraction can provide atomic imaging and reconstruction of crystal surfaces. TEM can provide images of atomic columns at interfaces based on Z^2 imaging as well as by mapping of EELS intensities at thresholds characteristic of specific elements.

Overall, electron beam techniques provide a wide array of tools to probe the atomic-scale geometrical structure of electronic material surfaces and interfaces.

9.6   Problems

1. What is the classical kinetic energy of a 100 eV electron in Joules? What is its momentum in kg m s^{-1}? From the de Broglie relation, what is its wavelength in Å?

2. Calculate the wave vector $k_{i,\parallel} =$ of a 100 eV electron incident on a surface at 45°. Calculate $k_{i,\parallel} =$ for a 400 eV electron. How does this compare with k at a Brillouin zone boundary of a cubic lattice with $a_0 = 5$ Å?

3. Derive the expression for Equation (9.3) from Equation (9.2).

4. For a hemispherical LEED screen located 5.0 cm from an ordered surface, 100 eV electrons produce a square (1×1) pattern of LEED spots with 1.05 cm spacing when projected as a flat pattern. What is the atomic lattice constant at the surface? Suppose the pattern becomes (2×1) with spots at twice the minimum spacing. What is the additional periodic spacing and the unit mesh? What are three possible reasons for this new periodicity?

5. Oxygen adsorption causes the (1×1) LEED pattern of a semiconductor to become (2×1) with half order spots. Describe two possible causes for this transition.

6. The transition in electron diffraction patterns is useful in calibrating the temperature of GaAs growth substrates. Identify a suitable reconstruction transition and calibration temperature.

References

1. Kahn, A. (1983) Semiconductor surface structures. *Surf. Sci. Rep*, **3**, 193.
2. Duke, C.B. (1977) Surface structures of compound semiconductors. *J. Vac. Sci. Technol*, **14**, 870.
3. Duke, C.B. (1992) Structure and bonding of tetrahedrally coordinated compound semiconductor cleavage faces. *J. Vac. Sci. Technol. A* **10**, 2032.
4. Takayanagi, K., Tanishiro, Y., Takahashi, M., and Takahashi, S. (1985) Structural analysis of Si(111)-7x7 by UHV-transmission electron diffraction and microscopy. *J. Vac. Sci. Technol. A*, **3**, 1502.
5. http://www.matscieng.sunysb.edu/leed/7x7.html
6. Takayanagi, K., Tanishiro, Y., Takahashi, M., and Takahashi, S. (1985) Structural analysis of Si(111)-7 × 7 by UHV-transmission electron diffraction and microscopy. *J. Vac. Sci. Technol. A*, **3**, 1502.
7. Pashley, M.D., Haberern, K.W., Friday, W. *et al.* (1988) Structure of GaAs(001) (2x4)-c(2 × 8) determined by scanning tunneling microscopy. *Phys. Rev. Lett*, **60**, 2176.
8. Herman, M.S. and Sitter, H. (1989) *Molecular Beam Epitaxy – Fundamentals and Current Status, Springer Series in Materials Science 7*, Springer-Verlag, Berlin.
9. Hasagawa, S., Daimon, H., and Ino, S. (1987) A study of adsorption and desorption processes of Ag on Si(111) surface by means of RHEED-TRAXS. *Surf. Sci*, **186**, 138.
10. http://en.wikipedia.org/wiki/Photoemission_electron_microscopy
11. Barrett, N., Winkler, K., Krömker, B., and Conrad, E.H. (2013) Laboratory-based real and reciprocal space imaging of the electronic structure of few layer graphene on SiC(000$\bar{1}$) using photoelectron emission microscopy. *Ultramicroscopy*, **130**, 94.
12. Escher, M., Winkler, K., Renault, O., and Barrett, N. (2010) Applications of high lateral and energy resolution imaging XPS with a double hemispherical analyser based spectromicroscope. *J. Electron Spectrosc. Relat. Phenom*, **178**, 303.

13. Xiong, G., Joly, A.G., Hess, W.P. *et al.* (2006) Introduction to photoelectron emission microscopy: Principles and applications. *J. Chinese Electron Microsc. Soc*, **25**, 15.
14. Hauser, A.J., Williams, R.E.A., Ricciardo, R.A. *et al.* (2011) Unlocking the potential of half-metallic Sr_2FeMoO_6 films through controlled stoichiometry and double-perovskite ordering. *Phys. Rev. B*, **83**, 014407.
15. Subrahmanian, A. and Marks, L.D. (2004) Surface crystallography via electron microscopy. *Ultramicroscopy*, **98**, 151.
16. Muller, D.A., Kourkoutis, L.F., Murfitt, M. *et al.* (2008) Atomic-scale chemical imaging of composition and bonding by aberration-corrected microscopy. *Science*, **319**, 1073.

Further Reading

Fadley, C.S. (2010) X-ray photoelectron spectroscopy: Progress and perspectives. *J. Electron. Relat. Phenom.*, **178–179**, 2.
Herman, M.A. and Sitter, H. (1989) *Molecular Beam Epitaxy*, Springer-Verlag, Berlin.
Brydson, R. (2001) *Electron Energy Loss Spectroscopy*, Royal Microscopical Society (Great Britain), Oxford.

10

Scanning Probe Techniques

A variety of scanning probe techniques are now available to image surfaces on an atomic scale including atomic force microscopy (AFM), scanning tunneling microscopy (STM), and ballistic electron energy microscopy (BEEM). These techniques are capable of imaging, measuring, and indeed manipulating surface constituents on a sub-nanometer and even atomic scale. Combined with other surface science techniques and theory, these scanning probe techniques provide many valuable tools to understand the electronic, chemical, and structural properties of surfaces. In each case, the microscope measures the response of a scanning tip to either current that passes between tip and surface or forces between them. Such forces include Van der Waals, electrostatic, magnetic, capillary, and other weaker forces.

10.1 Atomic Force Microscopy

AFM is now a widely used technique to measure surface morphology as well as tribological, electrical, and magnetic forces, all on a nanometer scale. Figure 10.1 illustrates three detection modes of operation for scanning probe microscopes. In Figure 10.1a, a pair of piezoelectric rods that expand or contract with applied voltage controls the distance between a probe tip and the surface. These rods respond to both the forces and tunneling current between surface and tip, causing the tip to either: (i) maintain a fixed position as the vertical distance to the surface varies or (ii) vary position to maintain a constant vertical distance above the surface features as the tip moves horizontally. Here the tip is affected by both surface morphology (dashed line) as well as charge density (dot-dashed line).

In Figure 10.1b, forces between the tip and surface deflect a cantilever spring, causing changes in distance and thereby capacitance between the spring and an electrode. In Figure 10.1c, the reflection of a laser off a force-sensitive cantilever into a position-sensitive photodetector measures the cantilever deflection. The force exerted on the tip in proximity

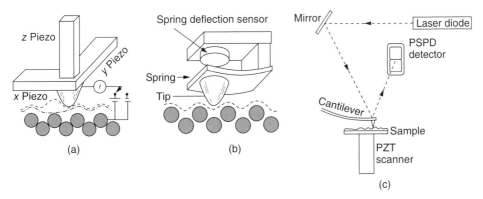

Figure 10.1 *Detection modes of scanning probe techniques (a) scanning tunneling microscopy, (b) capacitance-sensitive force microscopy, and (c) optical-reflection-sensitive force microscopy. (Brillson 2010. Reproduced with permission of Wiley.)*

to a surface is given by Hooke's Law, that is, $F = k_S \cdot d_C$, where k_S = the spring constant of the cantilever and d_C is the cantilever's deflection. The forces between tip and surface are measurable for all surfaces and since no electrical currents between tip and surface are needed, AFM is effective with all materials, both insulators and conductors.

AFM tips can be fabricated from crystalline silicon or other materials that can be etched by orientation-selective techniques. Tips may be coated with other materials, such as diamond for wear resistance or magnetic metals for magnetic force measurements. Metal tips sharpened electrochemically or by field emission can achieve near-atomic sharpness for high resolution tunneling microscopy.

10.1.1 Non-Contact Mode AFM

The AFM non-contact mode allows the tip to probe the surface morphology without disturbing the surface. Since the tip does not contact the surface, it is less likely to change during the scanning process. Without contact, the tip can also measure differences in electrical potential relative to the surface. For example, see Track II T10.1. AFM Electrostatic Force Imaging. For metals and semiconductors, such measurements yield the surface work function and Fermi level position relative to those of the tip.

For magnetically polarized tips, the non-contact mode yields the corresponding polarization of magnetic domains. Track II T10.2. AFM Magnetic Force Microscopy Imaging provides a magnetic force microscope (MFM) example. AFM systems are available to measure both morphology and electric potential simultaneously as the tip scans a surface area, sometimes revealing correlations between structure and electronic properties.

10.1.2 Kelvin Probe Force Microscopy

Kelvin probe force microscopy (KPFM) measures surface electric potential in non-contact mode. Figure 10.2 illustrates how one type of KPFM measures the electric potential between a metal tip and a semiconductor surface. In Figure 10.2a, the potential difference between tip (reference *REF*) and surface represents a capacitor with electrode separation

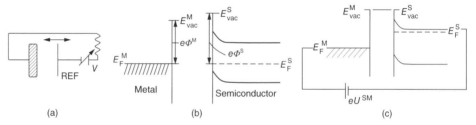

Figure 10.2 *(a) Vibrating Kelvin probe and capacitor with feedback circuit for measuring surface potential. (b) Metal and semiconductor with work functions $q\Phi_M$ and $q\Phi_S$, respectively, relative to vacuum levels E^M_{vac} and E^S_{vac}, respectively, with Fermi levels E_F^M and E_F^S aligned versus (c) with compensating voltage eU^{SM} that aligns E^M_{vac} and E^S_{vac} (Brillson 2010. Reproduced with permission of Wiley.)*

d. This potential difference is offset by a variable compensating voltage source U_{comp} and load resistor in the circuit connecting the two.

For a metal tip and a semiconductor surface with work functions $q\Phi_M$ and $q\Phi_S$, respectively, Figure 10.2b shows that these metal and semiconductor work functions can be expressed in terms of their vacuum levels E^M_{vac} and E^S_{vac} and Fermi levels, E_F^M and E_F^S, respectively, as $q\Phi_S = E^S_{vac} - E_F^S$ and $q\Phi_M = E^M_{vac} - E^M_{vac}$. With the Fermi levels aligned, the work function difference is defined as the *contact potential difference* U^{SM} such that

$$e(\Phi_S - \Phi_M) = E^S_{vac} - E^S_F - E^M_{vac} + E^M_F = E^S_{vac} - E^M_{vac} = U^{SM} \qquad (10.1)$$

In general, the charge Q on each capacitor plate is given by $Q = CV = C[-e(\Phi_S - \Phi_M) + U_{comp}]$. If the spacing of the reference probe from the semiconductor is modulated at a frequency ω such that $d = d_0 \sin(\omega t)$, the resultant vibrating capacitor generates a displacement current

$$I = dQ/dt = dC/dt[-e(\Phi_S - \Phi_M) + U_{comp}] \qquad (10.2)$$

and a voltage $I \cdot R$ across the resistor. This displacement current is typically small, for example, 10^{-10} A, requiring a high impedance preamplifier to generate significant voltages. Equation (10.2) shows that when compensating voltage U_{comp} is equal to $e(\Phi_S - \Phi_M)$, $I = 0$ and U_{comp} equals the contact potential difference U^{SM}.

Case Study: ZnO Nanomounds and Surface Potential

The combination of AFM and KPFM provide significant correlations between morphological and electrical properties of ZnO surfaces. Figure 10.3a illustrates an AFM map of nanometer scale features on a ZnO(000$\bar{1}$) surface [1]. These features correspond to clusters of "nano-mounds" hundreds of nm wide and tens of nm high separated by relatively smooth (1.27 nm rms) regions. The KPFM map in

Figure 10.3b shows that the nanoclustered regions have lower contact potential difference, corresponding to higher work function ($e\Phi_S$ in Figure 10.2b) for this particular AFM and tip. In turn, the work function changes reflect changes in the semiconductor band bending and/or electron affinity. Here the formation of clusters contributes to the formation of Zn vacancies, increasing the band bending and moving E^S_F lower in the ZnO band gap.

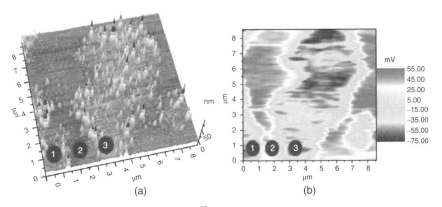

(a)   (b)

Figure 10.3 *(a) AFM map of ZnO(000$\overline{1}$) surface with two adjoining surface morphologies: a smooth surface region (2) bounded by surfaces with clustered nano-mounds (1 and 3). (b) KPFM map from the same area. Regions (1) and (3) show lower contact potential difference than (2). Reproduced with permission of Elsevier.) Color online: Track II TC10.3.*

10.1.3   Contact Mode AFM

Contact mode AFM has several applications including surface friction measurements, that is, *tribology*, and its relation to surface morphology and adsorbates. Here the deflection of the cantilever tip provides a gauge of surface friction and/or height variations. See Track II T10.3. AFM Surface Friction Imaging. However, care is needed to distinguish effects of both the tip and surface charging during the measurement. This mode also enables contacts to metal electrodes for electric measurements such as charge transport and capacitance at nanoscale Schottky barriers or heterojunctions.

10.2   Scanning Tunneling Microscopy

Scanning tunneling microscopy (STM) has had an enormous impact on the study of electronic material surfaces. It has helped resolve many of the semiconductor surface reconstructions described in Chapter 9 whose geometric structures have been unresolved by LEED measurements alone. STM can provide electronic structure information including densities of filled and empty states at semiconductor surfaces and in their bulk. It can also provide atomic-scale maps of surface charge density that reveal the positions of individual elemental constituents. It can detect native point defects, such as lattice vacancies, the elemental structure and imperfections of atomic layer steps, and the formation of quantum-scale objects such as nanoclusters, nano-"huts", and quantum dots.

Despite the similarities of AFM and STM involving a scanning probe tip, STM is based on the principle of quantum mechanical tunneling rather than local forces. This tunneling can occur for very small (<10 Å) tip-material distances d and involves small (pA to nA) tunnel currents I_T flowing across the gap between them. This tunnel current depends exponentially on d according to

$$I_T \propto (V/d)\exp(-kd\sqrt{\phi_{av}}) \tag{10.3}$$

where ϕ_{av} is the average work function, $V(<< e\phi_{av})$ is an applied voltage across d, and $k \sim 1.025 \text{ Å}^{-1} \text{ev}^{-1/2}$.

10.2.1 STM Overview

Figure 10.4 illustrates the key components of an STM and the conversion of piezoelectric signals from a scanning tip into a tunneling image. A feedback circuit controls the piezoelectric voltages to maintain a constant position of the tip above the sample surface.

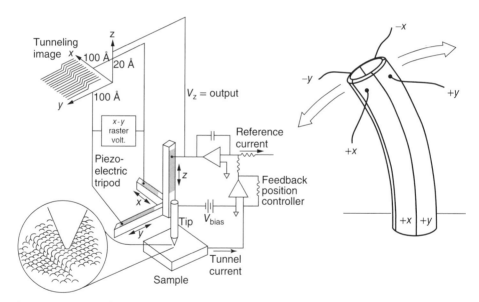

Figure 10.4 *(a) Schematic of the STM tip, sample surface, and electric circuit used to detect tunneling current. The current from tip to sample with an applied bias voltage V_{bias} is maintained constant by an electronic feedback system that controls the tip position normal to the sample by means of a piezoelectric tripod. The voltage applied to the z-leg of the tripod produces a tunneling image as the x − y tripod legs raster scan the tip laterally across the surface. (Quate, C. 1986 [3].) Reproduced with permission of American Institute of Physics (b) A piezoelectric scanner tube with electrodes attached to its side drives changes in tube length along different directions [2].*

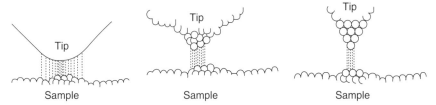

Figure 10.5 *Tunneling current visualization (dashed lines) between tip and sample surface. A blunt tip (left) yields only low-resolution images, whereas tips with only a few atoms (center) or ideally one atom (right) achieve proportionally higher spatial resolution. (Quate, C. 1986 [3]. Reproduced with permission of American Institute of Physics.)*

Because of the small tip–material distances required, the instrument must control d to a precision of $0.05 - 0.1$ Å – a formidable challenge! Furthermore lateral resolution must be controlled to within 1–2 Å in order to resolve individual atoms. The challenge then becomes to (i) reduce mechanical vibrations down to the sub-Å level and (ii) prepare tips with atomic sharpness.

Both challenges were overcome with ingenious designs of piezoelectric positioners and vibration isolation. Rods with sidewalls coated with electrically isolated piezoelectrics (Figure 10.4b) with one end fixed can expand or contract in length by atomic-scale increments with a combination of applied voltages, driving the tip to control both normal and lateral positions on this scale. Various techniques are used in order to reduce mechanical vibrations to this level. These include equipment tables supported by heavily loaded air legs, tip/sample stages mounted in UHV with long springs with small restoring force constants k, and heavy concrete floors separated from building floors that "float" in order to avoid building vibrations. Weak springs supporting large mass m frames are aimed at lowering the resonant frequencies ω of equipment coupled to the tip/sample stage since $\omega = \sqrt{(k_S/m)}$.

Ideally, the STM tip should consist of a single protruding atom in order to obtain images with atomic-scale resolution. Figure 10.5 illustrates the tunnel current between tip and sample for successively smaller numbers of atoms in proximity to a sample surface [3]. As the number of protruding atoms decreases, the tunneling current becomes increasingly localized and the spatial resolution of the tip increases accordingly.

With sufficient tip sharpness and vibration isolation, STM researchers have obtained beautiful atomic images of surfaces, such as that shown in Figure 10.6. This image of the Si (7×7) surface discussed in Chapter 9 illustrates the various superstructure atoms and the extended nature of its periodicity.

Careful inspection reveals changes in domain structure between the upper and lower halves of the topography. Furthermore, Figure 10.6 displays wavelike features spreading across the lower half of the topography, indicating a long-range coupling of the surface atoms. Images such as this illustrate the dramatically higher information content afforded by STM to the understanding of surface atomic structure and associated chemical and electronic properties.

Figure 10.6   *STM image of the Si(111) (7 × 7) surface. (Reproduced with permission of J. Pelz, Ohio State University.)*

10.2.2   Tunneling Theory

The tunneling current at the core of STM involves the wave functions of electrons that can move as traveling waves inside their solids but can also "leak" out of the surface into vacuum. See, for example, Figure 4.1. In quantum mechanical terms, these electrons have wave functions

$$\psi_S(z) = \psi_S{}^0 \exp(-kz) \tag{10.4a}$$

$$\psi_T(z) = \psi_T{}^0 \exp(-k(d-z)) \tag{10.4b}$$

for the surface and the tip, respectively, at a separation distance d and wave vector

$$k = [2m(V-E)/\hbar^2]^{1/2} = [2m\phi]^{1/2}/\hbar \tag{10.5}$$

where ϕ is the energy barrier for electrons with mass m and kinetic energy E relative to a local potential V. In physical terms, these two wave functions represent traveling waves with amplitudes that oscillate sinusoidally inside their solids but that decrease exponentially with distance from their respective surfaces. Figure 10.7a shows a simplified picture of electron "baths" in two solids with different work functions Φ_1 and Φ_2. Their wave functions $\psi_S(z)$ and $\psi_T(z)$ extend into vacuum but their electrons' probability $|\psi|^2$ at that distance decreases exponentially with increasing d. When these two solids are brought close enough together, $\psi_S(z)$ and $\psi_T(z)$ begin to overlap and there is a finite probability of tunneling between the two baths. The shape of the barrier between the two solids through

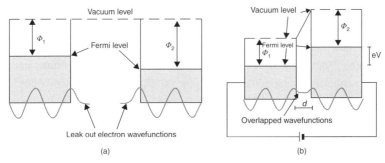

Figure 10.7 *Simplified illustration of electrons near the Fermi level of two metallic solids with work functions Φ_1 and Φ_2 before (a) and after (b) joining electrically. (Brillson 2010. Reproduced with permission of Wiley.)*

which tunneling occurs will depend on the distance d between the two surfaces, their work functions, and their work function difference.

The tunneling current depends on the probabilities of finding electrons whose wave functions extend out from the solids into the gap between them. It is proportional to the product of their individual probabilities, $|\psi_S^0|^2 |\psi_T^0|^2 e^{-2kd}$, which decreases with increasing tip–surface separation d and increasing energy barrier ϕ. For a spherical tip and position r, this tunnel barrier is found (Track II. T10.4. STM Tunnel Current and Barrier) to depend on the current according to $\phi \propto (\partial \ln I/\partial r)^2$.

In Figure 10.7b, the two solids are connected electrically and their Fermi levels align. A voltage eV between them shifts their Fermi levels by the amount shown. For a bath with Φ_1 and bath with Φ_2, electrons can tunnel from occupied states in bath 2 to unoccupied states in bath 1 so that a tunnel current I_T of electrons described by Equation (10.3) passes between them. For a tip at a surface, both the gap distance d and potential ϕ change locally so that a record of I_T versus position results in a density of states (DOS) "map" as the tip moves laterally across the surface.

Figure 10.8a shows tunneling from a filled DOS at a solid surface with negative bias and separation d into an empty DOS of a tip biased more positively. When the tip is biased more

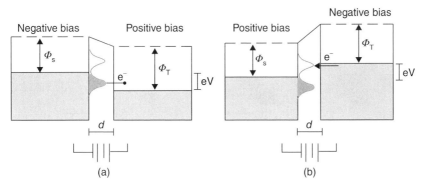

Figure 10.8 *Tunneling from filled to empty densities of states at a given energy and position as a function of applied bias (electron volts). (Brillson 2010. Reproduced with permission of Wiley.)*

negatively, Figure 10.8b shows that tunneling occurs in the opposite direction – from filled states in the tip to unoccupied states in the surface. By changing tip bias from positive to negative, the magnitude of tunneling current probes either the filled or the empty surface density of states. Tunneling from occupied states in the sample to empty states in the tip provides DOS distributions analogous to those of photoemission spectroscopy but based on a tunneling conductance averaged over an area defined by the tip. Conversely, tunneling from occupied states in the tip to empty states in the sample provides DOS distributions analogous to those of *inverse photoemission spectroscopy*. The local density of states (LDOS) at a given position and energy is directly proportional to the tunneling conductance at that position and energy. See Track II T10.4.

Case Study: Filled and Empty States on a GaAs Surface

Figure 10.9 illustrates the difference in spatial distribution between filled and empty states for a GaAs (110) surface [5], which contains equal numbers of Ga and As atoms. Density of states calculations indicate that the occupied states density is concentrated around the As atom while the unoccupied state density is situated around the Ga atoms, as described with Figure 9.4. Positive GaAs bias induces tunneling from filled states in the tip into empty states in the GaAs, while negative GaAs bias induces tunneling out of filled states in the GaAs into empty states in the tip. Here the contrast between images acquired with opposite bias voltages shown in Figure 10.9a and b reveals features due to filled versus empty states that show a shift in lateral position in tunneling intensity that is consistent with the unit cell structure in Figure 10.9c. The separation between occupied and unoccupied states is due to the charge transfer that occurs from Ga to As atoms at the reconstructed free (110) surface. This example shows that bias-dependent STM can detect the spatial distribution of filled and empty bond orbitals for atoms at semiconductor surfaces.

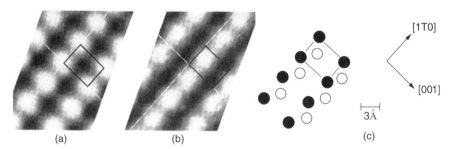

(a)   (b)   (c)

Figure 10.9 *Constant-current STM images for the GaAs(110) surface simultaneously acquired at (a) +1.9 and (b) −1.9 V sample voltages. Surface height in grey scale ranges from 0 (black) to (a) 0.83 and (b) 0.65 Å (white). (c) A top view of the surface atoms represents As (open circles) and Ga (filled circles) atoms with the unit cell indicated in all three figures. (Feenstra et al. 1987 [5]. Reproduced with permission of American Physical Society.)*

10.2.3 Surface Atomic Structure

STM has provided a wide range of three-dimensional atomic structures at surface and interfaces with unprecedented detail. Beside intricate unit cell patterns and extended wave-like patterns as already seen with the Si (7 × 7) surface shown in Figure 10.6, STM can image extended features, such as step edges, dislocations, and atomic clusters. Figure 10.10 illustrates a vicinal Si surface intentionally miscut 5° off the (001) axis, showing a succession of terraces descending from upper left to lower right and alternating planes of Si atoms with dimers oriented either parallel or orthogonal to the step edges. Depending on the dimer orientation, these step edges are rough or smooth. Deposition of Ge on a Si (001) surface can produce three-dimensional "quantum huts" that even have surface facets that are crystallographically oriented (Figure 10.10b).

Atomic-scale STM in cross-section can provide chemical information at electronic material interfaces. Figure 10.11 exhibits a cross-section of a cleaved AlAs/GaAs superlattice, revealing Al–Ga interdiffusion and preferential formation of As precipitates on the GaAs side of the heterojunctions. Annealing this superlattice promotes precipitate nucleation away from the interface and affects the apparent superlattice period.

On a larger scale, STM illustrates how dislocations at a semiconductor surface affect the atomic arrays extending hundreds of nm away. Figure 10.12a shows STM images of 6 nm Pt on GaN (0001) that display features of the underlying semiconductor surface. Besides the step features of the GaN lattice and pits where the threading or mixed threading/screw dislocations emerge, Figure 10.12a shows how the pits distort or "pin" the steps into "wing"-like features. Figure 10.12b shows how isolated screw dislocations produce growth structures with matching spiral arms, again over hundreds of nm.

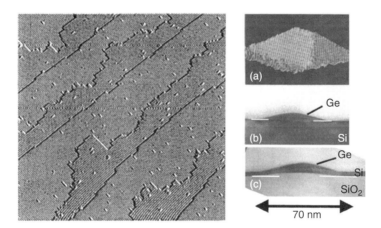

Figure 10.10 *(a) STM image of a Si(001) surface misoriented by 0.5°. (Swartzentruber et al. 1993 [6]. Reproduced with permission of American Physical Society.) Ge quantum dot hut crystal on Si(001) (a) in TEM cross-section on Si (b), and on silicon-on-insulator (SOI) consisting of 1.5 nm of Ge deposited by MBE (c).*

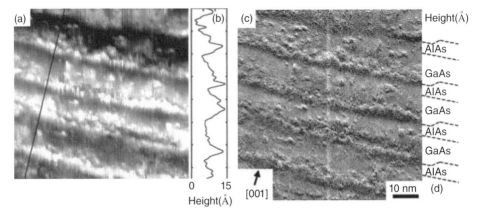

Figure 10.11 *Non-stoichiometric AlAs/GaAs superlattice (a) topographical map, (b) corresponding line profile along the [001] black line, (c) corresponding current image, and (d) identification of GaAs and AlAs layers. (Lita et al. 1999 [7]. Reproduced with permission of American Institute of Physics.)*

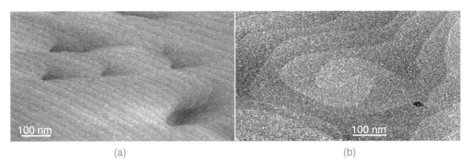

Figure 10.12 *STM image of GaN(0001) surface with a 6-nm thick Pt metal overlayer showing (a) monolayer steps and "wing" structures and (b) spiral growth structure. The granularity is due to nm-scale Pt clusters. (Im et al. 2001 [8]. Reproduced with permission of American Physical Society.)*

10.3   Ballistic Electron Energy Microscopy

Ballistic electron energy microscopy (BEEM) is a scanning probe technique that reveals electronic features of semiconductor surfaces and metal–semiconductor interfaces. The BEEM technique uses a third terminal in the tunneling process to control the kinetic energy of the tunneling electrons [9]. Figure 10.13a shows that a positive bias on the metal terminal relative to the tip raises the tip's Fermi level and draws *ballistic* (i.e., unscattered) electrons toward the metal and semiconductor. These ballistic electrons pass into a semiconductor conduction band if their kinetic energy entering the semiconductor exceeds the semiconductor's conduction band minimum. The onset of ballistic electron tunneling into the semiconductor occurs at a threshold voltage shown in Figure 10.13b and corresponds

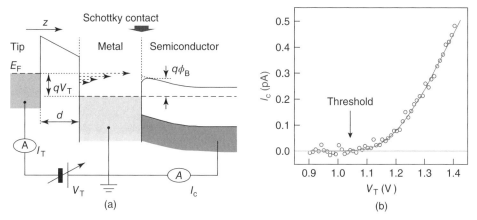

Figure 10.13 *(a) Energy level diagram of BEEM. (b) Representative BEEM spectrum and fitting curve to extract local Schotty barrier height. (Brillson 2010. Reproduced with permission of Wiley.)*

to the barrier height $q\Phi_B = E_C - E_F$. This threshold voltage is thus a direct measure of the Schottky barrier height on a local scale defined by the probe tip.

BEEM, $I-V$, and $C-V$ measurements provide considerable evidence for inhomogeneous Schottky barriers at conventional metal–semiconductor interfaces. Track II T10.5 STM and BEEM of Au on GaAs provides an example for a clean, smooth Au–GaAs interface prepared in UHV. Besides Schottky barriers, BEEM has been used to probe heterojunction band offsets, quantum wells and morphological features such as dislocations.

The advantages of BEEM include: (i) nm-scale real-space information about buried interfaces, (ii) good energy resolution ~5 meV at room temperature, (iii) direction information about local conduction bands, and (iv) direct information about local electric fields. Disadvantages of BEEM include that it: (i) uses a tip, which can, and often does, change with scanning, (ii) requires a "good" (i.e., non-leaky) barrier since a leaky barrier degrades signal-to-noise, (iii) requires a metal contact and conducting substrate, (iv) involves multiple interfaces at which hot-carrier scattering is not well known.

10.4 Atomic Positioning

The scanning probe tip is also capable of moving atoms across the surface to form patterns that display quantum mechanical features of the surface electrons. Figure 10.14 represents schematically a probe tip brought near a weakly bonded surface adsorbate, for example, a Xe atom. The van der Waals attraction between the adsorbate and the tip allows the tip to move laterally without contacting the surface underneath and drag the Xe atom with it, still adsorbed to the surface (Figure 10.14a). Once repositioned, the tip is retracted. For adsorbed atoms that exchange charge with the surface, the tip can be biased so that the tip–adsorbate attraction is stronger than the surface-atom bonding with the substrate. Once the tip moves the attached surface atom to its new position, the tip polarity is reversed and

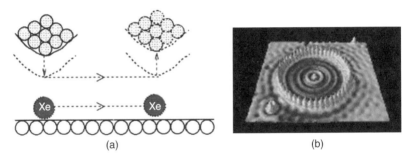

Figure 10.14   *(a) An STM tip brought close to an adsorbed atom can drag and reposition it on the surface before it is retracted. The atom remains in contact with the surface. (b) Quantum interference pattern from a Fe atom array on a Cu (111) surface. (Brillson 2010. Reproduced with permission of Wiley.) Color online: Track II TC 10.14.*

the atom rebonds with the substrate. This technique requires UHV conditions to prevent any residual air molecules from screening the forces between tip, adsorbate and surface.

Figure 10.14b displays an STM image set up by a ring of Fe atoms on a Cu(111) surface positioned using a low-temperature STM tip. Here each peak in the circle corresponds to a Fe atom. The wave pattern is due to scattering of Cu electrons off the Fe adatoms and point defects that produce an interference pattern.

10.5   Summary

Scanning probe techniques have opened a window into the structure and properties of atoms at surfaces. AFM can measure: (i) surface morphology, (ii) electrical potential, (iii) capacitance, (iv) carrier concentration, (v) magnetic polarization, and (vi) friction on a nanometer scale. AFM is highly versatile as a surface probe technique since it can measure properties of both conductors and insulators.

The BEEM technique provides: (i) Schottky barrier heights and their lateral inhomogeneity, (ii) the alignment of conduction bands at heterojunctions at the nm level, (iii) local electric fields, and (iv) "hot-carrier" effects – all with nanometer or better resolution.

The STM technique can measure (i) the geometrical arrangement of atoms within the surface unit cell, (ii) the extended distribution of atoms across tens or hundreds of nanometers and their relation to local strain and defects, (iii) the geometric positions of atoms adsorbed to a surface, (iv) the formation of extended terrace and cluster structures driven thermodynamically, (v) atomic segregation at interfaces, and (vi) the densities of filled and empty states within individual unit cells. The STM's tip can also be used to move atoms across surfaces and position them in arrays that display quantum interference and that can be used to construct electrical circuits at the atomic level.

10.6   Problems

1. Calculate the wavelength for an STM tip with a 5.0 eV work function across a vacuum gap. If the tip is initially positioned 2 Å from a surface and is moved 1 Å further away, does the tunneling current increase or decrease and proportionally by how much?

2. In BEEM, the beam of electrons is sharply forward-focused. Calculate the angle at which the transmitted intensity decreases by $1/e$ for a metal thickness of 50 Å, a sample work function of 4.3 eV, and an electron kinetic energy of 1 V. What is the corresponding spatial spread?

3. In Figure 10.14b the pattern inside the quantum "corral" is due to interference between electrons in the Cu surface. Calculate the wavelength of these electrons. Is it consistent with the wave pattern shown? Hint: use the separation between Fe atoms to establish a length scale.

References

1. Merz, T.A., Doutt, D.R., Bolton, T., Dong, Y., and Brillson, L.J. (2011) Native defect formation with ZnO nanostructure growth. *Surf. Sci. Lett.*, **605**, L20.
2. Golovchenko, J.A. (1986) The tunneling microscope: A new look at the atomic world. *Science*, **232**, 48.
3. Quate, C. (1986) Vacuum tunneling: A new technique for microscopy. *Phys. Today*, **39**, 26.
4. http://www.chembio.uoguelph.ca/educmat/chm729/STMpage/
5. Feenstra, R.M., Stroscio, J.A., Tersoff, J., and Fein, A.P. (1987) Atom-selective imaging of the GaAs(110) surface. *Phys. Rev. Lett.*, **58**, 1192.
6. Swartzentruber, B.S., Kitamura, N., Lagally, M.G., and Webb, M.B. (1993) Behavior of steps on Si(001) as a function of vicinality. *Phys. Rev. B.*, **47**, 13432.
7. Lita, B., Ghaisis, S., Goldman, R.S., and Melloch, M.R. (1999) Nanometer-scale studies of Al-Ga interdiffusion and As precipitate coarsening in nonstoichiometric AlAs/GaAs superlattices. *Appl. Phys. Lett.*, **75**, 4082.
8. Im, H.-J., Ding, Y., J.P., Heying B., and Speck, J.S. (2001) Characterization of individual threading dislocations in GaN using ballistic electron emission microscopy. *Phys. Rev. Lett.*, **87**, 106802.
9. Bell, L.D. and Kaiser, W.J. (1988) Observation of interface band structure by ballistic-electron-emission microscopy. *Phys. Rev. Lett.*, **61**, 2368.

Further Reading

Stroscio, J.A. and Kaiser, W.J. (eds) (1993) *Scanning Tunneling Microscopy*, Academic, New York.

11

Optical Spectroscopies

11.1 Overview

A variety of optical probe techniques that use light in the near-ultraviolet to near-infrared range can provide useful information about the physical properties of the surfaces and interfaces of electronic materials. Purely optical techniques (i.e., photon excitation and detection) include reflectance, ellipsometry, and Raman spectroscopy. Optical techniques that induce electrical changes include photocurrent and photovoltage spectroscopies. Conversely, electron excitation that induces optical emission can provide information specific to surfaces as well as interfaces below the surface.

Optical spectroscopies are advantaged in terms of their high energy precision both for excitation and detection, their non-destructive nature, and their ability to provide *in situ* or remote sensing. However, an inherent disadvantage of purely optical spectroscopies is their low surface sensitivity. This chapter reviews these various techniques, how they take advantage of different physical phenomena to achieve surface sensitivity, and the surface and interface information they can provide.

11.2 Optical Absorption

For optical techniques that involve photons incident on a sample, the optical absorption inside the material is a key parameter. The initial intensity I_0 of an incident light beam that penetrates a surface is attenuated by absorption with increasing path length d within the material according to

$$I = I_0\, e^{-\alpha d} \qquad (11.1)$$

and *optical absorption coefficient*

$$\alpha = 4\pi\, \kappa(\omega)/\lambda \qquad (11.2)$$

Here, ω is the *radiation angular frequency*, λ is the *optical wavelength*, $\kappa(\omega)$ is the frequency-dependent imaginary part and $\eta(\omega)$ is the real part of the *complex refractive*

An Essential Guide to Electronic Material Surfaces and Interfaces, First Edition. Leonard J. Brillson.
© 2016 John Wiley & Sons, Ltd. Published 2016 by John Wiley & Sons, Ltd.
Companion Website: www.wiley.com/go/Brillson/

index $n(\omega) = \eta(\omega) + i\kappa(\omega)$. Optical wavelength λ is related to *frequency* ν according to $\lambda\nu = c$, where c is the speed of light. See Track II T11.1 Optical Absorption Coefficient for derivation of these equations. Figure 11.1 illustrates the energy dependence of $\eta(\omega)$ and $\kappa(\omega)$ for InP, a typical III-V compound semiconductor, from below its fundamental band gap into the near-ultraviolet energy range. The energy E_0 at which $\kappa(\omega)$ begins to increase corresponds to the fundamental band gap transition, that is, electrons excited from the highest valence to the lowest conduction band. Both $\eta(\omega)$ and $\kappa(\omega)$ exhibit strong peaks at energies labeled E_1, E_2, and E_1' that correspond to transitions involving valence to higher energy conduction band transitions.

Absorption coefficient α is defined such that the incident light is attenuated by a factor $I/I_0 = e^{-(4\pi/\lambda)\kappa(\omega)d} = e^{-1}$ when $d = 1/\alpha$. α becomes significant for energies $E = h\nu$ above a semiconductor's band gap and varies strongly at higher energies. For example, $\lambda = 4416$ Å (2.80 eV) He-Cd laser excitation corresponds to $\kappa \sim 2$ in Figure 11.1 so that $d \simeq 4416$ Å$/[4\pi \cdot 2] \simeq 175$ Å. Lower energy (longer wavelength) excitation attenuates over correspondingly longer distances. Figure 11.2 shows the absorption coefficient of GaAs near its fundamental (i.e., lowest energy) band gap. Here $\alpha = 10^4$ cm^{-1} at its 1.43 eV threshold energy so that $d = 1/\alpha = 10^{-4}$ cm $= 10\,000$ Å. Thus probe depths for excitation with visible to near-UV light are in this range. Such extended probe depths highlight the challenge for surface studies using optical excitation. Since surface layers are typically 4–5 Å, the ratio of surface-to-bulk excitation is ~ 5Å$/10\,000$ Å $\sim < 10^{-3}$, so that the signal-to-background ratio is very small.

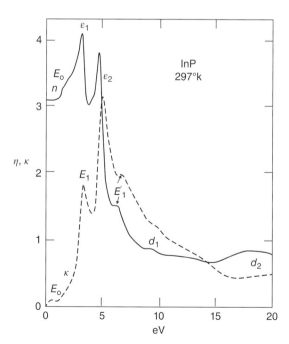

Figure 11.1 *Refractive index $\eta(\omega)$ and absorption index $\kappa(\omega)$ of InP at room temperature. (Cardona, M. 1965 [1]. Reproduced with permission of American Institute of Physics.)*

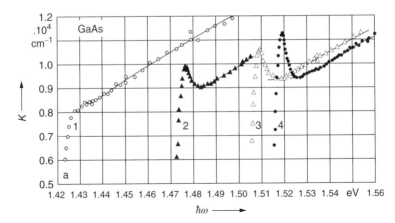

Figure 11.2 *Absorption coefficient versus photon energy in the region of the excitonic absorption edge for four temperatures: (i) 294, (ii) 126, (iii) 90, and (iv) 21 K. (Sturge, M.D. 1962. Reproduced with permission of American Physical Society.)*

Figure 11.2 exhibits two other characteristic semiconductor band gap features: (i) The band gap is temperature-dependent, increasing by nearly 0.1 eV between room temperature and 21 K for GaAs. Similar temperature variations are found for most other semiconductors. (ii) The absorption coefficient displays a peak feature at the absorption edge, due to exciton absorption several tens of meV below the conduction-to-valence band energy. Figure 11.2 also shows that absorption continues to increase with increasing energy, more so near *critical points*, energies between the valence band and higher energy conduction bands with high joint densities of states.

11.3   Modulation Techniques

Several experimental techniques are available to achieve high surface sensitivity with optical excitation: (i) Differential spectroscopy to remove the high background of bulk signals; (ii) multiple surface scattering including total internal reflection and multiple stacked interfaces to amplify the surface signal; (iii) surface or interface-specific features due to the addition or subtraction of adsorbates or other chemical processing; (iv) higher energy excitation to reduce the absorption length; (v) electrostatic sensitivity to detect small charge transfer changes electrically. Figure 11.3 illustrates a *differential reflectance spectroscopy* technique for detecting changes in surface adsorption. In Figure 11.3a, a light source is modulated in a square wave pattern and split in two so that half the light reflects off a specimen into an optical detector while the other half passes directly into the detector. Subtraction of the two beams enables detection of small intensity changes even though the intensities of the two beams separately are high. This method provides highly precise measurements of $\Delta R/R$, the *relative reflectance* change since any variations in light intensity with photon energy or experimental artifacts are cancelled out by the split beam. The grazing incidence geometry pictured in Figure 11.3a enables light polarized in the incident plane to couple strongly to adsorbate dipoles normal to the surface, since the light's

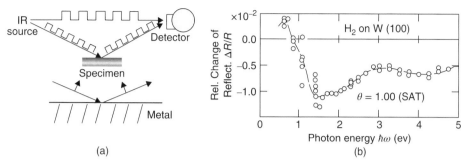

Figure 11.3 *(a) Split, modulated beam geometry for modulated reflectance spectroscopy. Polarization is in the plane of reflectance. (Brillson 2010. Reproduced with permission of Wiley.) (b) Spectral dependence of relative reflectance change due to saturation coverage of hydrogen ($\theta = 1$) taken to be two H atoms per W atom. The theoretical (solid) curve is obtained from an oscillator fit of ε_s to reproduce the $\Delta R/R$ data points. (Anderson et al. 1974 [2]. Reproduced with permission of American Physical Society.)*

electric field vector (indicated by arrows) is nearly parallel to any surface normal dipoles. Figure 11.3b shows how this technique is used to measure the $\Delta R/R$ spectrum for hydrogen molecules (H_2) on a clean tungsten (W) surface. An oscillator fit to the measured changes in $\Delta R/R$ yields changes in the imaginary part of the surface dielectric constant κ due to H_2 adsorption.

Besides differential reflectance, other modulation spectroscopies include (i) electroreflectance spectroscopy (surface electric field modulation), (ii) thermoreflectance spectroscopy (temperature-dependent changes in band structure features), (iii) wavelength modulations spectroscopy, and (iv) piezo-optical spectroscopies, that is, pressure-dependent changes in band structure detected optically.

11.4   Multiple Surface Interaction Techniques

Surface sensitivity is enhanced by multiple reflections and absorption of light with a surface. An example of this technique is the detection of surface states at a clean Ge surface.

Case Study: Ge–Oxygen reflectance

Figure 11.4 displays optical absorption spectra of Ge(111) obtained before and after exposure to an oxygen atmosphere. The clean Ge surface was prepared by cleaving a Ge single crystal in UHV with two opposing blades to expose a fresh cleavage plane, as shown in the Figure 11.4b insert. Since the Ge band gap is 0.67 eV, the optical absorption at 0.5 eV occurs in the band gap. Light in this energy range penetrates far into the crystal at normal incidence, but undergoes total internal reflection at glancing incidence. Here light with intensity I_0 is incident on a prism face of one of the cleaved

(continued)

(continued)

Ge sections and undergoes multiple reflections before exiting the crystal at another prism face with intensity I.

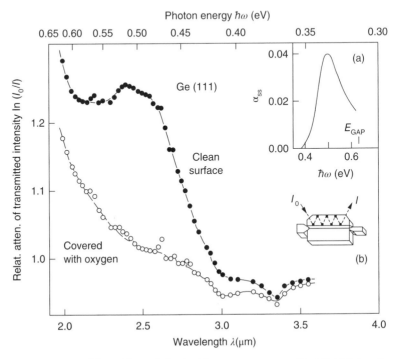

Figure 11.4 *Natural logarithm of the incident versus transmitted light intensity ratio I_0/I versus wavelength for the clean cleaved and oxygen-covered Ge(111) surface $(P(O_2) \sim 10^{-6}$ Torr). The surface absorption constant α_{ss} attributed to surface state transitions of the clean surface appears in insert (a), insert (b) shows the experimental configuration schematically. (Chiarotti et al. 1971 [3]. Reproduced with permission of American Physical Society.)*

With exposure to oxygen, the cleaved surface Ge atoms bond to O atoms, removing the 0.5 eV feature. Since only the surface has changed, the difference between these two spectra represents the absorption due to Ge surface states alone.

Here surface adsorption that removes an electronic feature provides a method to identify the clean surface feature, but signal amplification by multiple reflections is required for optical detection. A significant disadvantage of this approach is the difficulty in achieving this geometry in UHV and altering the surface chemical/electronic structure more than once. Nevertheless, multiple reflection techniques enable detection of the surface atoms' vibrational modes as well, providing considerable information about gas molecular reactions and bonding with semiconductors.

11.5   Spectroscopic Ellipsometry

Spectroscopic ellipsometry (SE) is a powerful and nondestructive technique to measure the optical properties of thin films and their surfaces. The SE technique involves light reflection from a surface and uses a broad band light source from which individual wavelengths can be selected and whose polarization can be controlled. SE measures both the change in light intensity and polarization as a function of wavelength and reflection angle. The surface sensitivity is due to the high precision of angular settings that determine the material's dielectric properties according to the ratio ρ of light intensity polarized parallel versus perpendicular to the plane of reflection expressed as

$$\rho = R_{||}/R_{\perp} = \tan \psi \; e^{i\Delta} \tag{11.3}$$

where Δ and ψ can be related to the real and imaginary parts of the dielectric constant, Re $\{\varepsilon\}$ and Im $\{\varepsilon\}$, respectively. The SE response can be interpreted using layer models with effective dielectric constants [4]. An important advantage of SE is its capability to measure films with thicknesses on a monolayer scale. Furthermore, with SE one can develop characteristic "fingerprints" of specific surface bond configuration for routine identification. Combined with its ability to probe surfaces remotely, SE can be used to adjust growth parameters during the growth of semiconductor surfaces. A disadvantage of SE for surfaces and interfaces is the requirement to model the dielectric response of the layers at a surface or interface. This can present difficulties for characterizing surfaces and interfaces that undergo changes in elemental composition and bonding such that their dielectric properties are not well known.

11.6   Surface Enhanced Raman Spectroscopy

Raman spectroscopy (RS) is a widely used technique to measure phonon frequencies of solid state materials. As with SE, RS involves reflecting a light beam off a surface. However, whereas SE measures changes in the specular (angle of incidence equals angle of reflection) reflected light beam, RS measures the scattered light away from the specular reflected beam and passes it into a monochromator, as shown in Figure 11.5a. Furthermore, whereas SE uses light with a broad range of photon energies from which to select a single incident energy, RS uses a laser source with a single energy $E = h\nu_0$ and narrow linewidth for excitation.

The scattered light spectrum contains a peak at $h\nu_0$ and peaks due to emitting or absorbing discrete vibrational modes termed *phonons*. The energies of phonon modes in semiconductors are typically in the range of hundreds of *inverse centimeters* (cm^{-1}), where 8065 cm^{-1} = 1 eV so that these modes are relatively close in energy to the laser energy. Nevertheless, these modes are observable even near the laser energy $h\nu_0$ because of the laser's narrow line width and the non-specular light collection.

The inelastic scattering of photons by phonons and phonons coupled to other lattice excitation modes is based on changes in the material's electric susceptibility. The Raman process can be modeled as a three-step process that involves absorption, scattering, and re-emission. The absorption process creates a free electron–hole pair that interacts with

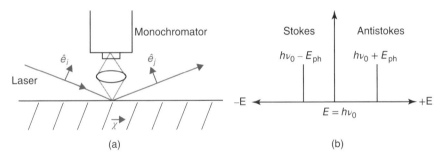

Figure 11.5 *Schematic illustration of Raman scattering spectroscopy (a) backscattering geometry with lens away from the incident and reflected beams to collect light into a monochromator. Here polarizations of both incident and reflected light,* **e**$_i$ *and* **e**$_{j'}$ *respectively, is in the plane of reflection. (b) Raman spectrum showing Stokes and anti-Stokes emission lines. (Brillson 2010, reproduced with permission of Wiley.)*

the polarization fields of the lattice. Scattering that produces a phonon with energy E_{ph} results in re-emission at an energy $hv_0 - E_{ph}$. These are termed *Stokes* emissions. Similarly, scattering that annihilates a phonon results in re-emission at an energy $hv_0 + E_{ph}$ and is termed *Antistokes* emission. The RS cross section can be expressed in terms of a scattering efficiency S_j.

$$S_j = dP_R/P_0 d\Omega = (\omega_s^4/c^4)VL \mid \hat{e}_i \cdot \chi^j \cdot \hat{e}_j \mid^2 \tag{11.4}$$

where P_0 is the *incident power*, $d\Omega = dA/R^2$ is the differential solid angle for a differential area dA at a distance R^2, V is the volume equal to AL, L is the scattering length in the dielectric medium, ω_s is the scattered photon frequency, and χ^j is the dielectric susceptibility of the jth excitation. Equation (11.4) shows how S_j depends strongly on frequency ω_s, increasing as ω^4. In general, *dielectric susceptibility* χ is defined for a given ε according to

$$\varepsilon = 1 + 4\pi\chi V \tag{11.5}$$

where χ^j is a three-dimensional tensor such that polarization $\boldsymbol{P} = \varepsilon_0\chi \cdot \boldsymbol{E}$ for electric field E and whose components contain the symmetry and strength of Raman scattering in three dimensions. The scattered light can be filtered through a polarizer to pass only one polarization into the monochromator. With both the laser and scattered light linearly polarized, RS can determine the relative magnitudes of the tensor components as well as confirming the effective crystal symmetry with any external forces applied.

Since the first step of the RS process involves absorption, one can increase signal intensity by using incident photons tuned to an absorption edge, as shown in Figures 11.1 and 11.2. This technique is termed *resonant Raman scattering* and it can enhance signal intensities by more than an order of magnitude. Another method to increase signal intensity is to magnify the electric fields induced by the light at the surface. A commonly used technique to achieve this magnification is to deposit metals such as Ag that form clusters on semiconductor surfaces and which couple strongly with incident light. The orders-of-magnitude signal enhancement with this technique makes possible studies of the top few monolayers of a semiconductor surface. Yet a third technique to monitor electrical properties at a semiconductor surface involves so-called *morphic effects*. An applied force

F_A such as strain or electric field can cause changes in χ^j that can be expressed in terms proportional to F_A for small F_A,

$$\chi_j = (\partial\chi/\partial \mathbf{Q}_j)\,\mathbf{Q}_j + (\partial^2\chi/\partial\mathbf{Q}_j\partial\mathbf{F}_A)\,\mathbf{Q}_j \cdot \mathbf{F}_A \text{ and higher-order terms in } F_A \qquad (11.6)$$

where $\mathbf{Q}_j$ could be a lattice displacement $\mathbf{u}_j$ or an electrical field $\mathbf{E}_j$ of a j th optical phonon mode. The first term in Equation (11.6) corresponds to the normal RS process, while the second term and higher order terms represent the effect of F_A. The electric field due to band bending in a semiconductor's surface space charge region represents a morphic effect that RS can probe.

Case Study: Electric-field-induced Raman Scattering

Raman scattering is sensitive to electric fields, either applied externally or built into the band bending of a surface space charge region. To illustrate how electric fields can induce RS, Figure 11.6 shows a portion of an RS spectrum for ZnSe grown epitaxially on GaAs, a heterojunction of interest for optoelectronics. The Raman spectra exhibit

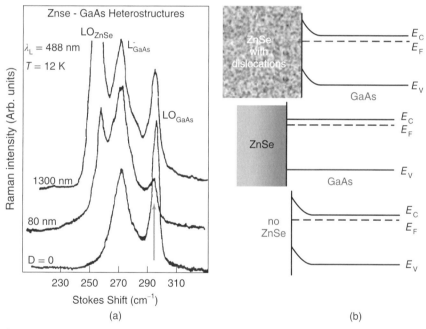

(a)   (b)

Figure 11.6 *First-order Raman spectra of ZnSe-GaAs heterostructures for thickness D = 0, 80, and 1300 nm, the latter two below and above, respectively, the ZnSe critical thickness. For D below the critical thickness, the LO$_{GaAs}$ phonon mode decreases relative to either the bare surface or the thick heterojunction. (Olega, D. 1987 [5]. Reproduced with permission of American Institute of Physics.)*

(*continued*)

(*continued*)

peaks corresponding to longitudinal optical (LO) vibrations in ZnSe and GaAs. RS detects band bending changes in the GaAs that depend on interface conditions. The LO_{GaAs} peak in particular is sensitive to the electric field below the heterointerface, increasing as Ej^2 in the GaAs band bending region. For the bare, air-exposed GaAs (100) surface prior to overlayer thickness ($D = 0$), negative charges at the free surface produce band bending, resulting in a strong LO_{GaAs} peak. With a deposition thickness $D = 80$ nm of ZnSe below the critical thickness to form dislocations, this phonon mode decreases strongly, indicating a band bending decrease. Above the critical thickness, the $D = 1300$ nm ZnSe overlayer forms dislocations that introduce band bending around the dislocation that in turn increases the field-induced LO_{GaAs} scattering. Such measurements are useful to measure the band bending and charge states introduced by dislocations, which can act as recombination centers that degrade the ZnSe/GaAs band gap emission. However, quantitative analysis of Ej requires Airy functions since the electric field varies with depth. Also, surface sensitivity is limited to the width of the surface space charge region, which can vary from tens of nanometers to a micron, depending on the doping concentration.

11.7   Surface Photoconductivity

The *surface photoconductivity* technique provides a measure of free carrier concentration near surfaces. Figure 11.7 shows a composite figure with light incident on a surface between two contacts as well as the typical band bending extending away from the surface into the bulk. Free carriers travel between the contacts in both the surface space charge region and the bulk. Photoexcitation that promotes charge out of states in the band gap to the conduction band in an n-type semiconductor increases the free carrier density such that conductivity $\sigma = ne\mu$ and $J = \sigma E$ increase. Likewise, photoexciting electrons from the valence band into gap states increases (decreases) p-type (n-type) conductivity.

Photoexcitation into and out of surface states changes the charge density at the surface, changing the band bending and conductivity as well. In this regard, surface

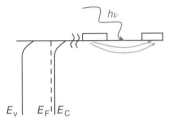

Figure 11.7   *Schematic experimental geometry for surface conductivity and associated band bending within semiconductor depletion region. (Brillson 2010. Reproduced with permission of Wiley.)*

photoconductivity can add or subtract free charge carriers and thereby identify the position of energy levels with respect to the band edges. In order to study surface states, however, this technique requires contacts, which make UHV cleaning or other surface preparation difficult. Another disadvantage is the limited surface sensitivity of photoconductivity, particularly for photon energies well below E_G, where excitation occurs primarily in the bulk. The resultant subsurface and bulk conduction in parallel pictured in Figure 11.7 can be reduced by narrowing the contact spacing so that most conduction occurs primarily in the surface region. Alternatively, photoconductivity studies of nanowires can be useful for wire diameters comparable to or smaller than the surface space charge width.

11.8 Surface Photovoltage Spectroscopy

Surface photovoltage spectroscopy (SPS) measures surface work function using a modulating capacitance technique. Instead of conductivity, SPS measures the change in work function versus incident photon energy. Figure 11.8a illustrates the vibrating capacitor pictured for KPFM in Figure10.2a but now with monochromatic light incident on the sample surface. The voltage difference between the surface and the reference probe of known work function $\Phi_M = E^M{}_{VAC} - E^M{}_F$ yields the semiconductor work function $\Phi_S = E^S{}_{VAC} - E^S{}_F$.

Figure 11.8b illustrates how photons with energy $h\nu_1$ provide the minimum or *threshold energy* to promote electrons from the valence band into a gap state. The additional negative charge on the surface increases the n-type (upward) band bending, lowering E_F and increasing $E_{VAC} - E_F$. Figure 11.8c shows the analogous depopulation of electrons from a mid-gap state into the conduction band with incident photon $h\nu_2$. The reduced negative charge on the surface decreases the band bending, raising E_F and decreasing $E_{VAC} - E_F$. Thus changes in work function at threshold photon energies provide measures of both changes in band bending as well as states within the band gap. With the sample and reference probe both grounded, their Fermi levels are aligned so that the voltage difference is equal to the difference U^{SM} in their vacuum levels, termed the *contact potential*.

$$e(\Phi^S - \Phi^M) = E^S{}_{vac} - E_F{}^S - (E^M{}_{vac} - E_F{}^M) = E^S{}_{vac} - E^M{}_{vac} = eU^{SM} \qquad (11.7)$$

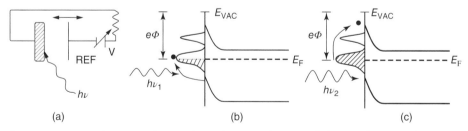

(a)   (b)   (c)

Figure 11.8 (a) *Vibrating Kelvin probe with monochromatic light stimulation and capacitor with feedback circuit for measuring surface potential in SPS. Surface state and/or bulk band transitions detectable by SPC and SPS spectroscopies for (b) filling and (c) unfilling optical transitions with work function* $q\Phi_S$, *semiconductor electron affinity* χ, *band bending* qV_B *and Fermi level* E_F *(dashed lines). (Brillson 2010. Reproduced with permission of Wiley.)*

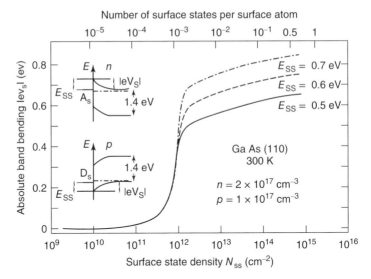

Figure 11.9   *Left: Calculated absolute band bending* $|V_S|$ *due to an acceptor surface-state level* A_s *and a donor level* D_s *for n- and p-type GaAs.* $|V_S|$ *is plotted versus the surface state density* N_{ss} *(lower scale) and related to the number of surface states per surface atom (upper scale). With the different definition of the energetic position* E_{ss} *for n- and p-type crystals (insets) the calculated curves for n- and p-type material are not distinguishable on the scale. (Lüth et al. 1977 [6]. Reproduced with permission of American Physical Society)*

By adding a compensating voltage to the circuit in Figure 11.8a, the charge on the capacitor plates $Q = CV$ for capacitance C becomes $C[-(\varPhi^S - \varPhi^M) + U_{comp}]$ and for a capacitor plate vibrating at frequency ω and spacing d,

$$I = dQ/dt = dC/dt \, [-(\varPhi^S - \varPhi^M) + U_{comp}] \tag{11.8}$$

With compensating voltage U_{comp} in the circuit, the *displacement current* through the resistor can be nulled out when $U_{comp} = \varPhi^S - \varPhi^M$. A feedback loop can supply a U_{comp} voltage that can be monitored continuously to provide SPS spectra versus $h\nu$. See Track II T11.2.Surface Photovoltage Apparatus provides an example of a home-built experimental system for SPS.

SPS has a much higher sensitivity to surface charge than photoconductivity since small changes in surface state population cause large changes in band bending. Consider, for example, the relationship between surface charge and bending in a common semiconductor, GaAs, with typical doping levels. Figure 11.9 illustrates the variation of band bending surface state charge density.

Case Study: GaAs Surface State Density versus Band Bending

From charge balance at a semiconductor surface, the charge in surface states Q_{SS} must be equal and opposite to the charge in the surface space charge region Q_{SC}. In turn, the charge in the surface space charge region W equals the width of the space charge region times the bulk doping n_B. With W from Equation (3.8) then,

$$- Q_{SS} = Q_{SC} = n_B \cdot W = n_B \cdot [2\varepsilon_S(V_0)/n_B]^{1/2} = [2\,\varepsilon_S(V_0)\,n_B]^{1/2} \qquad (11.9)$$

For GaAs with $\varepsilon_S = 13.2\,\varepsilon_0$, a typical n-type band bending $V_0 = 0.7$ eV and carrier density $n_B = 2 \times 10^{17}$ cm^{-3}, surface charge density is:

$$-Q_{SS} = [2\,(13.2)(8.85 \times 10^{-14}\ \mathrm{F\ cm}^{-1})\,(0.7\ \mathrm{eV}/(1.602 \times 10^{-19}\ \mathrm{C})$$

$$\times (2 \times 10^{17}\ \mathrm{cm}^{-3})]^{1/2}$$

$$= 1.81 \times 10^{12}\ \mathrm{electrons/cm}^2$$

Thus a density of just 2/1000th of an electron per surface atom produces band bending of 700 meV. Figure 11.9 illustrates how little charge density is required to produce band bending of hundreds of millivolts. Since the precision of contact potential measurements can be <1meV, this indicates that SPS sensitivity for this conventional semiconductor extends into the 10^{10} cm^{-2} range.

The oxidation of semiconductor surfaces often produces significant electronic changes. Figure 11.10 illustrates the measurement of surface states on a clean cleaved CdSe surface created by oxygen exposure.

Case Study: SPS of Oxygen-Induced States on a Semiconductor Surface

Arrows indicate changes in *contact potential difference* (cpd) that signify the onset of photostimulated population or depopulation of several states within the band gap. In addition to the 1.6–1.7 eV onset due to band-to-band transitions, the clean surface exhibits two gap-state transitions at 1.55 and 1–1.1 eV. With increasing θ, the surface develops several additional states that change with submonolayer oxygen coverage, reflecting different surface bonding sites for the adsorbed oxygen atoms. As expected, both the work function and band bending increase with increasing surface oxidation as the adsorbed oxygen atoms attract more negative charge to the surface. Since $e\Phi = e\chi + (E_C - E_F)$ at the surface, then

$$\Delta e\Phi = e\Delta\chi + \Delta(E_C - E_F) = e\Delta\chi + \Delta qV_B \qquad (11.10)$$

and the difference between work function and band bending in Figure 11.10 is due to changes in semiconductor electron affinity, that is, surface dipoles. UHV *in situ* comparison of the work function measured by KPFM versus band bending measured by photoemission spectroscopy provides a method to determine $e\Delta\chi$.

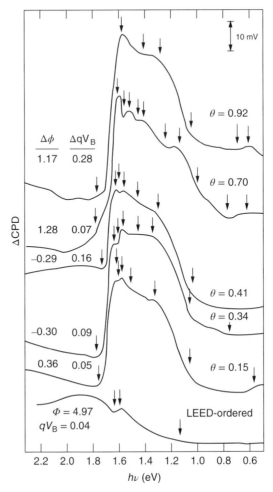

Figure 11.10   *Surface photovoltage spectra of clean, LEED-ordered (11$\bar{2}$0) CdSe with increasing molecular oxygen surface coverage θ, as determined by XPS. Arrows indicate energies of maximum slope change corresponding to onsets of photovoltage transitions. The changes in surface work function $\Delta\phi_s$ and band bending Δq_{VB} are given for each curve. (Brillson 1976 [7]. Reproduced with permission of American Institute of Physics.)*

Case Study: SPS of Metal-Induced States on a Semiconductor Surface : Au–GaAs

The adsorption of metal atoms on semiconductor surfaces also creates new electronic states. Figure 11.11 illustrates the formation of states within the band gap of a clean, cleaved GaAs (110) surface by the deposition of sub-monolayer Au atoms. The cleaved surface exhibits an SPS response only for $hv \geq 1.43$ eV, the GaAs band gap

(continued)

energy, indicating the near-absence of intrinsic surface states within the band gap. With the deposition of as little as 0.1 Å Au, additional features appear that correspond to SPS transitions into and out of states within the band gap. The energies of these photostimulated transitions change with increased Au deposition, finally exhibiting pronounced thresholds at 0.9 and 1.25 eV before being attenuated by the metallic Au overlayer. The 0.9 eV photo-depopulation transition from states 0.9 eV below conduction band edge E_C coincides with the commonly observed barrier height for Au on cleaved GaAs (110) surfaces. See also Chapter 15. These results are characteristic of many metal–semiconductor interfaces. They show that metals induce electronic states at semiconductor interfaces that can account for Schottky barrier heights. Indeed, for metals on these semiconductors, extrinsic rather than intrinsic states can account for Schottky barrier heights.

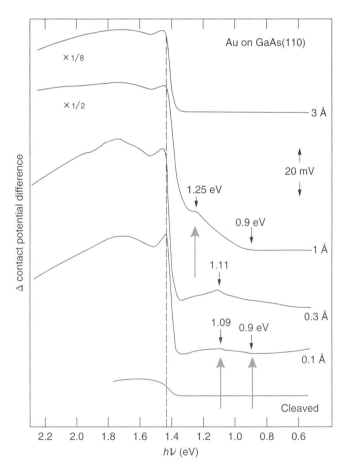

Figure 11.11 *SPS spectra of metal-induced interface states for Au on GaAs versus metal overlayer thickness. Arrows indicate slope changes. (Brillson, L.J. 1979 [8]. Reproduced with permission of American Institute of Physics.)*

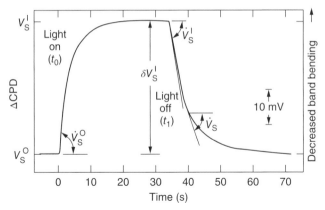

Figure 11.12 *Surface photovoltage transient for θ = 0.15 oxygen on CdSe (see Figure 11.10) and illumination I = 3.8 × 10¹⁵ photons cm⁻² at hv = 1.60 eV. (Brillson 2010.. Reproduced with permission of Wiley.)*

11.8.1   Transient Surface Photovoltage Spectroscopy

Besides providing surface state energies within the band gap and band bending, SPS can also yield information on the densities of these surface states. Here one measures the time-dependent cpd rate of response to known photon fluxes and energies corresponding to photo-induced population and depopulation of specific states. This *transient SPS* (t-SPS) method consists of monitoring Δcpd versus time for a sample initially at equilibrium in the dark, then switching light on to move Δcpd to a new equilibrium, and finally switching light off to allow Δcpd to return to its dark state. The transient photoresponse causes surface potential changes whose rates of change yield trap densities n^0_t according to [9]:

$$n^0_t = \frac{40 \times \delta V^1_s (2\varepsilon kTN_b)^{\frac{1}{2}}}{2q|V^0_s|^{\frac{1}{2}}(1 + \dot{V}^1_s/\dot{V}^0_s)} \tag{11.11}$$

where $\delta V_s^{\,1}$ is the light-induced surface potential difference, ε the dielectric permittivity, k is Boltzmann's constant, T the temperature (K), N_b the bulk electron density, $V_s^{\,0}$ the initial surface potential without light in dimensionless units (normalized to kT/q by a 40× factor), $\dot{V}_s^0$ is the rate of potential change with light on, and $\dot{V}_s^1$ is the rate of potential change with light off, as shown in Figure 11.12. See Track II T11.3 Transient SPS Response.

11.9   Photoluminescence Spectroscopy

Photoluminescence (PL) spectroscopy is an optical technique that provides a rich array of information about electronic states in bulk semiconductors. First, PL spectra exhibit emission that characterizes a semiconductor's band gap. Here the incident photon, usually from a laser, excites valence electrons into the conduction band, leaving behind valence band holes. These electrons and hole recombine, and the energy released during the recombination appears as band gap light. Electronic states due to crystal lattice imperfections and impurities produce emissions at lower energy, corresponding to energy levels within the band gap. Energy levels close to the valence or conduction band can be due to impurities that act as semiconductor dopants as well as electron–hole complexes with impurities and with each other. In general, these are termed *excitons*.

For many semiconductors, the very high energy resolution of laser-excited PL has enabled compilation of extensive data bases of gap state energies that are associated with impurity and excitonic states. PL can provide information about energy levels inside quantum wells and at their interfaces, particularly in semiconductor superlattices where multiple layers can contribute. The large absorption length of optical excitation at sub-band gap energies is a drawback for PL due to the low relative volumes of surface versus bulk excitation. Likewise, PL spectra are not depth dependent so that multiple layers of a junction provide only composite spectra. Finally, PL's spatial resolution is limited by a light diffraction limit so that nm-scale studies of many advanced electronic device structures are quite challenging.

11.10   Cathodoluminescence Spectroscopy

Cathodoluminescence spectroscopy (CLS) is another luminescence technique that is based on electron beam rather than laser excitation. Both CLS and PL involve creation of free electron–hole pairs, their subsequent recombination, and optical emission. CLS overcomes the depth and spatial resolution challenges of PL since one can control electron beam energies to excite the material at a continuum of depths on a nanometer-to-micron scale. Furthermore, one can focus the electron beam down to nm dimensions with an electron microscope, which enables studies on even a quantum scale. The lateral and depth resolution of CLS provides a wealth of information including: (i) luminescent center concentrations and spatial distributions, (ii) the concentration and distribution of extended defects such as dislocations, (iii) the composition of materials (from band edge emissions), (iv) surface recombination velocity (with transient CLS), (v) carrier diffusion lengths, and (vi) new interface compound formation.

11.10.1   Overview

From Figure 7.12, electron beams incident on a material create a cascade of secondary electrons that increase in number while decreasing in energy as they penetrate to increasing depths. These electrons can produce Auger electrons, X-rays, plasmons, and optical emissions at various depths. The penetration depths and the excitation energies of these cascading electrons depend on the energy loss mechanisms involved.

Figure 11.13 shows the calculated rates of energy per unit time dE/dt of electrons in a solid for the three primary energy loss mechanisms. At kinetic energies above $100\,kV$, energy loss occurs primarily by X-rays, whereas plasmon loss dominates at lower energies down to $\sim 10\,eV$. Energies below this point are sufficient to create only electron–hole pairs and optical phonons. The excitation of electron–hole pairs initiates the cathodoluminescence process.

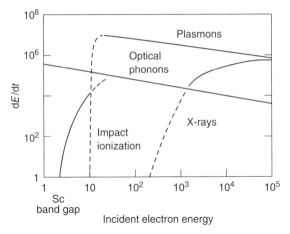

Figure 11.13 *Rate of energy loss per unit time* dE/dt *of electrons due to X-rays, plasmons, optical phonons, and impact ionization. (Rose, A. 1966 [10]. Reproduced with permission of Nature Publishing.)*

11.10.2 Theory

The scattering rate and energy loss of cascading electrons determines the depth and range of electron–hole pair creation and the CLS emission. In general, the scattering of incoming particles of mass M_1, charge Z_1, and energy $E = \frac{1}{2}M_1 v^2$ with a target of charge Z_2 and mass m has an energy loss per unit path length of

$$- dE/dx = (2\pi Z_1^2 e^4/E)\,(NZ_2)\,(M_1/m)\,\ln(2mv^2/I) \tag{11.12}$$

for an average excitation energy I and a target material of atomic density N, where $N = \rho/A$ for material density ρ and atomic weight A. See Track II T11.4 Energy Loss Rate/Stopping Power. Equation (11.12) shows that stopping power depends on (i) the ratio E/I of particle energy to average excitation energy, (ii) the atomic density N, and (iii) the number of target electrons NZ_2. For nuclear particles, nearly all the target electrons participate. The average excitation energy I for most elements is $\sim 10\,Z_2$ (in eV) for $Z > 12$. For example, $Z_2 = 13$ for Al so $I = 10\,Z_2 = 130$ eV. In the energy range 10–10 000 eV, the dominant mode of energy loss is the creation of plasmons with energy $I = \hbar\,\omega_p = \hbar(4\pi n e^2/m\varepsilon)^{1/2}$ See Chapter 7, Section 7.3.3. For $Z_1 = 1$ and $\varepsilon = \varepsilon_0 = 1$, Equation (11.12) then becomes

$$- dE/dx = (\omega_p^2\, e^2/v^2)\,\ln\,[2\,mv^2)/(\hbar\,\omega_p)] \tag{11.13}$$

For a single scattering event, $dE = \hbar\,\omega_p$ and $dx =$ the electron scattering length λ so that

$$(-dE/dx)/(\hbar\,\omega_p) = 1/\lambda = (\omega_p e^2/\hbar v^2)\,\ln[(2mv^2)/(\hbar\,\omega_p)] \tag{11.14}$$

This equation shows that λ decreases with decreasing kinetic energy $E = \frac{1}{2}mv^2$ but also increases at very low energies when $E \approx \hbar\,\omega_p$. Equation (11.14) yields $\lambda = 5.45$ Å at 350 eV, $\cong 2$ Å at 92 eV, and >4 Å at 4 eV, which is consistent with the escape depth minimum at 50–100 eV observed experimentally, as shown in Figure 6.2. The physical

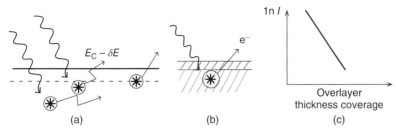

Figure 11.14   *(a) Schematic cross section of excited electron trajectories below a solid surface. Only electrons with the escape depth (dashed line) leave the solid without inelastic scattering. (b) Thin film deposition attenuates escaping electrons. (c) Decrease in electrons that escape without energy loss decreases logarithmically with increasing overlayer thickness. (Brillson 2010. Reproduced with permission of Wiley.)*

reason is that, at very low energies, there is much less energy and momentum to excite electronic transitions. The escape depth minimum on an Å scale for energies in this 50–100 eV range is the basis for the surface and depth sensitivity of techniques already described.

Experimental measurements of these scattering lengths at different energies consist of monitoring the attenuation of signals from substrates covered with increasing thicknesses of an overlayer material. Figure 11.14a illustrates one such method, where incident photons excite photoelectrons with a given kinetic energy inside a solid. These electrons can diffuse and scatter or reach the surface and escape into vacuum. Only those electrons at depths less than a scattering length contribute to the elastic photoelectron signal measured by an electron analyzer. In Figure 11.14b, an overlayer of known thickness attenuates the photoelectrons escaping a substrate. Consider a flux I_0 of electrons leaving the substrate with kinetic energy E_C, inelastic collision cross section σ, and density of scattering centers/$cm^3 N'$. I_0 decreases by $dI = \sigma I$ per scattering center so that $-dI = \sigma I N' dx$ per thickness increment and

$$I = I_0 e^{-\sigma N' x} = I_0 e^{-x/\lambda} \tag{11.15}$$

where mean free path $\lambda = 1/(N' \sigma)$. Thus the slope of the logarithmic decrease in I/I_0 versus known overlayer thickness yield λ^{-1}, assuming uniform overlayer thickness. Furthermore, by varying incident photon energy or selecting excitations with a different binding energy, one can obtain the dependence of λ on kinetic energy. Thus escape depths and their energy dependence are measureable and indeed are the basis for the values shown in Figure 6.2.

11.10.3   Semiconductor Ionization Energies

Besides scattering length, the electron cascade in semiconductors and insulators depends on the minimum energy to ionize atoms in the solid, that is, the band gap. The average energy to excite electron–hole pairs also increases with band gap but is higher than this *threshold energy* because of the requirement to conserve momentum as well as energy. Energies above the minimum energy allow a wider range of momentum space for electrons to scatter into. Figure 11.15 illustrates this average or *effective ionization energy*. To within

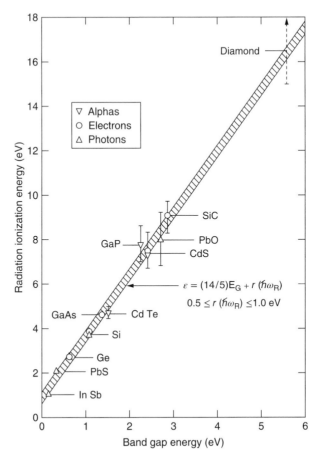

Figure 11.15 *Effective ionization energy E_{av} to generate electron–hole pairs in semiconductors as a function of bandgap. (Klein, C.A. 1965 [11]. Reproduced with permission of American Institute of Physics.)*

a close approximation, this effective ionization energy can be described by an expression

$$E_{av} = (14/5)E_G + r(\hbar\omega_R) \tag{11.16}$$

where $\hbar\omega_R$ is the highest-frequency zero-wave-vector phonon measured by Raman scattering and r is the average number of phonons per pair generated during the initial stage of the impact ionization process. Equation (11.16) spans a wide range of semiconductors and shows that incident electron energies must be several times the semiconductor band gap in order to produce cathodoluminescence emissions efficiently. This requirement of relatively high incident beam energies limits the use of atomic-scale scanning techniques to generate cathodoluminescence since currents into nanoscale areas must be severely reduced in order to avoid thermal damage. Likewise, while PL requires only band gap energy for efficient emission, it requires phonon scattering to enable indirect transitions, thereby lowering their probability of emission, whereas the momentum of incident electrons in CLS provides significant momentum transfer directly.

11.10.4   Universal Range–Energy Relations

The optical excitation depth of electron beams depends both on their energy and the material being probed. A variant of Equation (11.12) expresses the rate of energy loss along a path s of electrons inside the material.

$$- dE/ds = (2\pi N_A e^4)(Z\rho/A)[E^{-1} \ln (aE/I)] \qquad (11.17)$$

where the mean excitation loss energy $I = (9.76 + 58.8\, Z^{-1.19})Z$ (eV), constant $a = 1.1658$, $\rho = $ material density, $A = $ atomic weight, and $Z = $ atomic number of the material. [See Track II.T.11.4 Energy Loss Rate/Stopping Power.] Thus the energy loss rate increases with higher atom density, higher atomic weight, higher Z and therefore higher numbers of electrons to scatter with. As the incident electrons enter the material, they generate a cascade of secondary electrons. At each stage of this cascade, the number of electrons multiplies and their energy decreases. The final stage of energy loss involves impact ionization, that is, collisions between the incident electrons and atoms that detach valence electrons, resulting in free electron–hole pairs. These free electron–hole pairs can then recombine and emit light via conduction band-to-valence band recombination as well as between states within the band gap and the band edges. The maximum range over which the impact ionization occurs is termed R_B, the *Bohr–Bethe* range, which can be approximated by an expression of the form $R_B = C\, \xi^a$, where $\xi = aE/I$, $E = E_B$, the incident beam energy, C is a material-dependent constant equal to $9.40 \times 10^{-12} I^2 (A/Z)\, c'/\rho$, a is a material-independent constant $= 1.29$ for $\xi < 10$, c is a material-independent constant $= 1.28$ for $\xi < 10$. This approximation provides a 'universal' fit to experimental data for relatively low beam energies. A fit of experimental data to Equation (11.17) yields an expression [12] for R_B for energies above 1 keV.

$$R_B(E > 1 \text{ keV}) = 0.62\, \xi^{1.609} \qquad (11.18)$$

Likewise, the corresponding depth of maximum energy loss and impact ionization is

$$U_0\, (E > 1 \text{ keV}) = 0.069\, \xi^{1.71} \qquad (11.19)$$

[See Track II T11.5 Maximum Energy Loss Rate.] The U_0/R_B ratio varies between different materials but is relatively constant for a specific material with different E_B. Further approximations extend the energy dependence of R_B and U_0 to energies below 1 keV [12]. Figure 11.16a illustrates the rate of energy loss in GaN for an electron cascade with E_B in the low keV range, peaking at depths roughly one-third of the maximum cascade range. Figure 11.16b illustrates how R_B and U_0 vary in GaN with incident electron energy in the low keV range using the expression $R_B = C\, \xi^a$ with C and a indicated. These figures show that electron–hole pair creation increases with increasing beam energy and varies on a scale of tens of nanometers or less.

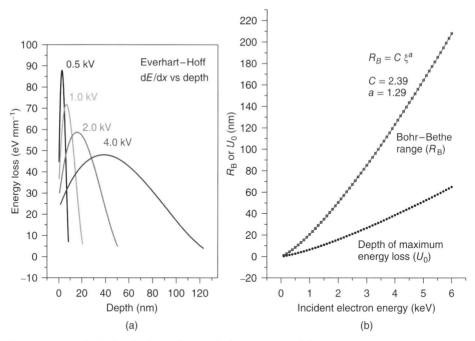

Figure 11.16 *GaN depth dependence of electron-excited luminescence spectroscopy. (a) Energy loss rate $-dE/dx$ versus depth. (b) Bohr–Bethe range R_B and depth of maximum energy loss rate U_0. (Everhart and Hoff 1971 [15]. Reproduced with permission of American Institute of Physics.)*

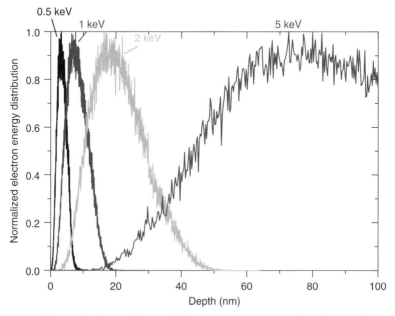

Figure 11.17 *Monte Carlo simulation of electron excitation in GaN for an electron beam incident at 45° to the surface. (Brillson 2010. Reproduced with permission of Wiley.)*

11.10.5   Monte Carlo Simulations

While the analytic expression in Equations (11.18) and (11.19) are useful for calculating excitation depths in a single material, they are less useful for multilayer structures with different materials. Monte Carlo simulations based on thousands of incident electron cascades can provide accurate models of electron penetration into solids with multiple material layers [13]. They take into account the stopping power of the material, the backscattered electron component, and the angle of incidence. Figure 11.17 provides an example of the agreement between this simulation and the calculated profile in Figure 11.16a for the same material and E_B. Since the simulation takes into account the density and stopping power of the material for each scattering event, it readily takes into account the changes in material with depth.

Case Study: CL vs. PL Comparison for ZnO

For studies of materials in bulk, both CLS and PL provide useful information. To compare their rates of excitation, consider these measurements for ZnO ($E_G = 3.39$ eV). The effective threshold energy for CLS with this band gap is $E_0 = 10$ eV. For an incident electron beam energy $E_B = 1$ keV, Bohr–Bethe range $R_B = 20$ nm from both Figures 11.16 and 11.17 and the number of secondary electrons produced is 1000 eV/(10 eV/secondary electron) = 100 secondary electrons. Assume a minority carrier lifetime $\tau = 10^{-10}$ s. For this E_B and nominal beam current $I_B = 10$ μA and area $A = 10^{-3}$ cm^2 with a glancing incidence electron gun, the number of free electron–hole pairs N (assuming 1 per secondary electron) is:

$$N = \text{generation rate } G \times \text{lifetime } \tau \text{ where } G = (I/A \cdot d)\,(E_B/E_0) \text{ so that}$$

$$N = [(10^{-5}A \cdot 6.24 \times 10^{18} \text{ electrons s}^{-1}/A)/(10^{-3} \text{ cm}^2 \cdot 2 \times 10^{-6} \text{ cm})]$$

$$\times (10^3 \text{ eV}/10 \text{ eV}) \cdot \tau(s)$$

$$= 3.12 \times 10^{24} \text{ electrons cm}^{-3} \text{ s}^{-1} \cdot \tau(s)$$

$$= 3.12 \times 10^{14} \text{ electrons cm}^{-3} \text{ within the top 20 nm.}$$

For comparison, a 10 mW incident He-Cd laser beam with photon energy $h\upsilon = 3.82$ eV produces $I = 6.242 \times 10^{16}$ eV s^{-1}/3.82 eV/photon = 1.63×10^{16} photons s^{-1} with a penetration depth of 83 nm based on an absorption coefficient $\alpha = 1.2 \times 10^5$ cm^{-1}. Note the power conversion: 1 W = 1 J s^{-1} = 6.242×10^{18} eV s^{-1}. The number of free electron–hole pairs produced by illuminating the same 10^{-3} cm^2 area with the same minority carrier lifetime is then:

$$N = (I/A \cdot d) \cdot \tau = 1.63 \times 10^{16} \text{ photons s}^{-1}/(10^{-3} \text{ cm}^2 \cdot 0.83 \times 10^{-5} \text{ cm}) \cdot 10^{-10} \text{ s}$$

$$= 1.96 \times 10^{14} \text{ electron-hole pairs/cm}^3 \text{ within the top 83 nm.}$$

Thus CL and PL generate comparable densities of free electron-hole pairs. However, the CLS densities are more than four times more localized. Furthermore, while continuous wave (CW) laser powers can be one order of magnitude higher without inducing thermal damage, electron beam densities per unit area can increase by several orders of magnitude, particularly with scanning electron microscope (SEM) spot sizes on a nanometer scale. A consideration for both CL and PL is sample heating by thermal dissipation of the beam energy. While similar heating can occur, a helpful indicator of temperature is the band edge energy, which decreases with increasing temperature. For most semiconductors, typical CL fluxes under the above conditions produce temperature rises of only a few °C.

In general, the advantages of CLS are: (i) nanoscale depth resolution, (ii) nanoscale lateral spatial localization (with an SEM), (iii) high excitation intensity, (iv) sensitivity to defects, impurities, and compound formation at the 10^{11} cm^{-2} and $<10^{15}$ cm^{-3} level, (v) above band gap energies to excite wide band gap semiconductors and insulators, and (vi) the ability to probe selectively below free surfaces. Potential limitations of CLS are: (i) a dependence of relative feature intensities on injection levels, that is, "saturation effects", (ii) dependence of the cross section for luminescence transitions on incident beam energy, (iii) possible thermal or photolytic beam damage. All three limitations are common to both CL and PL. However, electron beams are more likely to damage organics by bond breaking.

11.10.6   Depth-Resolved Cathodoluminescence Spectroscopy

The dependence of excitation depth on incident beam energy enables depth-resolved cathodoluminescence spectroscopy (DRCLS) to measure electronic properties at surfaces and at interfaces below the surface with nm-scale depth resolution. DRCLS has been effective in probing the nature and spatial distribution of electronic states within semiconductor band gaps due to surface states, bulk trap states, and interfacial compounds as well as band structure changes due to temperature and strain [15].

Case Study: Defect Segregation near ZnO Surfaces

ZnO and several other binary compound semiconductors exhibit an increase in defect emissions near their free surfaces. This near-surface segregation is noteworthy since these defects are electrically active and can significantly alter free carrier concentrations and hence the effective Schottky barriers heights at metal contacts to these surfaces. Figure 11.18a shows DRCLS spectra that include a 3.3 eV near band edge (NBE) peak (to which each spectrum is normalized) and lower energy emissions due to native point defects.

Here the broad peak centered at 2.5 eV, commonly attributed to oxygen vacancies, increases increases with proximity to the free surface. Figure 11.18b shows the normalized defect intensity at different depths and its increase starting at ~30 nm. Exposure of this surface to a remote oxygen plasma decreases the magnitude of this segregation, consistent with filling of oxygen vacancies by oxygen atoms activated by the plasma. This near-surface segregation phenomenon and the changes that surface oxidation produces on a nanometer scale are detectable because of the high surface sensitivity of DRCLS and its nanometer scale depth resolution.

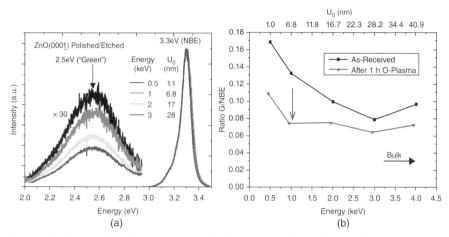

Figure 11.18 *(a) DRCLS spectra at 90 K of chemomechanically polished and etched ZnO (100$\bar{1}$) bare surface versus incident beam energy. (Brillson, L.J. 1979. Reproduced with permission of American Institute of Physics.) (b) NBE normalized 2.45 eV emission revealing segregation toward free surface and decrease (vertical arrow) with remote oxygen plasma treatment. (Mosbacker et al 2005 [16]. Reproduced with permission of American Physical Society.)*

11.10.7 Spatially-Resolved Cathodoluminescence Spectroscopy and Imaging

As with imaging of SIMS and TEM features, two-dimensional maps of DRCLS features are possible using hyperspectral imaging, that is, obtaining individual intensity versus photon energy spectra at each pixel of a two-dimensional array, then displaying a two-dimensional map of a selected emission feature. Such maps can reveal features on a mesoscopic scale that individual spectra might not. Nanoscale structures containing defects are well suited to illustrate this capability.

Case Study: ZnO nanowires

As in Figure 11.18, 3.3 eV NBE and 2.5 eV defect features are characteristic of many ZnO nanowires. Figure 11.19a presents a scanning electron microscope (SEM) image of 300–500 nm diameter ZnO nanowires.

Figure 11.19b displays a map of the normalized 2.5 eV intensity at each point in the same area. Here the light shaded (yellow) regions correspond to defect-rich regions

(continued)

(*continued*)

in the near surface regions of the nanowires while the darker (orange) regions correspond to the central core of each nanowire. Blue and green areas are from regions between nanowires. The radial extent of the defect-rich regions appears to be roughly independent of the nanowire radius. The apparent thickness of this surface segregated region appears to lie in the same range as exhibited for ZnO bulk crystals, that is, 30–50 nm. Images such as these can provide information on the driving forces that produce this segregation.

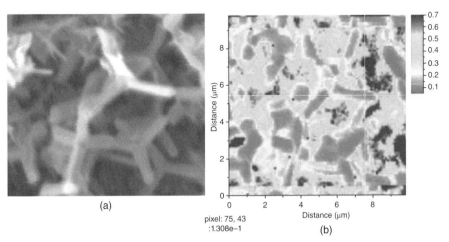

pixel: 75, 43
:1.308e−1

(a)

(b)

Figure 11.19 *(a) SEM image of ZnO nanowire array. (b) Hyerspectral image of I(2.5 eV)/I(NBE) for the same area as in (a). Color online: Track II TC11.19.*

11.11   Summary

The wide range of optical techniques described in this chapter makes use of many different physical effects to achieve surface sensitivity. Modulated excitation, background subtraction, and multiple reflections enable surface sensitivity for reflectance spectroscopies. Phase and polarization changes enable ellipsometry to achieve sensitivity to thin film and surface dielectric properties. Raman scattering yields lattice vibrations but also the magnitude of electric fields within the surface space charge region of semiconductors and insulators. Optical techniques, such as photoconductivity and photovoltage spectroscopies, that induce electrical changes take advantage of the latter's high sensitivity to charge and current near surfaces. Conversely, electronic techniques such as cathodoluminescence spectroscopy that induce optical emission take advantage of the strong depth dependence of electron scattering on electron energy inside solids. These techniques together provide researchers with many techniques and tools to explore electronic material surfaces and interfaces.

11.12 Problems

1. A useful conversion between optical wavelength and energy is $E(\text{eV}) = 1.24\,\text{eV}/\lambda(\mu\text{m})$. Thus a photon with $1\,\text{eV}$ energy has a wavelength of $1.24\,\mu\text{m}$. Derive this expression.

2. Calculate the excitation depth in InP for incident laser light of wavelength $6328\,\text{Å}$ (He-Ne laser) versus $325\,\text{nm}$ (He-Cd laser). Which of these laser energies would you choose to study: (i) a $50\,\text{nm}$ epilayer of InP grown on lattice-matched InGaAs, and (ii) the InP/InGaAs heterointerface? Explain your answers.

3. The semiconductor ZnO has a direct band gap of $3.39\,\text{eV}$ at $300\,\text{K}$ and an electron trap $0.9\,\text{eV}$ below the conduction band edge. Sketch the photoluminescence spectrum of photon intensity versus energy for $3.8\,\text{eV}$ laser excitation of this material using (a) a GaAs multiplier and (b) a Ge photodiode.

4. A $100\,\text{nm}$ thick GaN film on a sapphire substrate is to be probed optically using both a $325\,\text{nm}$ He-Cd laser line ($\alpha = 10^{-5}\,\text{cm}$) and electron beam excitation. Will photoluminescence be limited to the GaN? What is the maximum incident electron beam energy that will probe only the GaN film?

5. For a Si epilayer with doping $N_{\text{d}} = 10^{17}\,\text{cm}^{-3}$, calculate the film thickness needed to minimize competition from bulk conduction.

6. (a) For an InP crystal with $10^{16}\,\text{cm}^{-3}$ doping initially with flat bands, how many electrons per cm^2 does a surface photovoltage excitation add to a surface state in InP if the measured work function increases by $20\,\text{meV}$? (b) If the semiconductor doping is increased to $10^{17}\,\text{cm}^{-3}$, how much more incident light is needed to achieve the same band bending as in (a)?

7. A GaAs crystal with charged surface states has $10^{16}\,\text{cm}^{-3}$ doping and $0.8\,\text{eV}$ equilibrium band bending in the dark. Photostimulated depopulation of these surface states with 10^{15} photons $\text{cm}^{-2}\,\text{s}^{-1}$ intensity at $h\nu_{\text{d}}$ produces a $50\,\text{meV}\,\text{s}^{-1}$ transient toward decreasing work function. The capture cross section for photostimulated depopulation is given as $5 \times 10^{-17}\,\text{cm}^2$. What is the trapped surface electron density at this energy?

8. For an incoming electron ($Z_1 = 1$) scattered by an atom with Z_2 electrons, the distance d of closest approach is given by equating the kinetic energy E with the potential energy $Z_1 Z_2 e^2/d$. Calculate the distance of closest approach for a $2\,\text{keV}$ electron with Si. Compare to the Bohr radius.

9. Calculate the stopping power of a $2\,\text{keV}$ incident electron beam in Ge.

10. The incident electron beam used to excite cathodoluminescence can produce high electron–hole densities. Calculate the free electron density for an $n = 10^{17}\,\text{cm}^{-3}$ ZnO semiconductor with minority carrier lifetime $\tau_{\text{p}} = 10^{-10}\,\text{s}$, incident beam current $I_{\text{B}} = 10^{-9}\,\text{A}$, and beam diameter $20\,\text{nm}$ with an incident beam energy (a) $E_{\text{B}} = 1\,\text{keV}$ and (b) $5\,\text{keV}$. (c) Which is more likely to produce significant flattening of the semiconductor bands? [Hint: for large cascade distances, approximate the pear-shape as a sphere.]

11. Given an experimentally determined plasmon frequency ω_{p} corresponding to $15\,\text{eV}$, calculate the effective electron density within the metal.

12. Calculate the Bohr–Bethe range for a $2\,\text{keV}$ electron beam cascade in GaAs.

13. Calculate the Bohr–Bethe range for a $25\,\text{keV}$ electron beam cascade in ZnO. Use $a = 1.62$ and $c' = 0.68$ for $10 < \zeta < 100$.

References

1. Cardona, M. (1965) Infrared dielectric constant and ultraviolet optical properties of solids with diamond, zinc blende, wurtzite, and rocksalt structure. *J. Appl. Phys.*, **36**, 2181.
2. Anderson, J., Rubloff, G.W., Passler, M., and Stiles, P.J. (1974) Surface reflectance spectroscopy studies of chemisorption on W(100). *Phys. Rev. B*, **10**, 2401.
3. Chiarotti, G., Nannarone, S., Pastore, R., and Chiaradia, P. (1971) Optical absorption of surface states in ultrahigh vacuum cleaved (111) surfaces of Ge and Si. *Phys. Rev. B*, **4**, 3398.
4. Fujiwara, H. (2007) *Spectroscopic Ellipsometry: Principles and Applications*, John Wiley & Sons, Inc., New York.
5. Olega, D. (1987) Effects on ZnSe epitaxial growth on the surface properties of GaAs. *Appl. Phys. Lett.*, **51**, 1422.
6. Lüth, H., Büchel, M., Dorn, R. *et al.* (1977) Electronic structure of cleaved clean and oxygen-covered GaAs (11) surfaces. *Phys. Rev. B*, **15**, 865.
7. Brillson, L.J. (1976) Surface photovoltage and electron energy-loss spectroscopy of oxygen adsorbed on (1190) CdSe. *Surf. Sci.*, **13**, 325.
8. Brillson, L.J. (1979) Chemical reaction and charge redistribution at metal-semiconductor interfaces. *J. Vac. Sci. Technol.*, **16**, 1378.
9. Lagowski, J., Balestra, C.L., and Gatos, H.C. (1972) Determination of surface state parameters from surface photovoltage transients: CdS. *Surf. Sci.*, **29**, 203.
10. Rose, A. (1966) The acoustoelectric effects and the energy losses by hot electrons. *RCA Rev.*, **27**, 600.
11. Klein, C.A. (1965) Bandgap dependence and related features of radiation ionization energies in semiconductors. *J. Appl. Phys.*, **39**, 2029.
12. Brillson, L.J. and Viturro, R.E. (1988) Low energy cathodoluminescence spectroscopy of semiconductor interfaces. *Scanning Microsc.*, **2**, 789.
13. Hovington, P., Drouin, D., and Gauvin, R. (1997) *Scanning*, **19**, 1–14. The source code is available at Web site http://www.gel.usherbrooke.ca/casino/What.html. Wiley-Blackwell (United States) http://onlinelibrary.wiley.com/journal/10.1002/%28 ISSN%291932-8745.
14. Everhart, T.E. and Hoff, P.H. (1971) Determination of kilovolt electron energy dissipation vs penetration distance in solid materials. *J. Appl. Phys.*, **42**, 5837.
15. Brillson, L.J. (2012) Applications of depth-resolved cathodoluminescence spectroscopy. *J. Phys. D: Appl. Phys.*, **45**, 183001.
16. Mosbacker, H.L., Strzhemechny, Y.M., White, B.D. *et al.* (2005) Role of near-surface states in ohmic-Schottky conversion of Au contacts to ZnO. *Appl. Phys. Lett.*, **87**, 012102.

Further Reading

Kronik, L. and Shapira, Y. (1999) Surface photovoltage phenomena: theory, experiment, and applications. *Surf. Sci. Rep.*, **37**, 1–206.

12

Electronic Material Surfaces

This chapter describes the structural, chemical, and electronic properties of electronic material surfaces, the interplay of these properties, and how they affect charge transfer at their interfaces. These properties not only reflect the basic physical mechanisms involved, but they also highlight the various applications that these surfaces enable.

12.1   Geometric Structure

12.1.1   Surface Relaxation and Reconstruction

Chapters 9 and 10 addressed the atomic-scale geometrical and morphological structure of semiconductor surfaces. Here surface relaxation and reconstruction of the outer atomic layers can produce a variety of atomic arrays. Atomic bonding and geometric order at electronic material surfaces can differ significantly from those of the bulk crystal. These changes can involve changes in bond length, bond rotation, coordination, and even place exchange that are driven by the reduction of electrostatic and mechanical energy. In turn, these structural changes alter surface chemical and electronic properties that determine electronic structure and charge transfer.

12.1.2   Extended Geometric Structure

Domains, steps, and defects are morphological features that can dominate growth, epitaxy, and etching. These domains exist on a scale of tens of nanometers or less. As an example, Figure 9.10 illustrates an array of in-phase (IP) and anti-phase (AP) domains distributed across GaAs(001) (2×4)-c(2×8) reconstructed surface. The AP domain boundaries produce the (2×8) pattern while the IP domain boundaries produce the (2×4) boundaries. Here the domain size can be as small as a few nanometers or extend across the surface over much larger areas. This is an example of geometric structure determined by extended arrays of primitive cells. It provides a natural explanation for the range of stoichiometry often

An Essential Guide to Electronic Material Surfaces and Interfaces, First Edition. Leonard J. Brillson.
© 2016 John Wiley & Sons, Ltd. Published 2016 by John Wiley & Sons, Ltd.
Companion Website: www.wiley.com/go/Brillson/

observed for the same LEED pattern. Techniques to assess domain size include LEED, STM, and LEEM. Domain boundaries can move and areas can change at elevated temperatures. *In situ* monitoring of such movements shows the influence of impurities, elastic strain, and dislocations. For example, Figure 10.6 exhibits apparent ripples in the Si(111) (7×7) surface due to strain. Likewise, Figure 10.12a shows "wing" features due to dislocations that alter the epitaxial growth of GaN(001) surfaces. Steps are another major feature of electronic material surfaces. Step arrays are defined by a crystallographic direction and a misorientation angle. Figure 12.1 illustrates such a misoriented surface. Here the average step spacing is defined by the misorientation angle and the step height. Low-angle steps are desirable in crystal growth since they act as nucleation sites that increase the rate of growth. Figure 12.2 illustrates a stepped surface with adatoms that arrive on substrate terraces, then migrate to ledges where they can bond to the additional crystal sites and continue the lateral growth of the upper terrace. Regular step arrays can also introduce grating structures that make possible additional periodic chemical and electronic properties.

Steps also expose new crystal faces and different chemical bonds. For example, Figure 12.3 represents a vicinal GaAs(100) surface inclined 2° toward the [111]B direction. Here, step heights can be single-layer while terrace widths can extend from 50 to 120 Å. In addition to the exposed Ga atoms of the (100) surface, the vicinal step edge exposes As atoms with chemically-active dangling bond electrons. Energy levels

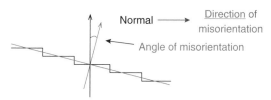

Figure 12.1   *Schematic view of a vicinal surface. Misorientation angle, misorientation direction, and step height determine the average step spacing. (Brillson 2010. Reproduced with permission of Wiley.)*

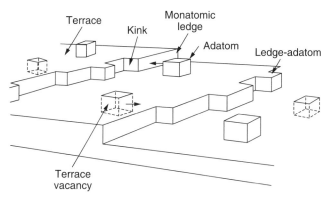

Figure 12.2   *Block illustration of adatoms, advacancies, and kinks on a stepped surface. (Brillson 2010. Reproduced with permission of Wiley.)*

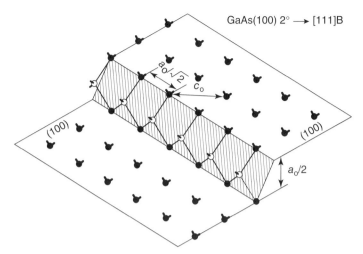

Figure 12.3 *Vicinal GaAs surface atomic geometry. Ga atoms (dark circles) terminate the (100) surface. The [111] surface (shaded) exhibits As (light circles) dangling bonds extending out of the plane. (Brillson 2010. Reproduced with permission of Wiley.)*

associated with these dangling bonds lie within the GaAs band gap and are observable by DRCLS. As the vicinal angle increases, both the number of exposed As sites and the intensity of deep level emission associated with these sites increase [1].

Vicinal step edges in III-V compound semiconductors such as GaAs also exhibit pronounced differences in morphology, depending on the orientation of vicinality. For example, the dimer rows in Figure 9.10 for GaAs(001) are parallel to steps edges along the [110] direction versus perpendicular along the [$\bar{1}$10] direction. As a result, STM shows that steps edges are relatively smooth along the [110] directions versus ragged along [$\bar{1}$10]. The energies associated with step atoms and long-range strain influence both step height distributions and step spacings. As an example, Si(001) single- versus double-layer step height distributions have a characteristic transition at a critical misorientation angle that is predicted theoretically based on energy minimization and entropy [2].

Step and kink energies can be extracted from distributions of kink separations and kink length where the energy for a kink of length n atoms is $E(n) = n\varepsilon + C$, and ε is the kink energy per atom with C a constant. Based on a Boltzmann distribution of independent kinks of length n, $N(n) \propto e^{-E(n)/k_B T}$. For vicinal Si, $E(n)$ are comparable to $k_B T$ at room temperature. Strain can also vary across semiconductor surfaces depending on the details of surface cleaning and can determine different metastable atomic surface structures that can form. In general, kink energetics are short-range and govern kinks and step fluctuations. Strain fields are longer range and control the average spacings between steps.

Facets are new crystal faces that become exposed on an otherwise uniform crystal surface. Facets have non-uniform morphologies that can be altered by impurities and adsorbates. On a yet smaller scale, point defects can form on electronic material surfaces. For example, anion and cation vacancies are electrically charged and can be imaged by STM. Figure 12.2 shows examples of these vacancies formed at a growing crystal surface. These vacancies may be mobile and recombine with a step edge, removing the defect.

12.2   Chemical Structure

For a clean semiconductor surface, chemical structure can involve (i) addition of atoms by, for example, crystal growth and the dynamics of deposition, (ii) atom movement across the surface as in surface diffusion, and (3) atom removal from the surface by, for example, desorption, evaporation, or etching.

12.2.1   Crystal Growth

Numerous techniques are available for bulk crystal growth including: (a) Czochralski growth from the melt, (b) horizontal Bridgman growth from the melt, (c) float-zone crystal growth, and (d) liquid phase epitaxy. These techniques involve molten material in contact with a seed crystal that can initiate further ordered atom growth of the same material. Several epitaxial crystal growth methods include: (i) vapor phase epitaxy (VPE), (b) liquid phase epitaxy (LPE), (c) molecular beam epitaxy (MBE), and chemical beam epitaxy (CBE or metal organic chemical vapor deposition, MOCVD). These methods require deposition on an ordered substrate with similar lattice constant. For gas phase methods of deposition, the various processes involved can be represented by a *Mass Balance Equation*. Thus for GaAs MBE, the change in surface concentration of Ga atoms during growth is:

$$(d\Theta/dt) = J_{Ga} - nJ_{As_n}S_{As_n} + 2R_{As_2} + D_{Ga} - D_{As} \tag{12.1}$$

where Θ = surface concentration of Ga atoms during growth, J_{Ga}, J_{As_n} ($n = 2$ or 4) are the incident fluxes of Ga and As_n, respectively, S_{As_n} is the As_n sticking coefficient, R_{As_2} = dissociation/desorption rate of As_2, while D_{Ga} and D_{As_n} are Ga and As diffusion rates from the sub-surface bulk to the surface, respectively. Here the Ga atoms are added to the surface by Ga deposition, Ga diffusion from the bulk, or As removal by desorption, freeing up Ga atoms. Ga atoms are removed from the surface by reaction with deposited As or diffused As from the bulk.

In MBE, neutral, monoatomic tetramers or dimers evaporate from Knudsen cells and deposit on a crystal surface, one monolayer at a time. The UHV environment insures that contamination levels in the deposited material are low. For MBE growth of GaAs, the 1:1 Ga-to-As ratio is maintained as long as $J_{As_n} > J_{Ga}$ since excess As is not bonded to Ga and re-evaporates. Layer-by-layer growth is possible under these growth conditions, which RHEED can monitor from the oscillation of diffracted streaks on a fluorescent screen. Figure 9.13 illustrates RHEED intensity behavior associated with laminar growth of epitaxial monolayers. As each layer reaches full coverage, intensities reach a maximum versus a minimum at half-layer coverage. Evaporation rates versus temperature measured by residual gas analyzers provide activation energies of sublimation in Equation (12.1).

The amplitudes of RHEED oscillations indicate how completely each monolayer forms for a given growth rate, which in turn provides the rate of surface diffusion. For diffusion lengths $L >$ terrace widths W, no growth occurs on the terrace, which would produce RHEED disorder. Diffusion coefficients are expressed as $D = D_o\,e^{-E_D/kT}$ so that $L = \sqrt{D\tau}$ where τ is the diffusion time. For example, E_D for Ga on GaAs(001) (2×4) is ~1.3 eV and $D \sim 10^{-5}\,cm^2\,s^{-1}$. RHEED spot intensities also increase during interrupted growth as the surface smooths at high temperature. The rates of surface smoothing yield an *activation energy* E_a – for MBE growth of GaAs, $E_a \sim 2.3$eV. Since the *cohesive energy* of GaAs is ~1.7 eV, this activation energy implies that some bond breaking must take place.

Overall, diffusion coefficient studies indicate that kinetics determines the smoothness of epitaxial interfaces, a major consideration for growth of multilayer structures such as quantum wells or superlattices.

12.2.2   Etching

Numerous etch processes are available to remove surface atoms.These include wet chemistry, reactive ions, plasma, thermal desorption, sputtering, electrolytic processes, ion bombardment, and melting. Among various electronic applications, etching is used to create gratings, mesas, highly anisotropic holes, grooves and vias, and selectively patterned heterostructures. Wet chemical etching provides a means to illustrate the role of atomic scale properties on macroscopic surface features. Besides cleaning surfaces in preparation for growth or the formation of specific device structures, etching can be used for defect identification and measurements of impurity distribution.

Primary factors that determine etching properties are: crystal orientation, the presence of defects such as dislocations, and the presence of impurities. Figure 12.4 illustrates the effect of crystal orientation on wet chemical etching. Here the left SEM image reveals a hexagonal pattern for GaAs(111) etched on a (111) surface, whereas the same etch process on the parallel ($\overline{111}$)B surface of the same GaAs(111) wafer produces much more pronounced etch features. Similar differences between A and B face etch rates are found for other III-V compounds.

One can understand the dramatic difference with orientation in these zinc-blende III-V compound semiconductors based on the density of dangling bond states. Figure 12.5a illustrates the zinc-blende crystal structure of a III-V compound semiconductor. The dashed line between AA and BB atomic layers represent planes that separate the two layers with a minimum of bonds, whereas the dashed line between AA and B'B' signifies a plane with higher bond density. Figure 12.5b illustrates this difference in bond density for InSb,

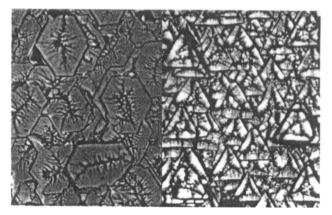

Figure 12.4   *SEM Micrograph of etched Ga- and As-terminated surfaces of a GaAs(111) wafer after 10 min in 0.2N Fe^{3+} in 6N HCl. The As surface exhibits much more pronounced etch features. (Gatos, H.C. 1994 [3]. Reproduced with permission of Elsevier.)*

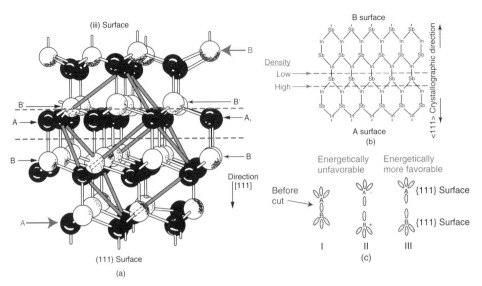

Figure 12.5 *(a) III-V compound semiconductor zinc-blende crystal structure. Dark circles are A (Group III) atoms. Open circles are (Group V) atoms. (b) Two-dimensional structure of the zinc-blende structure along the polar direction, showing the formation of an A atom and B atom surface plane by separation of bonds between AA and BB. (c) Distribution of covalent bond electrons before versus after separation of AA and BB. (Gatos, H.C. 1994 [3]. Reproduced with permission of Elsevier.)*

requiring lower energy to break bonds in forming the free A and B surfaces. This low bond density separation yields surfaces with just one type of atom. Figure 12.5cI shows the covalent bonding with two electrons shared between A and B atoms along the polar direction. Separation of these A and B planes in Figure 12.5cII to leave one electron in both A and B is energetically unfavorable compared to Figure 12.5cIII with both electrons paired on B and none on A. Hence the most energetically favorable configuration of the Group III polar surface has only six electrons, resulting in sp^2 hybrid bonding, whereas the Group V polar surface has eight electrons that result in sp^3 bonding. The missing electrons at the Group III surface atoms lead to a planar configuration and more strain.

The unshared pair of electrons at the Group V surface atoms produces dangling bonds that make this surface more chemically active. Indeed, surface chemical reactivity increases with dangling bond density. Measured etch rates in III-V compound semiconductors increase with orientation from A{111} (planar) to {110} with triply-bonded A and B surface atoms to {100} with an average of two electrons at A and B surface atoms to B{$\overline{111}$} with dangling bonds on all surface atoms. As a result of the higher strain, the planar A(111) surface exhibits more mechanical damage [3].

Edge dislocations at the surface also exhibit increased reactivity so that the creation and propagation of etch pits depends on the chemical bonding and structure of the atoms along the dislocation line. Thus chemical etching is useful for identifying dislocations and their density, but etch pit growth depends on the reactivity of the dislocation relative to the surface plane. Impurities are another factor that can alter the rate of chemical etching depending on their adsorption site. Finally, semiconductor doping has an effect on etching since excess carriers are involved in the chemical reaction process.

Photochemical etching is an example of how excess carriers accelerate the etch process. Chemical etching under UV illumination speeds up etching of GaN. However, the etched surface leaves behind nanowires that form as a result of slower etch rates. This slower etch rate results from gap state recombination of the free carriers generated by photoexcitation [4].

12.2.3   Adsorbates

The geometric structure of adsorbates on semiconductors is distinguished by (i) adsorbate site selectivity, (ii) the effect on surface reconstruction, and (iii) the degree of chemical reactivity and new chemical species that form. Thus in Figure 9.9, there are a number of characteristic sites on the Si(111) (7 × 7) surface, termed: corner hole, center, and rest atom locations, that adatoms prefer to bond to. For adsorbates on Si, STM measurements show that homoepitaxy such as Si on Si(111) exhibits a preference for (7 × 7) unit cell alignment with growing islands and a preference for a particular adsorption site with the (7 × 7) unit cells. For Si oxidation, O strongly prefers the "faulted" side of the unit cell as well as corner sites. For H_2O, NH_3, PH_3, and Si_2H_6 on Si, rest atoms are the most reactive and reactivity increases from corner to center to rest atom locations. For metal adsorbates on Si, Column III atoms such as Al and In substitute at on-top or $T4$ Si or Ge sites whereas B substitutes for Si in an $S5$ site under two layers of Si, inducing a change in surface reconstruction. Cs on Si as well as GaAs leads to periodic one-dimensional chains. Transition metals such as Fe, Co, Ni, as well as Ca produce silicides with epitaxial variants. One can rationalize these different bonding sites in terms of their energetics.

Adsorbates on GaAs also assume a variety of bonding sites. Si on GaAs(100) exhibits multiple inequivalent sites. On the GaAs(110) surface, Au prefers Ga surface sites, Cl prefers As sites, Sb produces Sb zig-zag chains at top layer anion and cation positions which eliminate the surface reconstruction, Sm also produces zig-zag chains but is chemically disruptive, and Al exhibits staged adsorption, reaction, and adsorbate incorporation. The initial adsorption of Al on the GaAs(110) surface leads to Al clusters, whose condensation energy leads to bond breaking and place exchange between top layer Ga and Al. This results in an energy gain since the AlAs heat of formation $H_F(AlAs) = -27$ kcal mol^{-1} versus -17 kcal mol^{-1} for GaAs. This replacement reaction can proceed to the next layer below but not further since there is no additional energy gain for Al diffusion. Indeed, the reacted layer now serves as a diffusion barrier against further Al diffusion. Track II Table T12.1 Adsorbates on GaAs summarizes these results.

12.2.4   Epitaxical Overlayers

In contrast to adsorption, there are relatively few examples of abrupt, elemental metal epilayers on silicon. An isolated example is Pb on Si, for which the metal and semiconductor are aligned as: (111)Pb || (111)Si and [1$\bar{1}$0]Pb || [1$\bar{1}$0]Si. Both a commensurate Pb/Si(111) (7 × 7) and incommensurate Si(111) ($\sqrt{3} \times \sqrt{3}$)R30°-Pb structure are observable depending on deposition temperature and coverage. Metals on Si can react to form silicides that are epitaxially ordered. Included are transition and refractory metals such as

Ni, Ti, V, Cr, Fe, Co, Zr, Nb, Mo, Pd, Ta, W, and Pt. A prime example is the $NiSi_2/Si(111)$ interface. $NiSi_2$ has a cubic CaF_2-type structure and a lattice constant a_0 only 0.44% smaller than $a_0(Si)$ at room temperature. $NiSi_2$ can grow exclusively in two variants, termed A and B and rotated by 180° around the surface normal, depending on surface preparation, template structure, and annealing conditions [5]. These variants are of particular interest since they are reported to exhibit different Schottky barrier heights.

There are multiple elemental metal systems that exhibit epitaxy on compound semiconductors. A common example is Al on GaAs, which aligns as either {100}Al||{100}GaAs or <010>Al||<011>GaAs with a 1.8% lattice mismatch. Another is Fe on GaAs where $a_0(Fe)$ aligns with $1/2a_0(GaAs)$ to within 1.2%. See Track II Table T12.2. Metals on GaAs for additional examples of metal epitaxy on GaAs.

Process-dependent epitaxy, chemical reaction, and diffusion are evident for most systems at elevated temperatures. Epitaxial metals on semiconductors can be useful in different applications if they can form conducting sheets inside semiconductor lattices. For example, buried III-V/metal/III-V epitaxial structures could form Bragg reflectors that increase absorption in active semiconductor layers without increasing semiconductor volume (unlike layers of semiconductor layers with different refractive indices), thereby minimizing volume defects. Similarly, buried metal sheets in semiconductors can enable permeable base transistors. In order to achieve such structures, the metals must meet one or more criteria for growth of stable epitaxial metal/III-V semiconductor heterostructures, namely: (i) single-variant epitaxy, (ii) smooth morphology, and (iii) phase stability. No single elemental film meets all these criteria. Instead, metal alloys offer greater flexibility to lattice match and achieve all these criteria. The lattice parameters of various transition metal (TM) – Group III phases with the CsCl(B2) structure are $\sim 1/2a_0$ for many III-V compound semiconductors. Likewise, the lattice parameters of the rare-earth (RE) metal-Group V phases with the NaCl(B1) structure are close to a_0 for these III-V compounds. These TM-III and RE-V compounds have high thermal stability since their phase equilibria include Ga, In, and Al in common with the semiconductor. They also have high morphological stability due to their high melting points so that alloy film agglomeration is low. From a fundamental standpoint, these systems have excellent interface structures for probing electronic properties since they are well ordered.

12.2.5 Growth Modes

The initial growth of atoms deposited on surfaces can proceed in three ways, as shown in Figure 12.6: (a) Frank Van der Merwe or layer-by-layer, (b) Volmer–Weber or island growth, (c) Stranski–Krastanov – combination of islands on monolayers. These modes of growth are determined largely by (i) the heat of condensation of the adsorbed atoms with the surface atoms and with each other and (ii) with the surface mobility of the adsorbed atoms, which depends on the chemical bond strength between adsorbate and substrate atoms.

Between the monolayer and multi-micron thickness regimes, metal overlayers on semiconductors exhibit several types of growth modes. Starting with (d) individual atoms on surfaces, adsorbed atoms can form epitaxial layers (e), depending on the adsorbate bond lengths and epitaxial relationship to the substrate, (f) intermixed layers of deposited and

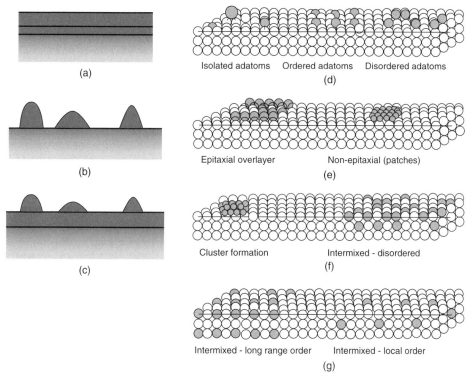

Figure 12.6 *Alternative growth modes of atoms deposited on surfaces: (a) Frank Van der Merwe, (b) Volmer–Weber, and (c) Stranski–Krastanov. (Brillson 2010. Reproduced with permission of Wiley.) Alternative pathways for interfaces to evolve from deposited atoms on a surface. Atomic distribution and bonding with increasing deposition and/or time elapsed for (d) individual atoms, (e) atomic groups, (f) sub-surface distributions, and (g) interface compounds or solutions. (Weaver, J.H. 1988. Reproduced with permission of Elsevier.)*

substrate atoms, or (g) a combination of short-range and longer-range intermixed layers, which could be the result of a chemical reaction and formation of new compounds.

Thermodynamic factors determine how the interface formation proceeds. Strong adatom–substrate bonding or a high density of adatom-defect bonding sites leads to low adatom mobility and more uniform, layer-by-layer growth. High heats of fusion or condensation promote strong adatom–adatom bonding and cluster formation. Strong bonding between dissimilar atoms can produce new chemical phases with more long-range order and well-defined boundaries due to the large energy gained. Nearly equivalent chemical interactions between atoms leads to entropy-driven diffusion, where the disorder contribution to the heat of formation becomes a deciding factor. Besides these thermodynamic equilibrium factors, energy processing can produce non-equilibrium conditions that alter interface chemistry. These processes include: (i) heat from thermal or rapid thermal annealing, (ii) light from laser annealing or UV excitation to break bonds, and (iii) energetic particles from, for example, ion bombardment.

12.2.6   Interface Chemical Reaction

The formation of reacted interface phases depends on both thermodynamics and kinetics. Thermodynamics determines the driving force for compounds to form based on the energy difference between substrate and compound heats of formation. Phase formation in thin ($\sim$100 nm) films is dictated by kinetics, that is, how fast the atoms move). Kinetics becomes a critical factor for interfaces whose constituents can form more than one compound. Classic examples of kinetically-driven phase formation in electronic materials include metal–Si silicides. For example, Ni can react with Si to form phases of Ni_2Si, Ni_5Si_2, Ni_3Si, $NiSi$, or $NiSi_2$, depending on reaction temperatures, times, and the availability of constituents. Track II Table T12.3.Metal-on-Si Reacted Phases shows interface phases formed with thermal annealing of metals on silicon substrates. These different phases are significant since they can have ohmic or rectifying characteristics. These phases fall into two categories of solid-state reactions: (i) laterally uniform growth with well-defined kinetics and temperature dependence and (ii) non-uniform growth whose nucleation and agglomeration depends critically on temperature. Thus, the resultant equilibrium phases depend not only on thermodynamics and rate of compound growth but also the identity of the moving species and its concentration at the reaction interface.

The diffusion velocity v of one constituent into the other can be expressed in terms of an effective species mobility μ_a under a driving force equal to the gradient of chemical potential ΔH_R over distance x.

$$v = \mu_a(-\Delta H_R/x) \tag{12.2}$$

For a given concentration C, the diffusion flux F of atoms is then

$$F = v \cdot C = (DC/k_B T)\,(-\Delta H_R/x) \tag{12.3}$$

See Track II T12.4.Diffusion Flux for derivation of v and F. The phases that form for a given time and temperature of anneal depend on the relative velocities of diffusing species through interfacial films of the various phases that can form. Furthermore, growth can be diffusion-controlled, in which case several phases can grow in parallel, or interface reaction-controlled, where multiple phases begin to grow only after it reaches a critical thickness determined by the relative diffusing velocities [6].

Besides laminar interactions between metals and semiconductors, non-uniform reactions can occur that are undesirable. For example, Al contacts to Si have a transition to liquid plus Si at 580°C and can form metal spikes that protrude unevenly into the semiconductor. Similar interactions occur at Al–SiO_2 and Al–Pd_2Si–Si junctions. Besides reactions between metals and semiconductors, the diffusion and reaction between multiple metals on the semiconductor can present challenges. Au on Al is an important binary system for microelectronics. Since Au is ductile and does not oxidize, Au wire can bond easily to conductor pads on a chip, whereas Al is a low work function metal for ohmic contact and good adhesion to Si. However, Au forms several intermetallic phases with Al, $AuAl_2$, that is brittle, has low conductivity, and changes volume to create cavities in the metal near the Au–Al interface. At clean interfaces, reactions can take place at room temperature under thermal compression so that bonds break under stress. Based on its bright purple color and the damage it produces, this intermetallic is termed "*purple plague*". Controlling such chemical reactions and diffusion becomes more pressing as microelectronics shrinks ever smaller and the need to increase thermal conductivity increases accordingly.

For multiple metal contacts to provide both low contact resistivity and desirable wire bonding properties, barrier layers that prevent diffusion between contiguous layers are

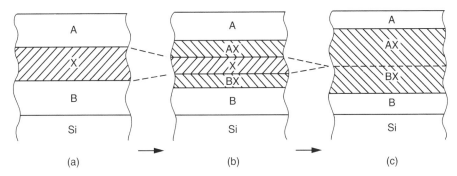

Figure 12.7 *Sacrificial diffusion barriers: (a) interfacial barrier layer between layers A and B, (b) intermediate compounds AX and BX formed at the A–X and B–X interfaces, and (c) full consumption of X into compounds AX and BX. (Nicolet, M.-A. 1978 [7]. Reproduced with permission of Elsevier.)*

necessary. These *diffusion barriers* must satisfy a number of conditions. For a metal A on metal B and a diffusion barrier X between them, Figure 12.7a: (i) transport of A and B across X must be low, (ii) the loss rate of X into A and B must be small, (iii) X must be thermodynamically stable against A and B, (iv) adhesion between X and both A and B must be strong, (v) the specific contact resistance of X to A and B must be small, (vi) X must be laterally uniform in thickness and structure, (vii) X must be resistant to mechanical and thermal stress, and (viii) X must be highly conductive, both electrically and thermally.

It is usually difficult to satisfy all these criteria with a single diffusion barrier. One approach is to use metal interlayers that form compounds with A and B in order to form the diffusion barrier. Figure 12.7 illustrates the formation of compounds AX and BX that consume *sacrificial layer* X almost completely for carefully controlled times and temperatures to avoid forming any additional compounds. Another method to inhibit interdiffusion is termed "stuffed barriers". Since each metal layer is polycrystalline, the boundaries between grains represent pathways for rapid diffusion. In order to eliminate these pathways that short-circuit the slower bulk diffusion, one can "stuff" these grain boundaries with impurities such as oxygen. As an example, Ti and Mo intermix but Ti–O–Mo do not. Alternatively, one can eliminate grain boundaries altogether by amorphizing the metal. In general, amorphous layers are metastable and tend to crystallize and single elements don't form amorphous phases at room temperature. One can select elements for amorphous alloys where the larger the difference in atomic size, crystalline structure, and electronegativity, the easier it is form metallic amorphous alloys. Examples include alloys of near-noble metals and refractory metals such as Ni_xW_{1-x} and Cu_xTa_{1-x} with nearly 50:50% composition ratio. Fast quenching during deposition also promotes amorphous alloys. Diffusion barriers are important for reducing Si consumed in silicide formation, especially as electrical contact dimensions continue to decrease in large-scale microelectronic integration.

12.3   Electronic Structure

The local atomic arrangement and the associated charge transfer that take place at semiconductor and insulator surfaces determine their electronic structure. This charge transfer depends on the specific arrangement of near-surface atoms or adsorbed molecules and determines the work function, the band bending, and the electron affinity. In turn, this electronic structure leads to a variety of fundamental properties and important applications.

12.3.1   Physisorption

Earlier chapters described the electronic structure of clean semiconductor surfaces. Atoms or molecules adsorbed on semiconductor surfaces can interact in a number of ways. The weakest form of such interaction is *physisorption*, based on *Van der Waals bonding*. Here the electronic structure upon adsorption is only slightly perturbed. Whereas Van der Waals forces between molecules scale with distance r as r^{-6}, the forces between adsorbates and surfaces scale as r^{-3}. The attractive force between an adsorbate is due to correlated charge fluctuations and mutually induced dipole moments. Figure 12.8 illustrates a physisorbed atom consisting of a positive ion and valence electron. Figure 12.8 shows how vibrations with displacement u of the ion and valence electron induce image charges within the solid. The resulting attractive potential is expressed as

$$V(r) \cong -(q^2 \mu^2)/(4r^3) \tag{12.4}$$

where u is the ion–electron separation and r is the adsorbate–surface distance. Because of the repulsion between outer adsorbate and surface electrons, the potential $V(r)$ increases with decreasing r so that the overall potential has a minimum at an equilibrium distance r_0 shown in Figure 12.9, which varies slightly with adsorbate but is $\sim \frac{1}{2} a_0$. The binding energy for physisorption is low (~ 10–100 meV) relative to $k_B T = 25.9$ meV at 300 K, and the separation from the surface is relatively large (3–10 Å). An example of physisorption on a semiconductor is Xe on GaAs(110), which requires cryogenic temperatures for Xe to adsorb.

12.3.2   Chemisorption

In contrast to physisorption, chemisorption involves strong chemical bonding and charge transfer between adsorbate and surface. As a result, the electronic structure of the adsorbate

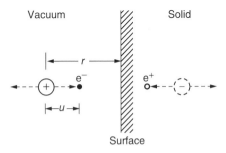

Figure 12.8 *Schematic illustration of a physisorbed atom consisting of a positive ion and valence electron. Oscillations normal to the solid surface (arrows) follow classical dynamics with attractive forces due to image charges. (Lüth, H. 2001 [8]. Reproduced with permission of Springer.)*

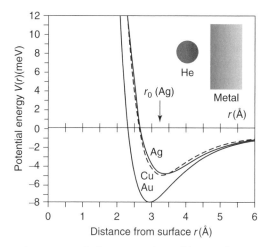

Figure 12.9 *Physisorption potentials for He on the noble metals Au, Cu, and Ag. Here the metal is described by a "jellium" model (sea of electrons in a mean density of positive ionic charge). (Lüth, H. 2001 [9]. Reproduced with permission of Springer.)*

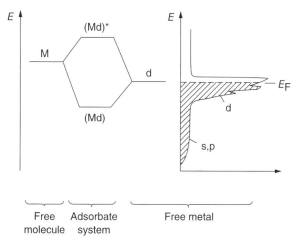

Figure 12.10 *Hybridized bonding (Md) and antibonding (Md*) electronic levels due to bonding between a molecule (M) and a d-band metal. (Lüth, H. 2001 [8]. Reproduced with permission of Springer.) (Lüth, H. 2001 [8]. Reproduced with permission of Springer.)*

and surface atoms can change significantly. As an example, consider how a molecule M with a partially-filled molecular orbital bonding interacts with the d level of a transition metal atom in a surface. Orbital overlap between molecule and metal surface should lead to rehybridization and formation of new Md orbitals (Figure 12.10). Charge transfer from molecule to metal or *vice versa* can be described by a linear combination of two new states. Energy minimization of this two-state

$$\widetilde{E}_\pm = \tfrac{1}{2}(H_1 + H_2) \pm \left[((H_1 - H_2)/2)^2 + H_{12}^2 \right]^{1/2} \qquad (12.5)$$

where ψ_1 represents a state corresponding to an electron transferred to the molecule and having energy $H_1 = \langle\psi_1|H|\psi_1\rangle$, ψ_2 and $H_2 = \langle\psi_2|H|\psi_2\rangle$ represent the state and energy, respectively, for an electron transferred to the metal, and $H_{12} = \langle\psi_1|H|\psi_2\rangle$ represents the interaction between the two "ionic" charge transfer states. See Track II T12.5. Energy Levels of Hybridized Molecule-Metal Chemisorption. The "+" solution to $\tilde{E}$ represents an "antibonding" state where electrons orbit primarily outside the nuclei of two atoms whereas the "−" solution represents the "bonding" state where electrons orbit primarily between the two nuclei. More sophisticated wave functions and approaches are available, such as cluster models with a finite number of substrate atoms. These apply methods of quantum chemistry to chemical bonding. Important examples of such bonding in solid state electronics include the oxidation of III-V and II-VI compound semiconductors. The rearrangement of electronic orbitals changes the adsorbate shape due to the new chemical bonds to the substrate. In turn, this can lead to new chemical species and molecular dissociation. An example of the latter is hydrogen molecules interacting with transition metal surfaces. [See Track II T12.6. Case Study: Hydrogen on d-Band Metal].

12.3.3　Surface Dipoles

The change in bonding for adsorbates on semiconductors produces changes in surface work function and surface dipoles. The work function $e\phi = E_{vac} - E_F$ is the energy to remove an electron from inside the bulk to a distance far enough ($\sim$1 micron) from the surface that image forces can be neglected. The work function can be defined thermodynamically. Thus the change in energy E_N of a system with N electrons as one electron is removed to the vacuum is:

$$e\phi = E_{N-1} + E_{vac} - E_N = E_{vac} - (E_N - E_{N-1}) \tag{12.6}$$

$(E_N - E_{N-1}) = (\partial F/\partial N)_{T,V}$ is the change in free energy $F = \mu N$ by removing one electron at constant temperature and volume so that

$$e\phi = E_{vac} - (\partial F/\partial N)_{T,V} = E_{vac} - \mu \tag{12.7}$$

where μ is termed the *electrochemical potential* of an electron in the bulk. For a semiconductor or insulator, charge transfer between adsorbate and substrate changes both band bending and surface dipole. As pictured in Figure 4.1, the wave function of charge decaying out of a substrate into vacuum represents charge separation and a surface dipole. Adsorption modifies such wave functions spatially, changing the bare surface dipole and hence the electron affinity. Likewise, Figure 3.1 illustrated how charge transfer between surface and bulk causes band bending within a surface space charge region. The overall work function $e\phi$ is then

$$e\phi = \chi + eV_S + (E_C - E_F)_{bulk} \tag{12.8}$$

For a semiconductor, charge transfer between adsorbate and substrate changes both band bending and surface dipole and hence the electron affinity so that

$$e\Delta\phi = \Delta\chi + e\Delta V_S = e\Delta\phi_{dipole} + e\Delta V_S \tag{12.9}$$

where χ is the electron affinity, $e\Delta\phi_{dipole}$ is the dipole change, and $e\Delta V_S$ is the band bending. Figure 12.11 illustrates how the $e\Delta\phi_{dipole}$ contribution changes both the electron affinity and the work function.

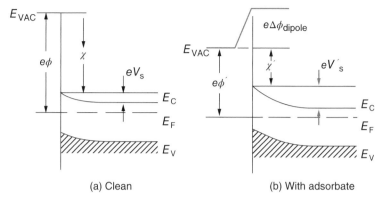

Figure 12.11   *Schematic energy band diagram of (a) clean and (b) adsorbed semiconductor surface. With adsorption, both band bending and effective electron affinity change. (Lüth, H. 2001 [8]. Reproduced with permission of Springer.)*

UV photoemission spectroscopy can detect changes in both $e\Delta\phi_{dipole}$ and $e\Delta V_S$. Figure 12.12a shows the filled valence and core level states of a semiconductor and the photoemission spectra for a photon energy $h\nu$ with the corresponding excited density of states superimposed on a background of secondary electrons. This secondary electron distribution drops to zero at E_{VAC} since electrons with kinetic energies less than E_{VAC} cannot escape the solid. For the positive dipole $e\Delta\phi_{dipole}$ (positive outward to vacuum) pictured in Figure 12.12b, both χ and ϕ decrease so that additional electrons can escape into vacuum. Hence the width of the *energy distribution curve* (EDC) increases (from dashed to solid lines). With additional band bending $e\Delta V_S$ (upward for n-type), all filled states move closer to E_F so that the photoemission energies from all the filled states increase as described in Section 7.1.5.3. While both effects increase the photoemitted electron energies in this case, the change in $\Delta EDC = e\Delta\phi_{dipole}$ determines the dipole contribution alone. Since $e\Delta\phi$ in Equation (12.9) equals the change in photoemitted electron energies measured, the difference $e\Delta\phi - e\Delta\phi_{dipole}$ yields the band bending $e\Delta V_S$ as in Equation (12.9).

$$e\Delta V_S = (E_F - E_V)_{measured} - (E_V - E_F)_{bulk} \qquad (12.10)$$

Photoemission spectroscopies can also measure the band bending $e\Delta V_S$ directly as described in Section 7.1.5.3 without $e\Delta\phi_{dipole}$ if one knows the semiconductor doping level. From the semiconductor doping, one can calculate $(E_C - E_F)_{bulk}$ (see Equation (3.18)), and from the semiconductor band gap E_G, one obtains $(E_F - E_V)_{bulk} = E_G - (E_C - E_F)_{bulk}$. As in Section 7.1.5.3, one measures $E_F - E_V$ where the highest energy emitted by a reference metal grounded to the analyzer determines E_F and the highest energy emitted by a semiconductor determines E_V. Then $e\Delta V_S$ is just the difference in $E_F - E_V$ between the surface and the bulk. An independent measurement of work function $e\phi$ from, for example, KPFM would then provide $e\Delta\phi_{dipole}$ from the measured $e\Delta V_S$ using Equation (12.9).

Figure 12.13 illustrates a simple model of a surface dipole as a parallel plate capacitor such that

$$e\Delta\phi = -q\mathcal{E}d \qquad (12.11)$$

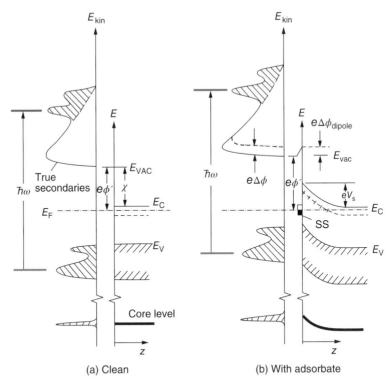

(a) Clean   (b) With adsorbate

Figure 12.12 *Adsorbate-induced changes in the photoemission spectra (a) before and (b) after adsorption. Electrons promoted to states above E_F exhibit core-level and valence band energies that are hν higher than their equilibrium values and a lower EDC cutoff at E_{VAC}. (Lüth, H. 2001 [8]. Reproduced with permission of Springer.)*

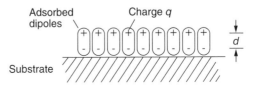

Figure 12.13 *Adsorbed dipoles on a surface, each with charges +q and −q separated by distance d. (Lüth, H. 2001 [8]. Reproduced with permission of Springer.)*

where electric field $\mathscr{E} = q_{dip} n_{dip}/\varepsilon_o$ from *Gauss's Law*, q_{dip} is the charge per dipole, n_{dip} is the density of surface dipoles, and d is the dipole charge separation. In general, the *dipole moment* $p = q_{dip} \cdot d$ is not well known since the effective polarization field $\mathscr{E}_{eff}$ is subject to depolarization, where

$$\mathscr{E}_{eff} = \mathscr{E} - f_{dep} \cdot \mathscr{E}_{eff} \tag{12.12}$$

and f_{dep} is defined as a *depolarization factor* that represents how the electric field at a dipole is reduced by the surrounding dipoles. For example, a square array of dipoles with

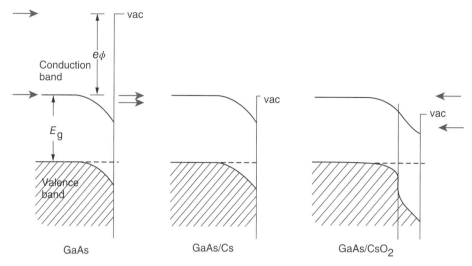

Figure 12.14 *Schematic energy band diagram of p-type GaAs before and after Cs and O adsorption. (Lüth, H. 2001 [8]. Reproduced with permission of Springer.)*

polarizability α produces an $f_{dep} \cong (9\,\alpha_{ndip}^{3/2})/(4\pi\varepsilon_o)$. However, adsorbate polarizabilities are also not well known due to their atomic scale and the effect that atomic scale probes can have on them. Track I T12.7 Case Study-Work Function Changes with Ordered Adsorption: H_2O on Cu illustrates how electron affinity changes with different LEED superstructures. Similarly, Track II T12.8: Case Study-GaAs Surface Work Function Changes with Reconstructions shows how the changes in surface dipole can be understood in terms of charge transfer between the outer atomic layers of Ga and As.

An important technological application of surface dipoles is *negative electron affinity* (NEA). Here, a positive surface dipole can effectively lower a semiconductor's vacuum level to below its conduction band minimum. This decreases or removes the barrier for electrons to escape the surface. Figure 12.14 illustrates this effect for a combination of Cs and O on a GaAs surface. Electron transfer from outer layer Cs atoms to O atoms at the intimate GaAs surface results in a large positive dipole and E_{VAC} below E_C in the band bending region below the surface. This NEA effect is useful for enhancing electron emission from metal filaments inside electron guns, such as in cathode ray tubes, from the electron multiplier stages of a photocathode, and even the charge injectors of a particle accelerator. In the case of GaAs, an additional NEA feature is that excitation of valence electrons to the conduction band minimum can be highly spin polarized due to spin–orbit splitting effects.

12.4 Summary

The macroscopic properties of adsorbates on semiconductors are sensitive to their geometric, chemical and electronic features at the atomic scale. The geometric structure of adsorbates on common semiconductors such as Si and GaAs shows that: (i) adsorption can be site selective, (ii) adsorbates can alter surface reconstruction, and (iii) adsorbate ordering on surfaces can enable epitaxial overlayer growth. The chemical structure of metals on clean semiconductor surfaces can include: (a) interface chemical reactions, leading to new chemical phases with properties different for either metal or semiconductor, (b) interface overlayer growth determined by kinetics and thermodynamics, (c) the opportunity to control such interface chemistry by surface science techniques. Overlayer growth depends on the specifics of chemical bonding, atomic structure and processing, which can control catalytic processes as well as the formation of electronic devices. Macroscopic thin film growth depends on atom surface mobility, the strength of adatom–adatom versus adatom–substrate bonding, surface defects, as well as the temperature and method of deposition. The electronic structure of adsorbates on semiconductors depends on local bonding and charge transfer, which determine the semiconductor χ, surface $e\Delta\phi$, and $e\Delta V_S$. While physisorbed atoms or molecules bond only weakly to surfaces and produce small electronic changes, chemisorbed atoms involve significant charge transfer that produce large electronic effects. Overall, local atomic structure at the atomic scale can have major effects on macroscopic electronic devices.

12.5   Problems

1. (a) For the metals Cu, In, Ni, Al, and Ti deposited on a GaAs (110) surface, which metal is most likely to produce uniform deposited overlayers within the first few monolayers? (b) Which is most likely to form islands? [Hint: Most thermodynamically stable metal-arsenide reaction products for Cu, In, Ni, Al, and Ti are: Cu_3As, InAs, NiAs, AlAs and TiAs with H_F values of -2.8, -57.7, -72, -116.3, and -149.7 kJ mol^{-1}, respectively.]

2. Calculate the driving force for an Al–In exchange reaction (a) at the InP(100) surface and (b) one monolayer below. Explain why this reaction produces a diffusion barrier to semiconductor outdiffusion from the bulk semiconductor and specifically to which element(s)?

3. Consider the diffusion of Si atoms into a 100 μm thick Al overlayer. At 327 °C, the diffusion length is 16.33 μm after 5 minutes. (a) What is the value of the diffusion coefficient D? (b) After annealing $D =$ for 5 minutes at 527 °C, the diffusion length is 1.67 times longer. What is the activation energy of diffusion?

4. If 1000 Å of Pt is deposited on a Si(100) surface and annealed to form a Pt_2Si low-resistance contact, (a) what thickness of Si is consumed by this much Pt in the formation of a low-resistance Pt_2Si contact? [Hint: atomic density $N = N_A \cdot \rho/A$ where $N_A =$ Avogadro's number and $A =$ atomic weight.] (b) What thickness of Pt is required to form one monolayer of Pt_2Si?

5. Consider a metal–semiconductor interface with a heat of reaction $\Delta H_R = -200$ kJ mol^{-1}. Assuming an activation energy for diffusion of 1.5 eV and an

activated hopping prefactor for diffusion $D_0 = 10^{-1}\,\text{cm}^2\,\text{s}^{-1}$, what is the initial velocity of diffusing interface atoms across the first 10 Å at room temperature? [Note: 1 eV/molecule = 96.48 kJ mol^{-1}.]

6. Ti is used between Al and Si to form a laterally uniform TiAl$_3$ interfacial layer that delays the formation of Al "spikes" into Si. The effective diffusion constant D for the formation of TiAl$_3$ is $0.15\,\exp(-1.85\,\text{eV}/k_{\text{B}}T)$. How thick an interfacial layer is required to prevent the onset of spiking at 500 °C after 1 hour?

7. How thick does an Au overlayer on a semiconductor have to be to prevent more than 3% interface oxidation at room temperature after 24 hours? How thick would an analogous film of In have to be? Assume a diffusion prefactor of $6.1 \times 10^{-16}\,\text{cm}^2\,\text{s}^{-1}$, an activation energy $E_{\text{A}} = b \cdot |\Delta H_0|$ where $b = 1.3 \times 10^{-5}\,\text{cal}^{-1}$ mole, and no diffusion through the edges of the film. [Hint: calculate the diffusion length and assume 100% oxidation of the free metal surface.] For In oxide, $\Delta H_0 = -220\,\text{kcal mole}^{-1}$.

8. Calculate the energy required to desorb a physisorbed molecule from a metal, where the molecule's equilibrium distance from the metal is 3.5 Å and the molecule has 0.5 electrons separated by 1 Å.

9. A UV photoemission spectroscopy experiment with $h\nu = 21.2\,\text{eV}$ of clean, cleaved n-type ZnS in UHV yields an EDC with $E_{\text{max}} = 13.78\,\text{eV}$ and $E_{\text{min}} = 0$ relative to E_{VAC} of the electron analyzer. Comparison with a clean Au foil in electrical contact shows that $E_{\text{F}} - E_{\text{C}} = 0.1\,\text{eV}$ at the surface. *In situ* adsorption of one-half monolayer of oxygen rigidly shifts the EDC such that $E_{\text{max}} = 15.76\,\text{eV}$. However, $E_{\text{min}} = 0.3\,\text{eV}$. (a) What is the change in band bending $\Delta q V_{\text{B}}$, work function $\Delta \Phi_{\text{SC}}$, and dipole $\Delta \chi$? (b) Neglecting depolarization and assuming a dipole length of 10 Å, calculate the adsorbed oxygen's dipole moment.

References

1. Chang, S., Brillson, L.J., Kime, Y.J. *et al.* (1990) Orientation-dependent chemistry and Schottky barrier formation at metal-GaAs interfaces. *Phys. Rev. Lett.*, **64**, 2551.

2. de Miguel, J.J., Aumann, C.E., Kariotis, R., and Lagally, M.G. (1991) Evolution of vicinal Si(001) from double-to single-atomic-height steps with temperature. *Phys. Rev. Lett.*, **67**, 2830.

3. Gatos, H.C. (1994) Semiconductor electronics and the birth of the modern science of surfaces. *Surf. Sci.*, **299/300**, 1 and references therein.

4. Youtsey, C., Romano, L.T., and Adesida, A. (1999) Rapid evaluation of dislocation densities in n-type GaN using photoenhanced wet etching. *Appl. Phys. Lett.*, **75**, 3537 (1999).

5. Tung, R.T. and Schrey, F. (1989) Topography of the Si(111) surface during silicon molecular-beam epitaxy. *Phys. Rev. Lett.*, **63**, 1277.

6. Mayer, J.W. and Lau, S.S. (1990) *Electronic Materials Science: for Integrated Circuits in Si and GaAs*, Macmillan, New York.

7. Nicolet, M.-A. (1978) Diffusion barriers in thin films. *Thin Solid Films*, **52**, 415.
8. Lüth, H. (2001) *Solid Surfaces, Interfaces, and Thin Films*, Springer-Verlag, Berlin, p. 503.
9. Zaremba, E. and Kohn, W. (1977) Theory of helium adsorption on simple and noble-metal surfaces. *Phys. Rev. B*, **15**, 1769.

Further Reading

Zangwill, A. (1988) *Physics at Surfaces*, Cambridge University Press, Cambridge.

13

Surface Electronic Applications

The charge transfer between semiconductor surfaces and adsorbates described in Chapter 12 has many useful applications. Changes in charge density at semiconductor free surfaces induce changes in band bending below the surface. As a result, the charge density within the semiconductor's surface space charge region changes, altering electrical transport across the surface which macroscopic measurements can detect. As Figure 11.9 showed, very small changes in surface charge density can manifest themselves as measureable voltage changes of surface potential or charge transport. Researchers have taken advantage of this high surface sensitivity to create high sensitivity sensors for gases, chemicals, pressure, and biological species.

13.1 Charge Transfer and Band Bending

The surface conductivity of metal oxide semiconductors is strongly dependent on specific gas adsorption. Figure 13.1 illustrates how exposure of a ZnO surface to oxygen versus hydrogen affects its band bending and electron concentration within the surface space charge region. Figure 13.1a shows how transient oxygen exposure produces a negatively charged surface with positive countercharge screened by the semiconductor's dielectric constant and extending across its surface space charge region. In contrast, transient hydrogen exposure produces a positively charged surface with negative countercharge. Figure 13.1b shows the corresponding band bending within the surface space charge region as described by Equations (3.7) and (3.8) for these two cases. The charge densities corresponding to these surface charges and band bending appear in Figure 13.1c.

13.1.1 Sheet Conductance

These changes in charge density have a direct effect on the sheet conductance of the semiconductor near-surface region. For a wire or solid, conductance $\sigma = ne\mu$, where n is free

An Essential Guide to Electronic Material Surfaces and Interfaces, First Edition. Leonard J. Brillson.
© 2016 John Wiley & Sons, Ltd. Published 2016 by John Wiley & Sons, Ltd.
Companion Website: www.wiley.com/go/Brillson/

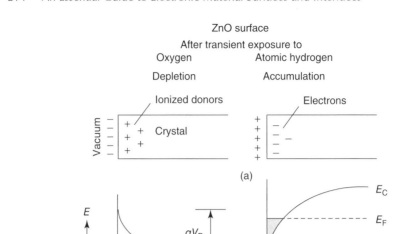

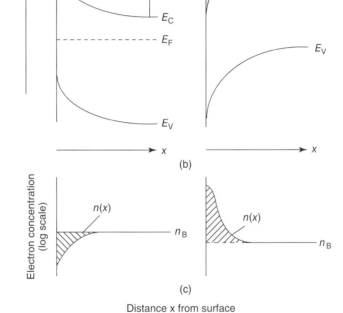

Figure 13.1 *ZnO surface (a) charge distribution, (b) band bending, and (c) electron concentration before and after transient exposure to oxygen and atomic hydrogen. Oxygen attracts electrons to the surface, depleting the sub-surface, while hydrogen donates electron, forming a surface accumulation layer. (Heiland, G. 1969 [1]. Reproduced with permission of Elsevier.)*

carrier density and μ is mobility. For the near-surface region pictured in Figure 13.1c, the sheet conductance $g_\square$ must account for the pathways parallel to the surface through which n varies. In general,

$$g_\square = e \int_0^l \mu n \; dz \qquad (13.1)$$

where l is the total layer thickness. For a semiconductor of space charge layer thickness d,

$$g_\square = e \int_0^l \mu_B n_B \ dz + e \int_0^d \mu_B [n(z) - n_B] \ dz - e \int_0^d (\mu_B - \mu_S) \ n(z) \ dz \qquad (13.2)$$

$$= \sigma_B l + \Delta\sigma \qquad (13.3)$$

where μ_B = bulk charge mobility, μ_S = surface charge mobility, n_B = bulk charge concentration, n_S = surface charge concentration $n(z)$ for $z < d$, σ_B = bulk conductivity, and $\Delta\sigma$ is the difference in surface versus bulk conductivity. Equation (13.3) neglects the third term in Equation (13.2), assuming $\mu_B = \mu_S$. Likewise, the excess charge in the space charge region is

$$\Delta N = \int_0^l [n(z) - n_B] \ dz \qquad (13.4)$$

13.1.2   Transient Effects

The adsorption of molecules produces new energy levels in the band gap that trap charge from the semiconductor. The electrical response to this adsorption depends on the rate of charge transfer between the semiconductor bulk and these new energy levels. In the case of oxygen adsorption, electrons move from below the surface to fill the new gap states with negative charge localized on the surface. As the surface charges negatively, the energy bands begin to bend, as shown in Figure 13.1b. Thus a dipole barrier begins to form between the negatively charged chemisorbed oxygen at the surface and the positively charged space charge region below (Figure 13.1c). Electron transfer over this barrier from the bulk to the surface is self-limiting: as adsorption proceeds, the band bending increases to a value that pinches off any further electron transfer. The transient response to the adsorption of gas on the semiconductor surface is temperature-dependent. As temperature increases, the electrical response becomes faster. The transient response also depends on the magnitude of band bending – the higher the barrier, the slower the response.

Case Study: Transient Response of ZnO to Oxygen Adsorption

Assuming that oxygen forms new acceptor levels on the surface at energies below the bulk E_F, calculate the rate of charge flow toward the surface at room temperature for $n_B = 10^{17}$ cm^{-3}. Assume a band bending of $qV_B = 0.3$ eV. For a surface trap density of $N_T = 10^{14}$ cm^{-2}, what is the time required to fill these traps? If temperature increases to 300 °C, how does the response change qualitatively? If the barrier increases to 0.6 eV, how does the response change?

Solution:

Using the band bending shown for oxygen in Figure 13.1b, the probability of conduction electrons reaching the surface is $P_S = n_s(T)/n_b(T) = e^{-qV_B/k_B T}$.

(continued)

(*continued*)

For $qV_B = 0.3$ eV at RT: $P_S = 9.3 \times 10^{-6}$. The *attempt frequency* for an electron to surmount the barrier is given by its thermal vibration frequency $v = k_B T/h = 6.25 \times 10^{12}$ s^{-1}.

The rate of charge flow toward the surface is $r = P_S \cdot n_B \cdot v$. For $n_B = 10^{17}$ cm^{-3} converted to charge density/cm^2, $r = 1.25 \times 10^{19}$ cm^{-2} s^{-1} so that for $N_T = 10^{14}$ cm^{-2}, the time to fill these traps is $t = N_T/r < 10^{-5}$ s. For a 300 °C temperature rise, the attempt frequency doubles but P_S decreases much more due to its exponential dependence on temperature. The net result is an increase in response time. Similarly, the exponential dependence of P_S on qV_B also increases the response time. Current measurements of ZnO exposed to H$_2$ exhibit this strong temperature dependence with response time decreasing from thousands of seconds to tens of seconds as the temperature increases from ~500 to 800 °C [1].

Response time can be a critical parameter for sensors that regulate other processes in real time, for example, engine feedback from automotive or aerospace exhaust gases.

13.2   Oxide Gas Sensors

The operation of oxide gas sensors is a two-step process: (i) the adsorption and creation of localized donor or acceptor states and (ii) the change in conduction band electron concentration due to these impurity states. Oxides such as ZnO or TiO$_2$ are particularly useful as gas sensors since (i) their surfaces are chemically inert compared to most non-oxides and (ii) their band gaps are large so that their thermally activated carrier concentrations are low and changes in current with gas adsorption are easily detectable.

Conductance changes of oxide films on insulating substrates can involve several possible mechanisms, including gas molecule adsorption, molecule dissociation, charge exchange with a semiconductor substrate or with defects at the surface, diffusion through grains or metal overlayers, and a change in free carrier transport between contacts on the substrate. In general, these multiple pathways for detecting gas phase molecules can be represented by an equivalent circuit with multiple sets of parallel resistances and capacitances linked in series. See Track II T13.1 Generic Sensor Design. The conductance will then have a response frequency dependence that has contributions from surface, bulk, contact and, in the case of granular films, grain boundaries.

An alternative gas sensor design involves Schottky barrier formation. For example, hydrogen gas diffusing through a Pt contact on ZnO alters the charge at the metal–oxide interface, which changes the band bending and conductance through the depletion region. Besides gas diffusion through the metal, however, the Pt itself may diffuse into the semiconductor and change the band bending region locally. Overall the detection of specific molecules requires the design of a three-dimensional architecture, substrate doping, contact geometry, operating temperature, and electrical frequency.

13.3   Granular Gas Sensors

A key advantage of granular gas sensors is their large surface area for gas adsorption and charge exchange. Furthermore, fabrication of polycrystalline films is considerably more cost effective than preparing single crystals. On the other hand, granular films involve grain boundaries, which have back-to-back Schottky barriers and add significant complexity to the charge transport. The conductance $G = \sigma A/L$ of such films depends strongly on the dimensions of these grains, the conductivity σ within the grains, the average path length L across the film, the average contact area A between grains, and the barrier heights at the grain boundaries. The intergranular resistance depends on the contact area and Schottky barrier qV_B. Gas adsorption changes qV_B and the carrier density in the space charge region, which changes σ both at the interface and within the grain itself.

13.4   Nanowire Sensors

Nanowires offer an exciting new architecture for sensors. Their sensitivity to external stimuli is enabled by their high surface-to-volume ratio, which can be large enough for the surface to affect their entire device volume. Charge transport in nanowires with radii comparable to the width of their space charge regions is highly sensitive to charge exchange into and out of localized states on the nanowire surfaces, which alters the width of the radial space charge region extending into the wire and thus the volume of nanowire that can conduct charge. This feature enables semiconductor nanowires to perform a wide range of functions to detect gas, biological, and chemical species [2].

Mechanical forces that bend, compress, or extend nanowires can also perturb charge transport through the wires or through their electrical contacts. The piezoelectric response of semiconductors such as ZnO and GaN enables them to serve as pressure or flexing sensors. Semiconductor nanowires can also be patterned into electronic circuits as optical detectors with high photon sensitivity. A number of techniques are available to probe states involved in charge exchange within semiconductor nanowires, including electron energy loss, photoluminescence, and cathodoluminescence spectroscopies [3]. Luminescence studies of ZnO nanowires with radii in the 50–300 nm range exhibit intensity variations consistent with the nanowire surface dominating the free carrier recombination and recombination occurring in a 30 nm surface annulus of characteristic thickness. Scanning tunneling microscopy also reveals a diameter dependence of electron injection into nanowire conduction bands that becomes evident for diameters below ~70 nm. Charge injection at these nanocontacts appears governed by a combination of band alignment at the contacts, electrostatics at dimensions comparable to space charge region widths, and surface-state dependent recombination time.

13.5   Chemical and Biosensors

Charge exchange at interfaces can be used to modulate current in a field effect transistor. Here target molecules, proteins, or other organic species adsorb and chemically link

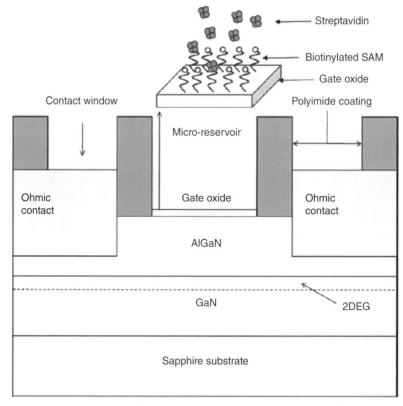

Figure 13.2 *Sensor design for detecting molecules in solution. Molecules that adsorb and transfer charge to a semiconductor alter conduction through a two-dimensional gas channel. (Gupta et al. 2008 [4]. Reproduced with permission of Elsevier.)*

to the semiconductor surface between a source and drain. The charge exchange alters the surface potential and thereby functions as the transistor "gate". The semiconductor surface can be functionalized with receptor molecules that only link to the target molecules so that the charge transfer is chemically or biologically selective. Such devices are termed *Chem-FET* or *BioFET* sensors. Figure 13.2 illustrates a sensor design for detecting molecules in solution. Here the 2DEG layer at an AlGaN–GaN interface forms the transistor channel, a micro-reservoir contains the physiological fluid above the gate region, and a semiconductor oxide forms an ultrathin barrier to ions in solution. A biotinylated self-assembled monolayer (SAM) on the semiconductor oxide forms receptor sites for linkage to protein molecules, in this case, streptavidin.

13.5.1 Sensor Sensitivity

Transistor designs such as in Figure 13.2 respond exponentially to the charge exchange at their interfaces since, from *Gauss's Law*, electric field

$$E = \sigma/\varepsilon \qquad (13.5)$$

where σ = surface charge density and ε = dielectric permittivity, increases linearly with surface charge density. The resultant voltage change in the diode equation

$$J = A^{**} T^2 \exp \ (-q \, \Phi_{SB}/k_B T)[\exp(qV/k_B T) - 1] \tag{3.10}$$

shows that current density increases exponentially with V. This effect enables higher sensitivities than sensors that are linearly proportional to detected species, such as changes in optical transmission, accumulated weight, or frequency changes. The ultrathin design and specific materials of both SAM and oxide serve two important functions: (i) to minimize the screening of charge transfer by counterions in solution that are normally needed for charge balance of the target molecules and (ii) to avoid diffusion of such ions into the oxide. For the design pictured in Figure 13.2, lower limits of detection are in the tens to hundreds of ng/ml. Recessed transistor gate designs can improve this sensitivity significantly.

13.5.2 Sensor Selectivity

Selectivity is especially important for biosensors, given the myriad of biological species that can give rise to false positive signals. Besides the chemical specificity of particular proteins or antibodies to their receptor counterparts, physical size restraints are useful to improve selectivity. Figure 13.3 illustrates an organic cage molecule designed to link with a Cl-bearing adsorbate molecule. This design restricts the available space for binding

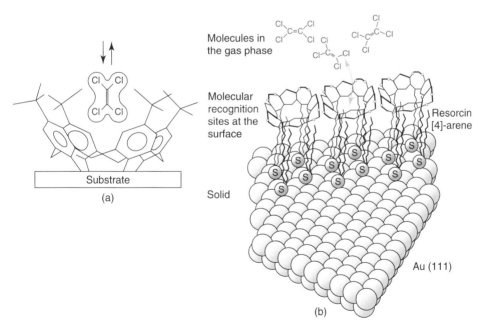

Figure 13.3 *(a) Organic cage molecule and Cl-bearing adsorbate molecule. The cage molecule restricts the size of the adsorbate [5]. (Heiland 1969 [1]. Reproduced with permission of Elsevier.) (b) Cage molecule array on Au(111) substrate. (Adapted from [6].)*

and charge transfer with the substrate. Similarly, capillaries and other restricted dimension screens are useful to filter molecules available for adsorption.

13.6   Surface Electronic Temperature, Pressure, and Mass Sensors

The resistance of electrically conducting materials is useful to detect changes in gas pressure via changes in heat dissipation due to current passing through comb designs. Inner "active" thermocouples exposed to controlled gas pressures measure temperature changes in comparison with external reference thermocouples. The conducting folds of a "comb" design increase the linear dimension for electrical measurements. Such devices are useful for low vacuum applications in the milli-Torr pressure range and can respond differently to different gases.

Mass-sensitive devices, such as the quartz crystal oscillator discussed in Section 5.4, are highly sensitive deposition sensors. For most elements, such oscillators coupled with frequency counters can discriminate thickness changes well below fractions of a monolayer. Besides oscillating plates, vibrating capacitors such as cantilevers fabricated lithographically produce capacitance changes that are sensitive to pressure, temperature or mass changes.

A wide range of electronic materials are available for sensors. These include relatively inert metals such as: (i) Pt, Pd, Au, Ag, Ni, Sb, and Rh, (ii) wide gap semiconductors such as CdS,ZnS, and GaN as well as micromachined Si, (iii) ionic compounds including the electronic conductors ZnO, SnO_2, TiO_2, Ta_2O_5, InO_x and $In_xSn_{1-x}O$, (iv) mixed conductors such as $SrTiO_3$, $La_{1-x}Sr_xCo_{1-y}Ni_yO_3$, and Ga_2O_3, (v) ionic conductors such as: ZrO_2, LaF_3, CeO_2, CaF_2, Na_2O_3, and β-alumina (Al_2O_3), (vi) molecular crystals such as the pthalocyanines $PbPc$, $LuPc_2$, and $LiPc$, (vii) Langmuir–Blodgett films, (viii) conducting polymers including polyacetylene, polysiloxanes, polypyrroles, and polythiophenes, and (ix) biomolecular functional systems such as phospholipids and enzymes.

13.7   Summary

Surface electronic effects lend themselves to a wide range of sensor applications, many of which are based on charge transfer between a semiconductor surface and a gas, liquid or otherwise mobile adsorbate. The band bending which results from this charge exchange alters volume charge densities and conductivity within thin films or structures with restricted dimensions.

The charge transfer is a time-dependent process with the rate of charge transfer and ultimate saturation increasing with increasing temperature. The magnitude of charge transfer and thereby the saturated change in surface conductivity depends on the energy and density of surface states within the band gap at the surface.

The reversibility of surface conductivity changes with atom or molecule desorption is an important feature of sensors that monitor adsorption changes continuously.

Granular gas sensors can magnify the electrical effects of gas adsorption due to their higher surface area and back-to-back Schottky barrier interfaces.

Transistor structures provide enhanced sensitivity due to their exponential dependence on charge transfer between adsorbate and semiconductor surface.

The selectivity of sensors can be engineered to permit charge transfer between target molecules or proteins and target-specific surface receptors.

Besides the field effect principle, electronic material sensors can monitor surface conductivity, capacitance, mass and temperature changes, converting them to chemical or electrical information.

A wide range of electronic materials are available for chemical and biological detection as well as temperature, pressure, and mass transfer.

13.8   Problems

1. A granular gas sensor consisting of polycrystalline grains is densified by applying high voltage that increases its conductivity. The increased current passing through the grains raises the temperature, decreasing the open volume between the grains. List at least three possible reasons why conductivity increases with this treatment.

2. Calculate the final rate of charge transfer between bulk and surface for O on ZnO with a carrier concentration of 10^{17} cm^{-3} at 200 °C. What is the final band bending for $E_C - E_T = 0.5$ eV. Assuming E_F stabilizes at this energy, how long does it take to reach equilibrium for an O trap density $N_T = 5 \times 10^{14}$ cm^{-2} at this rate?
   [Hint: intrinsic carrier concentration $n_i = 1.1 \times 10^{19} (T/300)^{3/2} \exp(-E_G/2k_B T)$ where $E_G = 3.34$ eV at 200 °C.

3. In Problem 2, how much charge does the calculated band bending correspond to for a ZnO(0001) surface? [Hint: use the depletion approximation.] Is the assumption of nearly full band bending throughout the charge transfer process reasonable?

4. Calculate the room temperature change in surface conductance of ZnO with $n = 10^{18}$ cm^{-3} assuming flat bands initially and that adsorption results in charge transfer such that $E_C - E_F = 0.5$ eV at the surface. $\mu_e = 50$ cm^2 V^{-1} s^{-1} for polycrystalline ZnO [Hint: Use $n_0 = n_i \exp[(E_F - E_i)/k_B T]$ at the surface to obtain the band bending.]
   Is this type of conductivity sensor more sensitive as a free standing crystal or a thin film?

5. GaN nanowires are to be grown for liquid sensors based on the field effect mechanism. Assuming a typical Fermi level stabilization with liquid contact of 0.9 eV below E_C, what is the optimal wire diameter for GaN doped $N = 5 \times 10^{17}$ cm^{-3}?

6. (a) Design a photosensor with a serpentine pattern. The semiconductor film is 5 μm thick with electron mobility $\mu_n = 50$ cm^2 V^{-1} s^{-1} $\gg \mu_p$, a donor density $N_d = 10^{14}$ cm^{-3}, electron lifetime $\tau_n = 2 \times 10^{-6}$ s, and a dark resistance of 10 MΩ. The photosensor must fit inside a 0.6 cm square area. Hint: $R = \rho L/A$ where L is length and A is cross-sectional area. (b) For band gap illumination that produces 10^{21} electron–hole pair cm^{-3} s^{-1}, what is the change in resistance?

7. An n-channel biosensor is made with source and drain on a p-type Si substrate with $N_a = 5$ 10^{15} cm^{-3}, $\mu_e = 1250$ cm^2 V^{-1} s^{-1}, and $Z/L = 10$. The SiO$_2$ thickness is 1 nm in the bare channel region. $\varepsilon_i(SiO_2) = 3.9$ ε_0. The effective interface charge is due to charge transfer between adsorbed molecules and the SiO$_2$. (a) Calculate the insulator capacitance C_i. (b) Given a threshold voltage $V_T = -0.284$ V, calculate the source-drain

current I_D for $V_D = 1$ V. (c) If I_D increases by 5% after adsorption of 10^{10} molecules cm^{-2}, calculate the charge transfer per molecule.

References

1. Heiland, G. (1969) Polar surfaces of zinc oxide crystals. *Surf. Sci.*, **13**, 72.
2. Patolsky, F. and Lieber, C.M. (2005) Nanowire nanosensors. *Materials Today*, April, 20.
3. L.J. Brillson, L.J. (2013) Surfaces and interfaces of ZnO. in *Semiconductors and Semimetals: Oxide Semiconductors*, Vol. **88** (eds B.G. Svensson, S.J. Pearton, and C. Jagadish), SEMSEM, Academic Press, Burlington, Ch. 4.
4. Gupta, S., Elias, M. Wen, X., *et al.* (2008) Detection of clinically relevant levels of biological analyte under physiologic buffer using planar field effect transistors. *Biosens. Bioelectron.*, **24**, 505.
5. Schierbaum, K.D., Wi-Xing, X., Fischer, S., and Göpel, W. (1993) Schottky barriers and ohmic contacts with Pt/TiO$_2$(110): Implications to control gas sensor properties in *Adsorption on Ordered Surfaces of Ionic Solids and Thin Films*, Springer Series in Surface Sciences, Vol. **33** (eds E. Umbach and H.J. Freund), Springer-Verlag, Berlin, pp. 268–278.
6. Schierbaum, K.D., Weiss, T., Thoden van Velzen, E.U. *et al.* (1994) Molecular recognition by self-assembled monolayers of cavitand receptors. *Science*, **265**, 1413.

Further Reading

Wilson, J.S. (ed) (2005) *Sensor Technology Handbook*, Elsevier, Inc, Oxford, UK.

14

Semiconductor Heterojunctions

Semiconductor heterojunctions offer unique electrical and optical properties not otherwise found in nature. Growth methods are now available to create new novel heterojunction structures with state-of-the-art physical properties for next generation electronics. Technologists can now select combinations of materials with band gaps and band offsets for desirable photon absorption and emission, carrier confinement and injection as well as tunneling. Figure 14.1 illustrates the three classes of semiconductor heterojunctions based on their energy band line-ups. These conduction and valence band offsets, ΔE_C and ΔE_V, respectively, and the differences in energy band gaps enable a host of opto- and microelectronic applications that depend on the particular straddling, staggered, or broken-gap alignments shown. For high crystalline quality, however, the growth of one semiconductor on another requires low defect densities and high thermodynamic stability. As epilayer dimensions shrink, such interface structure and defects near the junction ultimately control device performance. This chapter covers the unique interface features of semiconductor heterojunctions in terms of their geometrical, chemical, and electronic structure.

14.1 Geometrical Structure

14.1.1 Epitaxial Growth

High quality epitaxial growth requires a nearly precise lattice match near room temperature of the two semiconductors for: (i) a defect-free continuation of the lattice growth and (ii) minimum strain to prevent subsequent generation of imperfections. For a particular application, one chooses semiconductors or semiconductor alloys based on both their energy gaps and their good lattice match. See Track II Table T14.1 Energy Band gap versus Alloy Composition for the numerous III-V compound semiconductor alloys and their wide range of direct gaps.

An Essential Guide to Electronic Material Surfaces and Interfaces, First Edition. Leonard J. Brillson.
© 2016 John Wiley & Sons, Ltd. Published 2016 by John Wiley & Sons, Ltd.
Companion Website: www.wiley.com/go/Brillson/

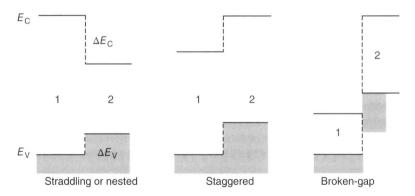

Figure 14.1   *Heterojunction classes of energy band line-ups.*

14.1.2   Lattice Matching

14.1.2.1   Alloy Composition and Lattice Match

Figure 14.2 illustrates lattice constants for ternary III-V crystalline solid solutions as a function of mole fraction [1]. Here the end components differ by <0.05 nm and Vegard's Law (lattice constant equal to the weighted average of end point values) is assumed to hold. Crossing points indicate where lattice match occurs between different alloys. Dashed lines indicate alloy ranges where miscibility gaps (phase segregation) is expected. Track II,

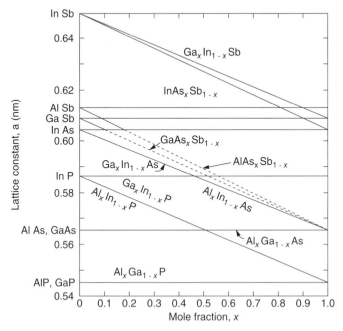

Figure 14.2   *Lattice match versus alloy composition. (Casey and Panish. 1978 [1]. Reproduced with permission of Elsevier.)*

T14.2. Lattice Constants vs. Band Gaps provide band gaps versus lattice constants for common III-V and II-VI compounds plus Ge, Si, and SiC. Lattice mismatch f is defined as $\Delta a / a_{\text{avg}}$ where $\Delta a = a_f - a_s$, a_f and a_s being the film and substrate lattice constants, respectively, and $a_{\text{avg}} = (a_f + a_s)/2$. For $f \leq 0.5\%$, the two materials are candidates for lattice-matched epitaxial growth. The relationships between composition, lattice constant, and energy band gap enable a design principle: selection of a semiconductor composition to achieve desirable optical and electronic properties while minimizing lattice mismatch and strain effects. Nearly identical lattice constants between GaAs, AlAs, and their alloys provide the basis for many light emitting devices in the visible range. Slightly mismatched GaN with its Al and In alloys yields numerous near-UV applications.

14.1.2.2   *Lattice Mismatched Interfaces*

The lattice mismatch between heterojunction constituents leads to strain and ultimately the formation of lattice dislocations. Figure 14.3a represents the different lattice constants a_f and a_s, respectively, of an epitaxial film overlayer and substrate. Figure 14.3b shows that the overlayer film conforms to the substrate lattice constant for thickness less than a

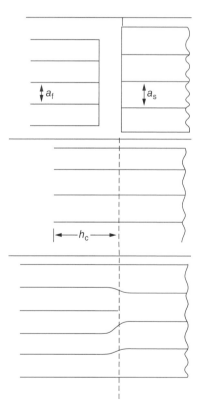

Figure 14.3   *(a) Epitaxical growth of epilayer on a substrate with dissimilar lattice constants. (b) Pseudomorphic growth below critical thickness. (c) Formation of misfit dislocations above critical thickness. (Mayer and Lau 1990 [2]. Reproduced with permission of American Physical Society.)*

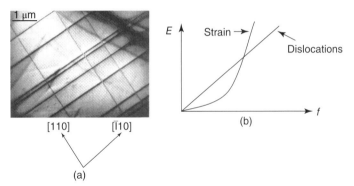

Figure 14.4 *(a) Transmission electron micrograph of partially relaxed $In_{0.008}Ga_{0.92}$ As on InP(100) showing array of misfit dislocations in the interface plane. (Raisanen et al. 1994 [3]. Reproduced with permission of American Institute of Physics.) (b) Schematic energy versus strain plot of energies due to strain or dislocation formation. Above the critical thickness h_c, the formation of dislocations is energetically favorable.*

critical thickness h_c. Here the lattice mismatch is accommodated by strain, and the strained overlayer is termed *pseudomorphic*. Figure 14.3c illustrates schematically the formation of misfit dislocations for thickness $h > h_c$. These misfit dislocations may act as electrically active sites – recombination centers or deep traps which strongly affect carrier lifetime and device properties.

Figure 14.4a illustrates a typical TEM micrograph of dislocations at a semiconductor heterojunction. Here lattice relaxation is greater in the [110] than the [$\bar{1}$10] direction.

14.1.2.3   Dislocations and Strain

The deviations from perfect lattice match lead to a build-up of strain with increasing epilayer thickness which ultimately results in dislocation formation. Figure 14.4b shows how the lattice energy associated with strain is initially less than that to form dislocations but increases faster than lattice mismatch f. Strain energy σ_s per unit area is defined as $\sigma_s = (Y/2)hf^2$ where Y is *Young's modulus* expressed in dynes or newtons per square centimeter. Thus strain energy builds up quadratically with lattice mismatch. In contrast, the energy of a grid of edge dislocations Γ_e can be expressed as

$$\Gamma_e \simeq \alpha Gb^2 \simeq \alpha Yb^2 \tag{14.1}$$

with the approximation $Y \simeq G$, the *shear modulus*. Here α is a geometrical factor ~ 1 and b is a *Burgers vector,* which expresses the magnitude and direction of the lattice distortion associated with a dislocation. In turn, the Burgers vector for an edge dislocation is perpendicular to the dislocation line and is in the slip (or glide) plane of the dislocations. When a dislocation slips on its slip plane, the atoms move in the direction of the Burgers vector, approximately the same distance as a_f and a_s.

Assuming a square dislocation grid with grid spacing $D = a_{avg}/f$, one can show that the *dislocation energy* is

$$\sigma_D = 2\alpha Yb^2 f/a_{avg} \simeq 2\,\alpha Ybf \tag{14.2}$$

where $b \simeq a_{avg}$. Since the energy to form dislocations increases linearly with f, the lattice energy due to strain exceeds that for dislocations above a critical thickness h_c so that dislocations become energetically favorable. At the intersection point, $\sigma_s = \sigma_D$, so that $h_c = 4\alpha b/f$, showing that the critical thickness increases with decreasing lattice mismatch. [See Track II, T14.3 Critical Thickness Derivation.]

A derivation of critical thickness that takes into account the orientation of slip plane and slip direction, as well as the *Poisson ratio* $\mu = -\varepsilon_X/\varepsilon_Z = -\varepsilon_Y/\varepsilon_Z$ of lateral to longitudinal strain under uniaxial elastic deformation is [4]

$$h_c = [b(1 - \mu \cos^2 \Theta) \ln((h_c/b) + 1)]/[2\pi f(1 + \mu) \cos \Phi] \tag{14.3}$$

where Θ = angle between dislocation line and b, Φ = angle between the slip direction and that direction in the film plane that is perpendicular to the line of intersection of the slip plane and the interface. For $f < 1$–2%, this curve can be approximated as

$$h_c = a/2f \tag{14.4}$$

For example, Equation (14.4) gives $h_c = 283$ Å for $a_0 = 5.65$ Å in GaAs and $f = 1\%$, similar to the $h_c = 300$ Å thickness indicated in Figure 14.5.

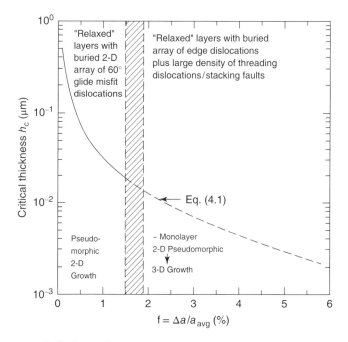

Figure 14.5 *Critical thickness h_c versus lattice mismatch $f = \Delta a/a_{avg}$ [4]. For $f \sim 1.5\%$, (dashed line), pseudomorphic two-dimensional growth continues up to a few monolayers, followed by three-dimensional growth. For $h > h_c$, dislocations and stacking faults form to relax the epilayer strain. (Woodall et al. 1988 [5]. Reproduced with permission of Materials Research Society.)*

14.1.3   Two-Dimensional Electron Gas Heterojunctions

The mechanical strain due to lattice mismatch can have practical value for electronic devices. Epitaxial overlayers of a piezoelectric material such as $Al_xGa_{1-x}N(x <\sim 0.4)$ on a GaN substrate are lattice mismatched, producing strain and a piezoelectric voltage normal to the epilayer plane. This voltage in turn can induce charge accumulation at their interface.

Figure 14.6 illustrates schematically the band structure of the wider band gap AlGaN on the lower band gap GaN, the electric field across the AlGaN, and the resultant charge accumulation at their interface. Increasing the Al content x and/or the epilayer thickness h increases the piezoelectric strain and the interface charge density for use in, for example, a high electron mobility transistor (HEMT). However, this charge accumulation is limited since increasing x and/or h decreases h_c so that dislocation formation is more likely. Strain would then decrease, reducing piezoelectric fields and charge accumulation. Hence, optimum values to maximize charge accumulation on polar face GaN are typically $x \sim 0.3$–0.4 and $t \sim 30$–50 nm. The accumulated charge at the heterojunction interface can have high mobility since the impurity atoms that donate the charge are physically separated from this interface. This high mobility in such nanoscale confinement structures has enabled observation of striking quantum phenomena, such as the *fractional quantum Hall effect*.

14.1.4   Strained Layer Superlattices

Strained layer superlattices (SLSs) are structures that can achieve higher critical film thicknesses and can have novel electronic properties. These superlattices consist of alternating epitaxial layers of two mismatched materials, each with thickness $h < h_c$. Layers with lower

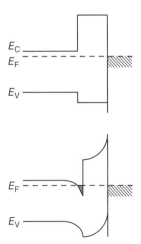

Figure 14.6   *Heterojunction interface between a wide gap insulating semiconductor and a narrower gap n-type semiconductor. Piezoelectric strain introduces electric fields within the wide gap material that bend the bands in both materials such that charge accumulates at the heterojunction. (Woodall et al. 1988 [5]. Reproduced with permission of Materials Research Society.)*

lattice constant are under biaxial tension whereas higher lattice constant layers are under biaxial compression. The resultant lattice has an average lattice constant between those of its constituents and, because of the alternating tension and compression, a critical thickness h_c that can increase by more than an order of magnitude. [See Track II, T14.4. Strained Layer Superlattices.]

14.1.4.1   Superlattice Energy Bands

The atomic layer by layer construction of superlattices enabled by epitaxial growth provides a new class of semiconductors with unique electronic properties that can be designed. These electronic properties depend on the nanoscale thickness of the individual layers as well as their physical properties in bulk. Quantum confinement at this nanometer scale leads to band gaps and energy levels that increase with decreasing layer thickness. As shown in Figure 14.7, allowed energy levels within these quantum-confined structures are analogous to quantum wells but extend throughout the solid due to the wave function overlap between layers. The bulk transport bands that form are termed *minibands* that define new conduction and valence bands. They are discrete in energy with bandwidths that increase and effective masses that decrease with decreasing layer width. In turn, the decreased effective masses in these man-made structures allow the design of higher carrier mobility materials.

14.1.4.2   Strain-Induced Polarization

Superlattice minibands in semiconductors with significant piezoelectric coefficients can exhibit useful optoelectronic effects. The piezoelectric fields induced by the lattice-mismatched strain cause a separation of electrons and holes within each layer of nm-scale thickness. In turn, the tilted bands lower the energy of the minibands' energy absorption edge. [See Track II, T14.5. Case Study: GaInAs-AlInAs Superlattice.] With above band gap illumination, however, these piezoelectric fields are screened by additional free carriers. As a result, the absorption edge increases. Photons with energy just below this increased absorption edge would then not be absorbed. This phenomenon provides a method for incident light to gate the passage of a second light beam through the material,

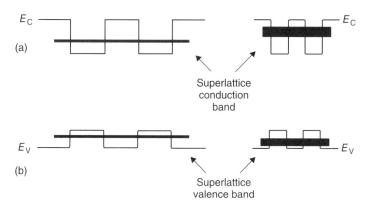

Figure 14.7   *Individual layer conduction and valence band edges versus energy bands of (a) thick and (b) thin superlattices.*

which represents a nonlinear optoelectronic switch. One can design the band gap to achieve such effects by tailoring: (i) the layer widths, (ii) the layer compositions, and (iii) the strain induced by lattice mismatch.

14.2   Chemical Structure

Semiconductor–semiconductor heterojunctions can exhibit a variety of chemical structures including interdiffusion, chemical reactions, and interlayer effects. These phenomena degrade the atomic layer abruptness of the heterojunction and can have strong electronic effects. Such chemical effects change device electronic properties and ultimately accelerate the rate at which devices fail.

14.2.1   Interdiffusion

At Column IV-IV element heterojunctions, such as Ge-Si, interdiffusion occurs at temperatures above ~200 °C, along with islanding and surface roughening at higher temperatures. Alloy layers between Ge and Si form at 350 °C. [See Track II T14.6. Case Study: Interdiffusion at IV-IV and III-V Heterojunctions.] For Ge-Si superlattices, the activation energy for this interdiffusion is 3.1 ± 0.2 eV in the range 640–780 °C. Within Ge-rich GeSi alloys, Ge segregation can occur with annealing.

At III-V–III-V compound heterojunctions such as Ge–GaAs, the interface is atomically abrupt at temperatures of 320–360 °C, while Ge diffuses hundreds of nm into GaAs at 650 °C, resulting in impurity doping of 10^{17}–10^{18} cm^{-3}. At AlGaAs–GaAs heterojunctions, interdiffusion is compositionally dependent, decreasing with increasing Al content. A similar compositional effect occurs in superlattices where the higher bond strength of Al in AlGaAs–GaAs and P in GaAsP–GaAs superlattices suppresses interdiffusion. At ZnSe–GaAs heterojunctions, the diffusion of Zn into GaAs and Ga into ZnSe produces dipoles that alter band bending at the heterojunction.

Case Study: III-V Compound Heterojunctions

Interdiffusion at III-V compound interfaces can be desirable for increasing doping. For example, Ge diffusion into GaAs can introduce p-type dopants. However, the time and temperature used for promoting this diffusion intentionally must be carefully controlled in order to produce such doped regions with very shallow thicknesses. The introduction of dopant impurities can also promote interdiffusion intentionally by changing the defect statistics inside the bulk semiconductor. For example, the patterned *impurity-induced diffusion* of Zn into strained GaAs–InGaAs–GaAs quantum wells produces disordered alloys with new band gaps and refractive indices. Such layers can act to confine electrically-generated laser light in vertical cavity surface emitting lasers (VCSELs).

14.2.2   Chemical Reactions

Chemical reactions can also occur at heterojunctions, altering the atomic abruptness and changing its electronic character. For example, heterovalent II-VI/III-V compound hetero-structures can lead to formation of III-VI compound interfaces. Thus growth of ZnSe on GaAs can produce a Ga_2Se_3 interfacial layer while CdTe on InP can lead to In–Te layers. Such compound formation can be suppressed in MBE by controlling the substrate stoichiometry and/or the order of deposition, for example, ZnSe/GaAs growth improves with As termination of the GaAs prior to ZnSe growth and Cd overlayers suppress In–Te reaction with Cd overlayers. Interlayers between two semiconductors can lower strain and improve growth, for example, a thin AlAs layer at the ZnSe–GaAs heterojunction.

14.2.3   Template Overlayers

With atomic layer control, a variety of template structures can improve heterojunction growth.

14.2.3.1   Bridge Layers

Crystalline alloy layers that "bridge" the difference in lattice constant between two lattices can enable epitaxial growth. For example, a $Ga_xIn_{1-x}As(0 \leq x \leq 1)$ pseudobinary alloy can bridge the 7% lattice mismatch between GaAs and InAs by gradually changing the alloy composition of the bridge layer over a thickness of tens to hundreds of nm from near-GaAs to near-InAs compositions.

14.2.3.2   Monolayer Passivation and Surfactants

Monolayer passivation can tie up substrate bonds in order to facilitate bonding at heterovalent interfaces. Thus GaAs grown on As-terminated Si(111) produces Si–As bonding (with atomic rearrangement at the intimate junction) to permit epitaxial overlayer growth, whereas Si on GaAs(100) produces mixed compositions in islands and interfacial layers. Monolayer surfactants can improve surface morphology. For example, As monolayers on Si promote improved morphology and an increase in the critical thickness of Ge overlayers. Thus, the As monolayers initially on Si(100) "float" on Ge as the overlayer film grows. Specific atomic structure and stoichiometry can control dipole fields

14.2.3.3   Orientation Dependence

Specific crystal orientations can yield preferred growth modes on vicinal surfaces, for example, laterally periodic changes in atomic composition, "tilted" superlattices and quantum wires. The choice of elements and their interface stoichiometry can also be used to control dipole fields.

   In general, it is important to avoid large electrostatic fields between heterovalent materials. For III-V and II-VI compounds, this is accomplished by: (i) graded or mixed interfaces with equal numbers of III-VI and II-V bonds, for example, Ga–Se and Zn–As bonds for ZnSe on GaAs. Specific planar reconstructions may be needed to achieve the same mixed bonding. Here, careful substrate temperature and flux ratio are required.

14.3   Electronic Structure

Chapter 3 introduced the electronic structure of heterojunctions, electrical techniques to measure their band offsets, and the importance of heterojunctions for next generation opto-electronic and microelectronic devices. This section covers both experimental and theoretical work to understand the nature of heterojunction band offsets as well as atomic-scale methods to control them.

14.3.1   Heterojunction Band Offsets

The heterojunction band offset is a central feature for the creation of novel band structure and electronic device features. Figure 14.1 introduced the three types of semiconductor heterojunctions – nested, straddling, and broken-gap. Here the nested heterojunction provides a model with which to consider the role of dipoles in forming heterojunction band offsets. Consider the band diagram for two semiconductors with nested band gaps. Here the difference in band gap $\Delta E_g = E_g{}^A - E_g{}^B$ and

$$\Delta E_g = \Delta E_C + \Delta E_V \tag{14.5}$$

where ΔE_C and ΔE_V are conduction and valence band offsets, respectively. Figure 14.8a shows a schematic energy band diagram for one of the semiconductors with an electron affinity χ consisting of both a surface dipole V_S and an internal potential S_S defined as shown. Dipoles are present at all surfaces as discussed in Section 12.3.3. Their sign and magnitude depend on both the intrinsic wave function "tailing" and extrinsic surface features such as surface reconstruction and adsorbate charge exchange.

Figure 14.8b shows the energy bands for two semiconductors with their (n-type) Fermi levels aligned in thermal equilibrium. The band alignment between the two semiconductors involves: (i) the dipoles V_S at the surfaces of the individual semiconductors, (ii) the voltage drops V_b across each semiconductor's band bending region and (iii) an interface dipole V_i

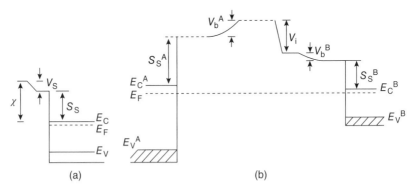

(a)   (b)

Figure 14.8   *Schematic energy band diagram illustrating the voltage drops across (a) an individual semiconductor surface and (b) the entire heterointerface. (Mailhiot and Duke 1986 [6]. Reproduced with permission of American Physical Society.)*

between the two semiconductors after contact. Figure 14.8a shows that $\chi = V_S + S_S$ and

$$S_S^A = \chi_A - V_S^A \tag{14.6a}$$

$$S_S^B = \chi_B - V_S^B \tag{14.6b}$$

Here the internal potential S_S allows us to express the surface dipole in terms of the electron affinity χ, a macroscopic measureable parameter. Equations 14.6a and b are represented schematically in Figures 14.8a and b. S_S is characteristic of the bulk, while V_S depends on the surface atomic composition and bonding. With E_F as a common reference level, one can use the internal potentials to equate the voltage drops on either side of the interface [6]. Thus,

$$S_S^A + V_b^A + E_C^A - E_F = S_S^B + V_b^B + E_C^B - E_F + V_i \tag{14.7}$$

Rearranging terms and using Equations 14.6a and b,

$$\Delta E_C = E_C^A - E_C^B = (\chi_B - \chi_A) + (V_i + V_b^B - V_b^A + V_S^A - V_S^B) \tag{14.8}$$

And finally at their surfaces,

$$\Delta E_C^S = E_C^A + V_b^A - (E_C^B + V_b^B) \tag{14.9}$$

$$= (\chi_B - \chi_A) + (V_i + V_S^A - V_S^B) \tag{14.10}$$

Equation (14.10) expresses the conduction band offset in terms of the difference between the two semiconductors of their bulk conduction band plus their surface band bending. This difference has the two components shown in Equation (14.10) – the difference in electron affinities plus a term that is just the difference between the interface dipole and the two surface dipoles of the separated semiconductors. The electron affinities and their difference are measureable quantities. However, the dipoles are not. One extracts their values only from their effects on the measureable quantities χ_B, χ_A, V_b^A and V_b^B.

   For the case in which the dipole term is zero, Equation (14.10) reduces to just the difference in electron affinities. This relation is known as the *electron affinity rule* [7] and is commonly used to approximate band alignments. However, experimental measurements of band offsets show that this classical relation does not predict band offsets accurately. Therefore, we can conclude that the dipole term in Equation (14.10) makes a significant contribution to the heterojunction band offset.

14.3.2   Band Offset Measurements

Many of the optical, scanned probe, and surface science techniques described in previous chapters are useful for characterizing semiconductor heterojunctions.

14.3.2.1   Electrical and Optical Techniques

Section 3.3.1 described the capacitance–voltage (C–V) technique for obtaining conduction band offsets from built-in voltages $V_{bi} = V_b^A \pm V_b^B$: plus for n-type facing p-type band bending and minus for same band bending type facing each other. C–V measurements are complicated by impurity gradients or near-interface trapped charge. Current–voltage (I–V) measurements can also provide V_{bi} but carrier recombination, tunneling, and shunt currents can affect their interpretation. Optical methods include internal photoemission spectroscopy (IPS) and photoluminescence spectroscopy (PL). IPS can extract band offsets from measurements of both valence-to-conduction band transitions between the two semiconductors or conduction-to-conduction band transitions. [See Track II, T14.7 Internal Photoemission at III-V Heterojunctions.] Given the spatial separation of the initial and final states and the weak transition and thus the low transition probabilities involved, both types of IPS measurements require intense, monochromatic light, particularly in the 0.1–0.5 eV region for the conduction-to-conduction band method.

PL can derive band offsets from sub-band transitions inside quantum wells since the sub-band energies are determined by the well width and heights. However, this method is indirect since it requires a self-consistent calculation of the quantum-confined states and, by induction, the well depths and hence the band offsets.

14.3.2.2   Scanned Probe Techniques

Several methods can gauge heterojunction band offsets by comparing electron excitation or injection across the microscopic heterojunction using scanned probes. The secondary electron threshold (SET) technique employs a UHV scanning electron microscope to excite electrons from the cleaved cross section of a heterojunction. The vacuum level E_{VAC} determines the minimum energy of electrons that escape the solid and reach an electron analyzer. This minimum energy is evident in the cutoff of the secondary electron distribution measured by the analyzer.

Figure 14.9 illustrates how electrons at E^S_{VAC} in the semiconductor appear with an energy ΔE above the vacuum level E^A_{VAC} in the analyzer. This ΔE represents the difference in work function $\varphi_S - \varphi_A$ between semiconductor and analyzer since $E^S_F = E^A_F$ with the two connected electrically. Likewise, the difference in ΔE between the two sides of the heterojunction yields the difference in their work functions. Since their Fermi levels are equal, SET measurements across the heterojunction provide the spatial variation of vacuum level between the two semiconductors. Assuming the electron affinities of

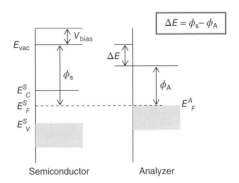

Figure 14.9   *Principle of secondary electron threshold (SET) measurement.*

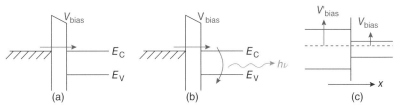

Figure 14.10   *STM methods for measuring heterojunction band offsets. (a) Electron injection versus bias voltage measurements $E_C - E_F$. (b) Electron injection induces light emission. (c) A scan across the interface yields difference in thresholds for light emission and hence the relative E_C positions.*

both semiconductors are known, then $\chi = E^S{}_{VAC} - E_C$ provides the difference in their conduction bands ΔE_C. For example, SET measurements of E_{VAC} for an InGaAs/InP junction grown under difference anion soak conditions as the growth transitions from one semiconductor to another shows sizable (0.25 eV) ΔE_C differences due to interfacial broadening with increasing soak time [8].

Ballistic energy electron microscopy (BEEM) using a scanning tunneling microscope (STM) tip provides an alternate transmission approach to measuring band offsets. Figure 14.10a shows that electrons injected from a metallic tip exhibit a current–voltage onset when the bias V_{bias} between tip and semiconductor increases above a threshold value. Comparison of the onsets for the two sides of the heterojunction in cross section yields ΔE_C directly.

Injected electrons can also recombine with holes inside the semiconductor to emit light, as Figure 14.10b illustrates. Here the difference in bias-dependent onset of light emission between the two sides provides ΔE_C, as illustrated in Figure 14.10c.

14.3.2.3   *Photoemission Spectroscopy Techniques*

Photoemission spectroscopy provides a combined core level–valence band approach to measuring band offsets between two semiconductors in contact. Here one first measures the energy difference from the valence band edge E_V to a core level E_{CL} for each of the two semiconductors separately. Then one measures the energy difference between the same two core levels with the two semiconductors in contact. In order to detect both core levels in the same XPS spectrum, photoelectrons must be able to escape from both sides of the heterojunction. This requires measurement of only a few monolayers of one semiconductor on the other, that is, during the initial stages of heterojunction formation.

Figure 14.11 illustrates the energy band line-ups for two semiconductors in contact. XPS measurements of the individual clean semiconductor surfaces yield $E_{CL}{}^A - E_V{}^A$ and $E_{CL}{}^B - E_V{}^B$. Measurement of $E_{CL}{}^A$ and $E_{CL}{}^B$ in spectra from the thin overlayer heterojunction yield the core level difference ΔE_{CL}. The valence band offset then follows from

$$\Delta E_V = (E_{CL}{}^B - E_V{}^B) - (E_{CL}{}^A - E_V{}^A) + \Delta E_{CL} \qquad (14.11)$$

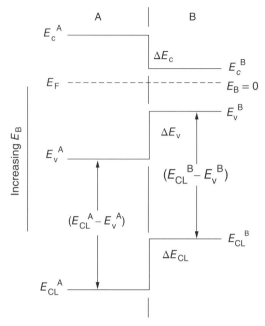

Figure 14.11   *Schematic energy band diagram for a semiconductor heterojunction. XPS spectra provide the valence-band offset from core-level binding energies and valence band edge energies.*

Precise ΔE_V measurements require determination of valence band edges from lineshape fits of the valence band spectra to a theoretical density of states calculation along with a deconvolution of core-level lineshapes to distinguish bulk from chemically shifted components.

A second valence band offset method involves XPS measurements of the two valence bands in the same spectrum. Figure 14.12a shows the bent band diagram and initial density of states for the semiconductor before overlayer deposition with the density of states DOS replicated in the excited XPS spectrum raised in energy by $h\upsilon$. With the deposition of the second semiconductor, Figure 14.12b shows an additional density of states in the energy band diagram and valence band spectrum. Note the possible band bending component BBC due to the overlayer.

Figure 14.12c illustrates this valence band change for a CdS ($10\overline{1}0$) surface with increasing thicknesses of amorphous Si. The additional valence band contribution appears with as little as 0.5 Å Si deposition, eventually dominating the spectrum by 20 Å. At an intermediate coverage of 3.5 Å, there is a well-defined ΔE_V between the two extrapolated valence band edges.

While this method provides the most direct measurement of ΔE_V, the two valence band edges are measurable only when the two valence band edges are distinct and when ΔE_V is positive. Otherwise, their individual features become difficult to resolve.

In general, XPS measurements of valence band offsets are limited by the X-ray line width resolution of 0.1–0.2 eV compared with $\Delta h\upsilon = 0.01$ eV or less with optical techniques.

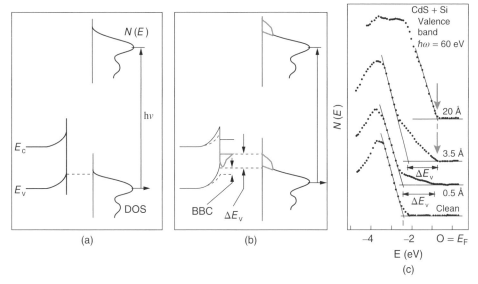

Figure 14.12   *Schematic energy band and density-of-states diagram for photoemission measurement of ΔE_V (a) before and (b) after deposition a second semiconductor over the clean surface of the initial semiconductor [9]. (c) A double-edged valence-band spectrum for amorphous Si on CdS ($10\bar{1}0$). (Katnani and Margaritondo 1983 [10]. Reproduced with permission of American Physical Society.)*

However, photoemission energy shifts can be measured with significantly higher accuracies than nominal XPS resolution.

14.3.3   Inorganic Heterojunction Results

Researchers have used photoemission (PH), C–V(CV), photoluminescence (PL), and other techniques (OT) to obtain a wide variety of heterojunction ΔE_C and ΔE_V band offsets. [See Track II Table T14.7 Heterojunction Band Offsets.] These experimental values form a database on which various theoretical models of heterojunction band offsets and the contribution of interface dipoles can be evaluated. Photoemission measurements using the coupled valence band-core level method provide numerous energy band alignments for semiconducting oxides. [See Track II, T14.8. Energy Band Alignments for Semiconducting Oxides.]

One can approximate band alignments for wide band gaps such as AlN, GaN, and SiC based simply on their measured electron affinities and band gaps. [See Track II, T14.8. Band Alignments for Wide Band Gap Semiconductors.] Conduction band discontinuities are large between AlN, GaN, and most other semiconductors because of AlN's extremely large (6.2 eV) band gap. However, energy band line-ups based on the electron affinity rule should be viewed with caution since the dipole contribution to heterojunction band offsets in Equation (14.10) introduces large deviations between measured values of ΔE_C and the $\Delta\chi$ values from electron affinity differences. [See Track II, T14.9. Test of Electron Affinity Rule.]

14.3.4   Organic Heterojunctions

Organic semiconductors pose an additional challenge for measuring heterojunction band offsets. Organic carrier densities are typically low so that their surface space charge regions can extend throughout the organic films with little or no band curvature. Furthermore, charge transfer at organic heterojunctions can involve a wide variety of mechanisms for the adsorbed organic molecule on a metallic surface.

Figure 14.13 illustrates how charge transfer produces localized dipoles since bulk conduction is low. Besides cation or anion formation pictured in Figure 14.13(a1) and (a2), respectively, dipole molecules can induce image charges in metallic substrates (b), surface charge rearrangement (c), chemical reaction (d), as well as interface states (e) and permanent dipoles (f). Existing charge forms a "pillow" effect that reduces the electron cloud and the dipole charge tailing out of the substrate.

For an organic substrate, image forces and wave function tailing are low because of the low organic charge density. Air exposure can be another challenge since surface oxidation changes the electronic structure, including the interface dielectric permittivity, E_{VAC}, and band bending qV_B near the interface.

14.3.5   Heterojunction Band Offset Theories

Beyond the electron affinity model, several different physical mechanisms provide a basis for theoretical models of band alignment at semiconductor heterojunctions. Leading mechanisms include: (i), tunneling approaches involving charge neutrality levels, (ii) local bonding approaches based on tight-binding or pseudopotential calculations of electrostatic potentials at the atomic scale, and (iii) empirical deep-level schemes relating the band

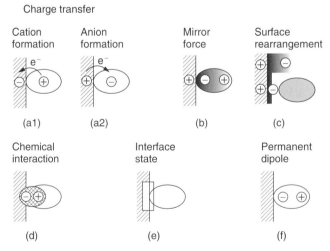

Figure 14.13 *Dipole formation mechanisms at organic heterojunctions. (a) Charge transport at the donor or acceptor molecule–substrate interfaces. (b) image force, (c) "pillow" effect, (d) chemical bonding, (e) interface states, and (f) oriented polar molecules or functional groups. (Ishii et al. 1999 [11]. Reproduced with permission of Wiley.)*

edges in the two semiconductors to transition-metal impurity levels. This section provides brief descriptions of each.

14.3.5.1   Charge Neutrality Levels

One approach to describe band offsets involves the alignment of charge neutrality levels in the two semiconductors. As Figure 4.1 illustrated, the wavefunction of charge at a semiconductor surface can tunnel into the bulk at energies within the band gap. For a semiconductor with free-electron-like bands, the energy at which states cross over from conduction to valence band-derived is termed the *branch point*. Based on the complex band structure of the semiconductor in one dimension, attenuation of this tunneling reaches a maximum at this energy E_B, at which the bulk E_F stabilizes between filled and empty states, termed the *charge neutrality level* (CNL). [See Track II T14.10 Branch Points and Band Alignments.] Alignment of branch points between two semiconductors therefore aligns their band structures, providing values for ΔE_C and ΔE_V at their interface. Calculated branch points derived theoretically display moderately good agreement with experiment values for various heterojunctions to within a precision of several hundred meV. [See Track II, T14.11 Branch Points and Alignments.]

14.3.5.2   Local Bond Approaches

In contrast to the CNL bulk band structure approach, local bond approaches take interface-specific local potential differences into account explicitly. For example, tight-binding pseudopotentials determine the semiconductor band edges from hybridized atomic orbitals (HAOs) such that the valence band energy maximum is given by

$$E_V = \frac{1}{2}(\varepsilon_a + \varepsilon_c) + \left[\frac{1}{4}(\varepsilon_a - \varepsilon_c)^2 + (4E_{xx})^2\right]^{1/2} \tag{14.12}$$

where ε_a and ε_c are the anion and cation orbital energies, respectively, and $E_{xx} \approx -1.28\ \hbar/md^2$ is an energy term that depends on the nearest anion–cation distance d and the free electron mass m. [See Track II,T14.12. Hybrid Atomic Orbital Approaches for Valence Band Offsets.] Both this HAO model *per se* as well as one corrected for lattice mismatch with an average lattice constant display 0.1–0.2 eV agreement with experimental measurements over a $\Delta E_V > 2$ eV range.

14.3.5.3   Empirical Deep Level Schemes

A completely different physical approach to band offsets involves the alignment of impurity energy levels within the semiconductor band gap. Transition metal impurities exhibit deep levels within the band gap of many III-V and II-VI compound semiconductors. Deep donor and acceptor energies referenced to semiconductor vacuum levels exhibit binding energies that are constant to within 0.1–0.2 eV for a given charge state, independent of the host semiconductor. [See Track II.T14.13. Empirical Deep Level Schemes.] The common binding energies between different semiconductors suggest relatively low bond differences for transition metals inside their host lattices.

The corresponding energy levels within these semiconductors are also nearly constant for a given transition metal for common III-V and II-VI compound semiconductors. Hence, such universal binding energies for a given impurity represent another approach to align the band gaps of two semiconductors. As with charge neutrality levels, this approach is based on bulk properties and ignores interface interactions.

For theoretical approaches overall, several methods show agreement with experiment to within ≤ 0.2 eV. However, this comparable agreement is based on conceptually different physical phenomena. Furthermore, higher experimental and theoretical precision to within ~ 0.01 eV is required to distinguish between various models and to be useful in electronic device design. Needed are models that *predict* ΔE_V and ΔE_C as well as heterojunction systems where ΔE_V and ΔE_C can be varied.

14.3.6   Interface Effects on Band Offsets

Another strategy to distinguish between theoretical models and assess the role of the interface is to measure ΔE_V and ΔE_C as they vary with interface composition, structure, and growth conditions. This approach is enabled by epitaxial growth and surface science analysis techniques for measurements during the initial stages of heterojunction formation. Interfaces can be studied as a function of: (i) growth sequence, (ii) crystallographic orientation, (iii) surface reconstruction, (iv) interface composition, and (v) interfacial layers. These studies reveal how specific features of the microscopic interfaces introduce the dipoles that account for the deviations from classical theory. Likewise, control of these interface features provides an avenue to engineer band offsets to improve electronic device properties.

14.3.6.1   Growth Sequence

Studies of band offsets as a function of growth sequence can gauge the extent to which dipoles contribute to band offsets. Dipole contributions alter the commutativity of band offsets, that is, whether growing semiconductor A on semiconductor B produces an equal and opposite band offset. For valence band offsets, commutativity is expressed as

$$\Delta E_V(A - B) + \Delta E_V(B - A) = 0 \tag{14.12}$$

Similarly, dipole contributions alter band offset transitivity with a third semiconductor, that is,

$$\Delta E_V(A - B) + \Delta E_V(B - C) + \Delta E_V(C - A) = 0 \tag{14.13}$$

Deviations from these linear expressions indicate the extent to which interface dipoles contribute.

Valence band-core level XPS studies of MBE-grown III-V compound semiconductors such as GaAs with AlAs establish that isovalent heterojunctions are linear to within ≤ 0.1 eV. In contrast, heterovalent interfaces are linear only to ≥ 0.2 eV, that is, the sequence of growth alters the energy band line-up. Thus $\Delta E_V(\text{Ge/ZnSe}) + \Delta E_V(\text{ZnSe/Ge}) = 0.23$ eV between Group IV and II-VI compounds, and

$\Delta E_V(\text{Ge/GaAs}) + \Delta E_V(\text{GaAs/Ge}) = 0.2$ eV between Group IV and III-V compounds. Such differences are attributed to sequence-dependent chemical reactions, for example, Ge on CuBr(110) leads to a chemical reaction whereas CuBr on Ge grows epitaxially. Hence chemical reactivity is an important factor in heterojunction dipole formation.

14.3.6.2 Crystallographic Orientation

High precision (0.01–0.02 eV) XPS measurements of Ge on GaAs showed only a 0.17 eV variation in ΔE_V with different crystal orientation and reconstruction [12]. These measurements were based on the electropositive substrate terminations of (111)Ga (2×2) and (100)Ga $c(8 \times 2)$, electronegative terminations of (100)As and (111)As (1×1), and the stoichiometric (110) (1×1) termination. These results indicate that the initial surface geometric arrangement of substrate atoms is at most a secondary factor in determining band offsets.

In contrast, surface reconstruction can significantly affect the band bending of Ge on GaAs. Thus Ge epilayer growth on GaAs $c(4 \times 4)$, $c(8 \times 2)$ and (4×6) surface exhibit >0.3 eV variations in band bending qV_B but nevertheless the same band offset [13]. Likewise, Ge overlayers doped with Al to alter the overlayer work function produce the same band offset but different qV_B. These results indicate that band offsets and band bending have different physical origins.

14.3.6.3 Interface Bonding

Heterovalent compound junctions can show a strong dependence of interface states on the chemical bonding at the interface. C–V measurements of interface traps D_{it} for ZnSe epilayers on GaAs substrates show a >100× variation in D_{it} versus RHEED reconstruction [14]. These localized trap densities and their effect on capacitance are highest for As-rich reconstructions, decreasing with increasing Ga–Se bonding. As with the effect of growth sequence on band offsets, interface atomic bonding has an important effect on band bending. Indeed, the introduction of new interfacial layers to alter this bonding can have an effect on both band offsets and band bending, as shown below and in Chapter 15.

14.3.7 Theoretical Methods

Unlike most material systems, epitaxial growth of semiconductor heterojunctions provides theorists with models of well-defined atomic structures from which band offsets can be calculated. However, such calculations are challenging since there exists no absolute reference potential for a semiconductor lattice V^{lattice} due to the long-range Coulomb potential. Thus one can only define V^{lattice} to within a constant. Despite this limitation, several methods have been developed that reflect the variations in heterojunction band offset with interface atomic structure and composition. See, for example, Franciosi and Van de Walle [15].

14.3.7.1   First Principles Calculations

Self-consistent, first-principles calculations that require no experimental input are based on ground state energies from single-particle solutions of Schroedinger's equation $(h^2/2m)\nabla^2\psi + V\psi = E\psi$. Analogous to the interface state calculation in Chapter 4, interface boundary conditions fix the particle wave function ψ, and the potential V consists of a lattice potential between the electron and the atom plus an exchange-correlation between electrons that depends on local charge density. The system's energy is calculated as a function of atomic position in order to determine stable structures that include relaxation and reconstructions.

14.3.7.2   Mathematical Approach

The mathematical approach to such calculations involves a self-consistent solution of a variational problem. The large number of lattice atoms involved represent a computational challenge so that periodic boundary conditions are used for isolated interfaces as well as for actual superlattices. Here one takes the average potential $\overline{V}(z) = (1/S) \int V(x, y, z)\, dx\, dy$ for each semiconductor, where S is a unit cell area in planes parallel to the interface. The variation of plane-averaged potential across the interface then yields a valence band offset. [See Track II T14.15. Planar Averaged Si-Ge Heterojunction Potentials.]

All-electron (core plus valence) calculations can mimic XPS valence band-core level offsets. Pseudopotential (averaged core plus valence) calculations provide no direct ΔE_V but obtain offset values from average potentials. Within each atomic layer parallel to the interface, planar average potentials $\overline{\overline{V}}(z) = (1/a)_{(z-a/2)}\int^{(z+a/2)}\overline{V}(z')\, dz'$ can provide macroscopic averages of charge and their deviation from average charge values. Here $a = $ length of one period perpendicular to the interface with unit cell volumes. In turn, deviations from average charge values near the interface provide a measure of the dipole moment contribution to the band offset. [See Track II T14.15. Planar Averaged GaAs-AlAs Heterojunction Potentials and Charge Densities.]

Other mathematical approaches for homovalent heterojunctions include linear response theory, simplified Hamiltonians, and simple models based on bulk properties. Linear response theory involves modeling the difference between two semiconductors across an interface in terms of their average, for example, $Al_{0.5}Ga_{0.5}As$, across the GaAs–AlAs interface. Simplified Hamiltonians include self-consistent tight binding calculations with parameters fitted to bulk properties rather than the interface. As a result, however, these calculations cannot extract energies. Simple models based on bulk properties include the *Electron Affinity Rule* and charge neutrality levels with either intrinsic or extrinsic interface dipoles.

14.3.7.3   Heterovalent Interfaces – Polarity and Interfacial Bonding Dependence

For heterovalent interfaces, crystal orientation determines the local bonding and charge transfer between atoms across the plane of the heterojunction. For nonpolar interfaces, there is no change in $\overline{\overline{V}}(z)$ at the interface. Figure 14.14 illustrates a Ge–GaAs(110) heterojunction, which has equal numbers of Ga and As atoms facing Ge atoms across the interface.

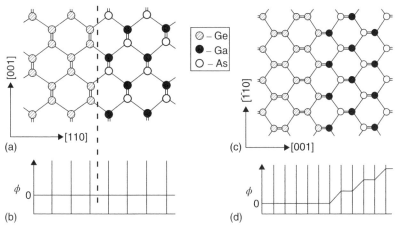

Figure 14.14   *(a) Schematic lattice representation of a Ge–GaAs(110) heterojunction, viewed along the [110] direction with the [001] direction vertical. Dashed line signifies the interface plane. Every atomic plane parallel to the junction is neutral on average, corresponding to a nonpolar junction. (b) Potential versus lattice plane. (c) Ge–GaAs(001) heterojunction viewed along the [110] direction with the [110] direction vertical. The first atomic plane to the right of the junction consists of only negatively charged Ga atoms. (d) Potential versus lattice plane increasing from left to right due to nonzero electric field. (Harrison et al. 1978 [16]. Reproduced with permission of American Physical Society.)*

All atoms are tetrahedrally coordinated with the double bonds shown representing two tetrahedral bonds separated by the usual 109° angle and projected onto the plane of the figure. Every atom in Figure 14.14.4a is neutral on average, corresponding to a nonpolar junction [16].

For the ideal polar interface shown in Figure 14.14c, the first atom to the right of the junction is composed entirely of Ga, which is negatively charged (since less negative charge is transferred to its Ge rather than to an otherwise As neighbor).

The potential averaged over planes parallel to the junction is calculated by integrating Poisson's equation from left to right. Figure 14.14d shows that this negatively charged layer leads to a non-zero average electric field in the GaAs region. As a result, electric potential increases from the interface into the GaAs, leading to what is termed a "polar catastrophe". Atomic mixing can reduce or eliminate the net charge, electric field, and potential build-up at polar interfaces. For example, replacing half the Ga atoms with Ge at the Ge–GaAs(001) heterojunction shown in Figure 14.14c, eliminates the average electric field. [See Track II T14.16. Interface Dipoles with Ge-GaAs Interface Atomic Mixing.] Nevertheless, this replacement introduces a dipole shift that affects the band offset.

With a double layer mixing such that the first layer past the all-Ge layer consists of three Ge and one As while the second layer consists of one Ge and three Ga atoms, there is no change in the average potential across the interface. Finally, two Ge layers at a GaAs(001) homojunction produces an opposite charge exchange between Ge and GaAs on either side of the interlayer and a net negative dipole. See Figure 14.15a, below.

14.3.8 Band Offset Engineering

The calculated dependence of band offsets on local atomic bonding show that "engineering" of heterojunction band offsets is possible with interface atomic control. Such control is now possible using UHV growth techniques that include: (i) atomic interlayers and (ii) local nonstoichiometry.

14.3.8.1 Atomic Interlayers

As shown in Figure 14.15a, the different charge exchange on either side of the GaAs–Ge–GaAs homojunction gives rise to a dipole that produces a large (0.74 eV) band offset. The sign of this offset depends on the GaAs atomic termination on which the Ge is deposited. The sandwiched Ge layer with oppositely charged GaAs interface is thus a built-in capacitor. Band offsets of this magnitude are measured at homojunctions of GaAs with Si (−0.38 eV) [9].

Similarly, homojunctions separated by compound semiconductor interlayers exhibit sizable dipoles: 0.35–0.45 eV for Ge homojunctions separated by Ga-As or Al-As interlayers as well as 0.4 eV for Si homojunctions separated by Ga-P or Al-P interlayers.

Elemental interlayers at heterojunctions include: Si at GaAs(100)–AlAs, as well as Al or Au at CdS–Ge, ZnSe–Ge, GaP–Si, and GaAs–Ge interfaces. Figure 14.15b shows the variation in ΔE_V at the AlAs–GaAs(100) heterojunction as a function of ordered Si interlayer thickness. Control of the Si coverage, growth sequence, and As flux alters ΔE_V continuously in the range 0.02–0.78 eV around its intrinsic $\Delta E_V = 0.40 \pm 0.07$ eV value. This figure illustrates that the AlAs–GaAs band offset can be controlled over a 0.8 eV energy range. The magnitude and sign depend on the growth sequence with maximum values clearly indicated at 0.5 monolayers of Si. Experimental results for all these systems

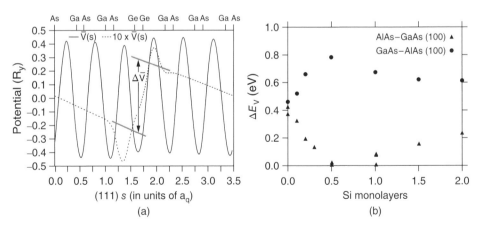

Figure 14.15 *(a) Planar-averaged potential variation of a GaAs-2-layer Ge-GaAs homojunction along the (111) direction (solid line) and the unit cell-averaged potential showing the band offset (dotted line). (b) Valence band offset for a GaAs-multilayer Si-GaAs heterojunction for AlAs-GaAs(100) (Δ) and GaAs-AlAs(100) (●) heterostructures as a function of ordered Si interlayer thickness. Band offset varies as a function of Si interlayer thickness. (Sorba et al. 1991 [17]. Reproduced with permission of American Physical Society.)*

show that: (i) heterovalent ΔE_V are larger than isovalent junctions, (ii) nonpolar (110) heterovalent junctions exhibit only $\Delta E_V \sim 0.2$ eV variations, and (iii) polar (100) heterovalent ΔE_V variations exceed 0.5 eV.

14.3.8.2   *Local Nonstoichiometry*

Band offsets can be "engineered" by controlling stoichiometry at the intimate (1–2 mono-layer) interface. By varying the Zn:Se beam pressure ratios during the growth of ZnSe on GaAs(100), one can achieve large variations in ΔE_V [18]. XPS measurements of the Zn3d versus Se 3d core level intensities provide Zn:Se ratios R versus ZnSe overlayer thickness starting at the first deposited monolayer. The corresponding ΔE_V versus Zn/Se R exhibits systematic variations from 0.55 to 1.2 eV, increasing with increasing R. [See Track II T14.17. ZnSe-GaAs Band Offset vs. Interface Stoichiometry.] This striking variation exhibits only a minor dependence on initial GaAs reconstruction and doping. These results suggest that slight stoichiometry variations at heterojunction interfaces can alter measured ΔE_V significantly. The ΔE_V control versus local stoichiometry also has practical applications. For example, the reduction of ΔE_V from $\sim$0.9 eV for 1:1 stoichiometry to 0.55 eV with Se-rich stoichiometry permits more efficient hole transport from GaAs to ZnSe.

In terms of practical band offset engineering, one seeks to: (i) minimize unintentional doping and dislocation density formation during growth and (ii) maximize chemical stability against deep level formation. For example, while the 0.55 eV ΔE_V at Se-rich ZnSe/GaAs(001) heterojunctions is desirable, it is less stable against intermixing at elevated temperatures due to the diffusion of Ga atoms into Zn vacancy sites [19]. Therefore, a tradeoff exists between control of ΔE_V and chemical stability at this and possibly other heterojunctions.

A final method for band offset engineering involves *delta doping*. With MBE growth techniques, one can create near monolayer-thick sheets of ionized donor and acceptor charges within a few monolayers of and on opposite sides of an interface. The dipole formed at such interfaces produces a voltage change $\Delta \Phi = \sigma d$ between the two sides, where σ is the sheet charge density per unit area and d is the sheet separation. Delta-doped layers separated by only a few atomic layers produce sharply bent bands that are nearly triangular in shape and enable tunneling to effectively reduce band offsets. For both the variable stoichiometry and delta-doped methods, atomic-scale control is required to engineer interface charge states and dipoles.

14.4   Conclusions

(i) It is now possible to create lattice structures with a wide range of band structures and physical properties not found in nature using atomically controlled growth techniques. These semiconductor heterostructures are the basis for many of today's optoelectronic and microelectronic devices. Achieving these heterostructures requires joining materials with similar lattice structures and controlling the chemical bonding at their atomic-scale interfaces. (ii) Lattice mismatch can be used as a design parameter in order to create carrier confinement channels at heterointerfaces with both high carrier density and mobility. (iii) Strained layer superlattices enable growth of semiconductor composites with novel band structures and properties. Band gaps in these structures can be engineered over wide ranges by controlling layer composition, layer width, and strain. (iv) Pseudomorphic template layers can bridge between semiconductors with mismatched lattice structures, providing surface passivation and introducing electric fields and dipoles that can alter heterojunction band offsets. (v) XPS measurements of band offsets use core level–valence band edge comparisons with a precision sufficient to gauge the contributions of growth sequence, crystallographic orientation, surface reconstruction, interface composition, and interfacial layers to ΔE_V. (vi) Band offsets measured experimentally are consistent with several theoretical models, notwithstanding their different physical bases, and require precision in the 0.01 range to achieve predictive value. (vii) Failure of the *electron affinity model* to account for band offsets quantitatively emphasizes the importance of interface dipoles in band structure alignment. (viii) Band offset engineering can produce ΔE_V variations as large as 1 eV, but may involve tradeoffs between electronic structure and thermodynamic stability.

14.5   Problems

1. Semiconductor heterojunctions are commonly used to build visible light emitting diodes and quantum well lasers. Describe the advantages and disadvantages of (a) GaAs/(Al, Ga)As versus (b) GaN/(In, Ga)N heterojunctions in terms of their growth and physical properties.

2. Consider mixed valence heterojunctions other than common arsenide or nitride systems that exhibit good lattice match. (a) Identify two binary semiconductors with good lattice match to Si along with their potential advantages and disadvantages. (a) Repeat for one binary semiconductor to GaN.

3. Using Vegard's Law, calculate the critical thickness of $Al_{0.4}Ga_{0.6}N$ on GaN. Why do AlGaN/GaN heterojunctions for high electron mobility transistors typically have Al compositions in this range?

4. (a) Calculate the critical thickness of GaN grown epitaxially on a ZnO wafer. (b) In terms of lattice matching, is ZnO better than SiC? (c) Besides lattice match, what other factors should be considered in using ZnO wafers as a large area template for GaN growth?

5. Sketch the band diagram for a heterojunction consisting of a p^+ semiconductor A with a band gap $E_G = 2.7$ eV and an undoped n-type semiconductor B with band gap $E_G = 1.4$ eV. Assume $\Delta E_C = 2/3 \ \Delta E_G$, $\varepsilon_A = \varepsilon_B$ and, from continuity of the electric flux density at the interface, band bendings $V_{0A}/V_{0B} = N_B \varepsilon_B / N_A \varepsilon_A$ for doping densities N_A and N_B. Show ΔE_C, ΔE_V, the band line-ups, the band bending, and relative depletion widths in each semiconductor.

6. Design a thin high mobility channel layer to be grown epitaxially on an InP substrate. What is the band gap of this layer?

7. Design an epilayer structure grown epitaxially on a GaAs substrate that emits light at 5900 Å.

8. The space charges within the band bending regions of a heterojunction are equal and opposite in sign. Assuming the depletion approximation for semiconductors A and B, show that the ratio of band bending V_{0A} in semiconductor A with doping N_A versus V_{0B}

in semiconductor B with doping N_B is $V_{0A}/V_{0B} = N_B\varepsilon_B/N_A\varepsilon_A$. [Hint: Use Equation (3.7)]

9. Prove Equation (3.31) for a heterojunction comprised of semiconductor 1 with ε_1 and doping N_1 versus a semiconductor 2 with ε_2 and doping N_2. Assume the depletion approximation for the two semiconductors.

10. Calculate the required sheet doping to increase the valence band offset by 0.1 eV across four monolayers at a GaAs–AlGaAs heterojunction. Assume $\varepsilon = \varepsilon(\text{GaAs})$.

11. The gain in an p-n-p bipolar junction transistor (BJT) depends in part on the *emitter injection efficiency* γ, the ratio of majority versus majority plus minority carriers injected under electrical bias from an n-type emitter into a p-type base region. Minority carriers lower γ depending in part on the relative conduction versus valence band offsets. (a) Give two examples of heterojunctions with offset differences that could be used to increase γ. (b) Calculate the ratio of thermal injection for electrons versus holes across one of these heterojunctions. (c) Draw a heterojunction band diagram to illustrate this concept.

References

1. Casey, H.C. Jr. and Panish, M.B. (1978) *Heterojunction Lasers, Part A: Fundamental Principles, Materials and Operating Characteristics,* Academic Press, New York, Part B.

2. Mayer, J.W. and Lau, S.S. (1990) *Electronic Materials Science: for Integrated Circuits in Si and GaAs*, Macmillan Publishing Company, New York, p. 424.

3. Raisanen, A., Brillson, L.J., Goldman, R.S. *et al.* (1994) Strain-induced deep levels in $In_xGa_{1-x}As$ Thin Films. *J. Vac. Sci. Technol. A,* **12**, 1050.

4. Matthews, J.W. and Blakeslee, A.E. (1974) Defects in epitaxial multilayers: I. Misfit dislocations. *J. Cryst. Growth,* **27**, 118.

5. Woodall, J.M., Kirchner, P.D., Rogers, D.L. *et al.* (1988) Semiconductor device research using non-lattice matched structures. *Mater. Res. Soc. Symp. Proc.,* **126**, 3.

6. Mailhiot, C. and Duke, C.B. (1986) Many-electron model of equilibrium metal-semiconductor contacts and semiconductor heterojunctions. *Phys. Rev. B,* **33**, 1118.

7. Anderson, R.L. (1962) Experiments on GeGaAs heterojunctions. *Solid-State Electron.,* **5**, 341.

8. Smith, P.E., Goss, S.H., Gao, M. *et al.* (2005) Atomic diffusion and band lineups at $In_{0.53}Ga_{0.47}As$-on-InP heterointerfaces. *J. Vac. Sci. Technol. B,* **23**, 1832.

9. Margaritondo, G. and Perfetti, P. (1987) in: *Heterojunction Band Discontinuities: Physics and Device Applications*, (eds. F. Capasso and G. Margaritondo) North-Holland, Amsterdam, ch. 2.

10. Katnani, A.D. and Margaritondo, G. (1983) Microscopic study of semiconductor heerojunctions: Photoemission measurement of the valence-band discontinuity and of the potential barriers. *Phys. Rev. B,* **28**, 1944.

11. Ishii, H., Sugiyama, K., Ito, E., and Seki, K. (1999) Energy level alignment and interfacial electronic structures at organic/metal and organic/organic interfaces. *Adv. Mater.,* **11**, 605.

12. Grant, R.W., Kraut, E.A., Waldrop, J.R., and Kowalczyk, S.P. (1987) *Heterojunction Band Discontinuities: Physics and Device Applications*, (eds F. Capasso and G. Margaritondo), North-Holland, Amsterdam, ch. 4.
13. Chiaradia, P., Katnani, A.D., Sang, H.W. Jr.,, and Bauer, R.S. (1984) Independence of Fermi level position and valence-band edge discontinuity at GaAs-Ge(100) interfaces. *Phys. Rev. Lett.*, **52**, 1246.
14. Qian, Q.-D., Gunshor, R.L., Kobayashi, M. *et al.* (1990) Influence of GaAs surface stoichiometry on the interface state density of as-grown epitaxial ZnSe/epitaxial GaAs heterostructures. *Appl. Phys. Lett.*, **56**, 1272.
15. Franciosi, A. and Van de Walle, C. (1996) Heterojunction band offset engineering. *Surf. Sci. Rep.*, **25**, 1–140.
16. Harrison, W.A., Kraut, E.A., Waldrop, J.R., and Grant, R.W. (1978) Polar heterojunction interfaces. *Phys. Rev. B*, **18**, 4402.
17. Sorba, L., Bratina, G., Ceccone, G. *et al.* (1991) Tuning AlAs-GaAs band discontinuities and the role of Si-induced local interface dipoles. *Phys.Rev. B*, **43**, 2450.
18. Nicolini, R., Vanzetti, L., Mula, G. *et al.* (1994) Local interface composition and band discontinuities in heterovalent heterostructures. *Phys. Rev. Lett.*, **72**, 294.
19. Raisanen, A.D., Brillson, L.J., Vanzetti, L. *et al.* (1995) Atomic diffusion-induced deep levels near ZnSe/GaAs(100) interfaces. *Appl. Phys. Lett.*, **66**, 3301.

Further Reading

Capasso, F. and Margaritondo, G. (eds) (1987) *Heterojunction Band Discontinuities: Physics and Device Applications*, North-Holland, Amsterdam.
Franciosi, A. and Van de Walle, C.G. (1996) Heterojunction band offset engineering. *Surf. Sci. Rep.*, **25**, 1–140 (1996).
Brillson, L.J. (1992) in *Basic Properties of Semiconductors*, vol. 1, ed. P.T. Landsberg, North-Holland, New York, ch. 7.

15

Metal–Semiconductor Interfaces

15.1 Overview

Contacts between metals and semiconductors are the most extensively studied of electronic material interfaces. This activity reflects the importance of metal contacts to semiconductors throughout all of solid state electronics, materials science, and physics as well as the wide range of phenomena that researchers have observed at these interfaces. In Chapter 4, we introduced the variety of phenomena that can give rise to interface states that affect Schottky barriers. Subsequent chapters described the array of experimental and theoretical techniques used to examine these interface states. Here we examine interface states and their effect on Schottky barriers in greater depth. First, we consider the role of interface dipoles at the metal–semiconductor interface, the different models that have been used to account for experimental observations, and then the experimental results used to evaluate and distinguish between these models.

Described next are the extrinsic phenomena that have observable effects on Schottky barriers. The physical framework that finally emerges includes both intrinsic and extrinsic factors. Contributing in varying degrees are: (i) intrinsic properties of the bulk metal and semiconductor, (ii) crystal quality, (iii) chemical reaction and diffusion near the interface, and (iv) chemical and thermal conditions under which the interface is formed and processed. Each of these factors can significantly affect the charge transfer across the metal–semiconductor interface and the effective Schottky barrier height.

15.2 Metal–Semiconductor Interface Dipoles

As described in Figure 4.1, the wave function of charge decaying away from a semiconductor surface into vacuum produces charge separation and a surface dipole. Dipoles are present at metal surfaces as well and are required to confine electrons inside the solid. Figure 15.1 illustrates the amplitude of wave function probability for charge at a metal or

An Essential Guide to Electronic Material Surfaces and Interfaces, First Edition. Leonard J. Brillson.
© 2016 John Wiley & Sons, Ltd. Published 2016 by John Wiley & Sons, Ltd.
Companion Website: www.wiley.com/go/Brillson/

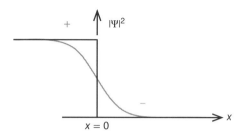

Figure 15.1 *Wave function probability function (grey) extending out from the semiconductor or metal surface (black) (x = 0) into vacuum. Dipoles are required to confine electrons to the solid. (Brillson 2010. Reproduced with permission of Wiley.)*

semiconductor. The electron density extending away from the surface and the corresponding electron-deficient region just below the surface represent a dipole that acts to confine near-surface electrons to the solid.

Figure 15.2a illustrates the surface dipole potentials V_M and V_S, respectively, for metal and semiconductor surfaces before contact. Here, the metal work function Φ_M is

$$\Phi_M = V_M + S_M \tag{15.1}$$

and the semiconductor electron affinity χ_{SC} is

$$\chi_{SC} = V_S + S_S + \zeta \tag{15.2}$$

As in Figure 14.8, S_S is defined with respect to the lowest unoccupied conduction band states, taking into account the Fermi level position relative to the conduction band edge $\zeta = E_F - E_C$ for the case of a degenerate conduction band. For E_F below E_C, χ_{SC} is just $V_S + S_S$.

Upon metal–semiconductor contact, Figure 15.2b shows that the two Fermi levels align, analogous to Section 14.3.1. Now the surface dipoles V_S and V_M are replaced by an interface dipole V_i. As shown,

$$S_M = V_i + V_b + S_S \tag{15.3}$$

where V_b is the semiconductor band bending and λ the width of the surface depletion region. V_b is the primary contribution to the Schottky barrier height Φ_{SB} since

$$\Phi_{SB} = V_b + (E_C - E_F) \tag{15.4}$$

and $E_C - E_F$ is typically $0.1 - 0.2$ eV or less. Solving for V_b and expressing the internal potentials in terms of measurable parameters,

$$V_b = S_M - S_S - V_i \tag{15.5}$$

$$= (\Phi_M - V_M) - (\chi_S - V_S - \zeta) \tag{15.6}$$

$$= (\Phi_M - \chi_S - \zeta) + (V_S - V_M - V_i) \tag{15.7}$$

As with semiconductor–semiconductor junctions, only the first term in Equation (15.7) is observable. The dipole term is inferred from the difference between $(\Phi_M - \chi_S - \zeta)$ and V_b. This difference represents the contribution of interface states.

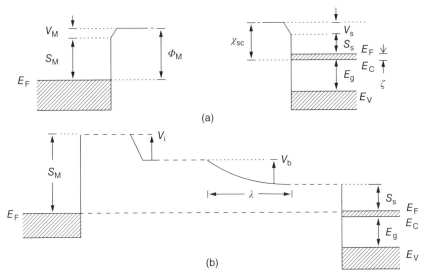

Figure 15.2   *Potential distribution for (a) separated metal and semiconductor surfaces and (b) the metal–semiconductor interface. The observable metal work function Φ_M consists of a calculated internal potential S_M plus a surface dipole V_M. The observable semiconductor electron affinity consists of a calculated internal potential S_S plus a surface dipole V_S. The local interface dipole V_i plus the dipole V_b of the surface space charge region account for the difference in internal potentials when the metal and semiconductor are joined. After Duke and Mailhiot. (Duke, C.B. and Mailhiiot, C. 1985 [1]. Reproduced with permission of American Institute of Physics.)*

15.3   Interface States

Chapter 4.1 introduced the various types of localized charge states that can contribute to dipoles at metal–semiconductor interfaces, including intrinsic surface states, metal-induced states, extrinsic states due to lattice imperfections/impurities, and chemical reactions or interdiffusion. This section addresses how well each of these mechanisms can account for the wide array of Schottky barrier measurements now available.

15.3.1   Localized States

Figure 15.3a illustrates wavefunction tunneling and charge localization at a semiconductor interface with vacuum, analogous to Figure 15.1. These *intrinsic surface states* are associated with the discontinuity of the lattice potential at the interface. For junctions without imperfections or chemical interactions, metals deposited on semiconductor surfaces with intrinsic surface states would yield Schottky barriers that are relatively insensitive to the metal work function, depending on the extent to which the surface disrupts the bulk lattice potential. Here, one expects a weak Φ_{SB} dependence on Φ_M for metals on high dielectric permittivity ε semiconductors, such as GaAs and Si, versus a stronger dependence for low ε semiconductors, such as CdS and SiO_2, due to their different strength and lattice bond localization.

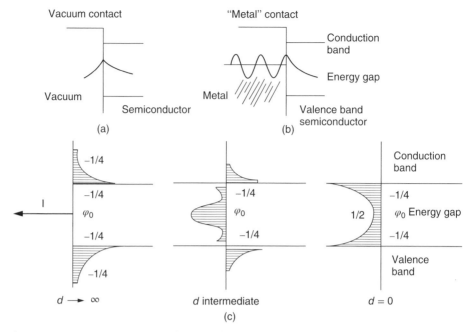

Figure 15.3 (a) Gap state wavefunction decay into vacuum and semiconductor. (b) Propagating metal wavefunction decay into semiconductor at gap state energies. (c) Metal-induced gap states (MIGS). The exponentially decaying behavior leads to the continuous density of interface states shown for a one-dimensional model of a covalent semiconductor–metal interface, where φ_0 defines the charge neutrality energy of the semiconductor and d is the metal–semiconductor separation. (Tejedor et al. 1977 [2]. Reproduced with permission of Institute of Physics.)

15.3.2 Metal-Induced Gap States

Figure 15.3b illustrates wavefunction tunneling from a metal into a semiconductor at energies within the band gap. The exponential wavefunction decay occurs over distances of only ~ 10 Å. Figure 15.3c shows that wave function tailing produces a new density of states within the band gap that is compensated by decreased valence and conduction band densities as the distance d between metal and semiconductor decreases.

Below energy φ_0 in the band gap, interface and valence band states compensate each other locally. This effective mid-gap energy is defined as the *charge neutrality level* or CNL. It determines E_F and the local charge distribution required to maintain charge neutrality. The dipole introduced by charge transfer and new gap states is proportional to the difference $\Phi_M - CNL$ so that

$$E_F - CNL = S\,(-CNL + \Phi_M) \tag{15.8}$$

where S is a parameter related to the semiconductor dielectric properties. The interface dipole D is then

$$D = (1 - S)\,(\Phi_M - CNL) \tag{15.9}$$

and

$$S = 1/[1 + (4\pi e^2 D(E_S)\delta/A)] \qquad (15.10)$$

for density of interface states $D(E_S)$ at energy E_S, state penetration distance δ into the semiconductor, and area A. Thus,

$$\Phi_M - CNL = D + (E_F - CNL) \qquad (15.11)$$

Finally, adding and subtracting Φ_M from $E_F - CNL$ in Equation (15.8) yields $E_F - CNL = (\Phi_M - CNL) + (E_F - \Phi_M) = S(\Phi_M - CNL)$ so that

$$(\Phi_M - CNL) = (\Phi_M - E_F)/(1 - S) \qquad (15.12)$$

Consider the limiting cases for D. If $D = 0$, then $S = 1$ and $E_F = \Phi_M$, which is just the Schottky or classical limit. If D is large, then $S = 0$ and $E_F = CNL$, which is just the limit of Fermi level pinning. Both limiting cases are consistent with the expected Schottky barrier behavior.

Case Study: Metal-Induced Gap State Calculation

Given the work functions for Au and Cu, $\Phi_M(Au) = 5.1$ eV and $\Phi_M(Cu) = 4.65$ eV and corresponding p-type barrier heights $\Phi_{BP}(Au) = 0.34$ eV, and $\Phi_{BP}(Cu) = 0.46$ eV, calculate (a) S and (b) CNL for Si.

Answer: (a) $S = \Delta\Phi_{BP}/\Delta\Phi_M = (0.46 - 0.34)/(5.1 - 4.65) = 0.27$. (b) For the p-type barrier height given for Au, $\Phi_{BP}(Au) = 0.34$ eV. From Figure 15.4, the Fermi level position relative to the vacuum level E_{VAC} follows from the *ionization potential*, that is, the valence band maximum E_V relative to E_{VAC}, which is just $\chi + E_G$, where E_G is the semiconductor band gap. From Appendix D, $E_{VAC} - E_V = 4.05 + 1.11 = 5.16$ eV so that $E_F = E_V - \Phi_{BP} = 5.16 - 0.34 = 4.82$ eV. From

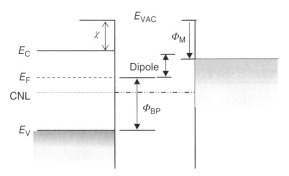

Figure 15.4 *The CNL represents the crossover energy between the weighted average of conduction and valence bands. Charge tunneling into the semiconductor forms a dipole that accounts for the difference between the metal work function and the Fermi level at the semiconductor–metal interface.*

(*continued*)

(*continued*)

Equation (15.12) and $E_F = E_V - 0.34$ eV, $(\Phi_M - CNL) = (5.1 - (E_V - 0.34))/(1 - 0.27) = (5.1 - (5.16 - 0.34))/0.73 = 0.28/0.73 = 0.38$ eV. Then the CNL $= \Phi_M - 0.38 = 5.1 - 0.38 = 4.72$ eV, which is $E_V - CNL = 5.16 - 4.72 = 0.44$ eV above the valence band.

Note that the CNL is sensitive to S, which depends on a linear fit to the Φ_M versus Φ_{SB} data. For $S \sim 0.1 - 0.3$ and $E_F - CNL \sim 0.1$ eV, typical for many studies, the interface dipole $D = (1 - S)(\Phi_M - CNL)$ can be on the order of 1 eV. Calculations of metal wave function tunneling yield large local dipole charge densities of $\sim 10^{14} - 10^{15}$ cm^{-2} eV^{-1}. Calculated energies where the gap states cross over from valence-to-conduction-band character in the complex band structure appear in Table 15.1.

15.3.3   Charge Transfer, Electronegativity, and Defects

For charge neutrality levels and theoretical models in general, the magnitude of charge transfer and the dipole that results depend on the boundary conditions assumed, whereas the local atomic bonding measured experimentally can differ significantly from ideal interface structures. For metals on both Si and GaAs, experimental barrier heights exhibit a monotonic dependence on the electronegativity of adsorbates and hence the charge transfer expected. The *Miedema electronegativity* used derives from a semiempirical fit with thermodynamic solubility data. For Si, the Φ_{SB} variation extends over half the band gap and includes both metals that react to form silicides and metals that do not react. [See Track II T15.1. Φ_{SB} versus Miedema electronegativity.] For GaAs, Φ_{SB} follow a similar trend with significant deviations attributed to defects, implying a significant effect of extrinsic factors on the otherwise intrinsic charge transfer.

Additional E_F "pinning" mechanisms involve dangling bond hybrid orbitals, electron correlation effects, and band gap narrowing analogous to Figure 15.3c. In general, theoretical models all require abrupt interfaces with no additional chemical phases. However, only a few such interfaces are known, and all indicate a dependence on atomic structure.

Table 15.1 *Charge neutrality levels (CNL in eV) relative to the valence band* E_V *from [3] and references therein*

Semiconductor	CNL	Semiconductor	CNL
Si	0.36	AlSb	0.45
Ge	0.18	GaAs	0.07
AlP	1.27	InSb	0.01
GaP	0.81	ZnSe	1.70
InP	0.76	MnTe	1.6
AlAs	1.05	ZnTe	0.84
GaAs	0.50	CdTe	0.85
InAs	0.50	HgTe	0.34

Tersoff 1984 [3]. Reproduced with permission of Elsevier.

15.3.4   Imperfections, Impurities, and Native Defects

Extrinsic interface states include lattice imperfections and contaminants. Studies of chemisorbed species provide evidence for atomic-like donor/acceptor states. For sub-monolayer adsorbates on clean cleaved p-type GaAs, E_F moves to energies consistent with the adsorbate's first ionization potential, then stabilizes at a different "bulk" value beyond two monolayer coverages [4]. Chemical contamination has strong effects on E_F stabilization and contact stability with annealing. Structural imperfections, such as lattice steps produced with crystal cleavage or with misorientation, can lead to strong pinning and even increased chemical activity. Point defects such as interstitial atoms and lattice vacancies can form due to surface cleaning, ion bombardment, and crystal cleavage.

Deep level defects can be electrically-active and are well studied in semiconductors such as Si, GaAs, GaN, and ZnO. For GaAs, the method used to grow the crystals has a major effect on the type and density of these defects. Melt-grown crystals have deep levels with densities in the mid-10^{16} cm^{-3} range, crystals grown by vapor-phase epitaxy have multiple defects but <10^{14} cm^{-3} densities, liquid-phase epitaxy results in densities <10^{13} cm^{-3}, while MBE-grown epilayers possess even lower defect densities. These native point defects are sensitive to the growth temperature, stoichiometry, and free carrier concentration. For lightly doped n- and p-type GaAs, dislocation density exhibits a dramatic dependence on melt stoichiometry, decreasing by orders of magnitude with slight As-rich composition. [See Track II T15.2. GaAs Dislocation Density vs. Growth Stoichiometry.] However, the As excess produces As antisite defects.

Dopant atoms within the semiconductor lattice produce energy levels near E_C or E_V that serve to increase electron or hole densities. Impurity atoms that produce deep levels can alter free carrier densities, lifetimes, and the surface space charge region. Examples include C, As clusters, and Au in GaAs.

Native point defects, such as lattice vacancies, interstitials, and antisites, can produce multiple deep levels with densities varying by orders of magnitude, depending on the growth process. These defects can be mobile and segregate to surfaces or interfaces. Extended surface or bulk imperfections, including misfit dislocations, antiphase domain boundaries, and defect complexes, can develop during growth and subsequent processing.

Misfit dislocations can incorporate states that trap charge and produce depletion regions that surround the dislocation core. At sufficiently high density, depletion regions between dislocations can overlap, effectively "pinching off" current transport through those regions. At metal–semiconductor interfaces, this effect can change the effective Φ_{SB}. Together, these results highlight the importance of both surface and bulk crystal quality.

15.3.5   Chemisorption, Interface Reaction, and Interfacial Phases

Chemisorption of metal atoms on semiconductors can release energy that can break surface bonds. The surface defects formed can have deep levels with densities high enough to "pin" E_F at an energy within the bandgap. XPS measurements of E_F versus metal adsorbate coverage display a rapid E_F movement to a pair of deep levels in GaAs with submonolayer metal or oxygen coverage [5]. [See Track II T15.3. Unified Defect Model.] Significantly, these gap state energies within the band gap are nearly identical to the energy positions of As$_{Ga}$ antisite levels found by photo-spin resonance experiments, indicative of a specific native point defect common in melt-grown, As-rich GaAs that stabilizes E_F, independent of the adsorbate [6].

Section 4.3 introduced phenomenological evidence for extrinsic states due to metal–semiconductor bonding and interdiffusion. Φ_{SB} variations of transition metal silicides on Si suggested a correlation with the thermodynamic bond strength of an interfacial layer. Correlations between Φ_{SB} and the heat of reaction ΔH_R displayed the same chemical transition, regardless of semiconductor ionicity, suggesting that chemical stability factors into the Φ_{SB} dependence on metal. See Figure 4.6. Indeed, semiconductors with the least chemical stability exhibit the lowest Φ_{SB} dependence on Φ_M, suggesting that interfacial reaction plays a dominant role in E_F stabilization. See Figure 4.2.

Besides interface reactions, atomic outdiffusion can occur at metal–semiconductor interfaces. For III-V compounds such as GaAs and InP, anion-rich versus cation-rich outdiffusion correlates with interface reactivity. Unreactive metals lead to anion-rich outdiffusion and high Φ_{SB} while metals that react with anions result in cation-rich outdiffusion [7]. The resultant outdiffusion produces electrically-active sites that stabilize E_F, providing a chemical basis for E_F pinning that involves native point defects.

Yet another chemical mechanism to account for E_F "pinning" involves changes in semiconductor stoichiometry at the metal interface. Here, an accumulation of anion atoms at the junction can alter the effective work function of the contacting metal. The resulting anion-rich interphase dominates the interface charge transfer, resulting in Φ_{SB} that appear relatively insensitive to Φ_M. [See Track II T15.4 Interfacial Phase Model.]

15.3.6   Organic Semiconductor–Metal Interfaces

As with inorganic semiconductors, the organic semiconductor–metal interface plays a dominant role in the charge injection and transport through these materials. Many of the interface features already presented can produce dipoles at organic–metal interfaces as well [8]. Depending on the organic's crystallinity, thermally-activated intermolecular hopping and orbital band shifts become additional parameters affecting charge injection and transport. Likewise, organics have typically low dielectric constants, nearly charge-depleted materials with high trap densities that require space-charge-limited, trap-limited, or injection-limited models to describe charge movement across their interfaces. Because they are soft, the processes used to form their junctions can strongly affect interface morphology, chemistry, and electrical properties.

As an example, Figure 15.5 illustrates how the sequence of deposition alters the interface morphology. Deposition of Alq_3 on Mg (Figure 15.5a) results in exponential attenuation of the substrate XPS Mg 2p peak, indicating an abrupt interface, whereas Mg deposition on Alq_3 (Figure 15.5b) attenuates the C 1 s substrate peak much more slowly, indicating a diffuse junction. Nevertheless, both interfaces exhibit similar core level shifts, defect states in the Alq_3 band gap, and E_F position relative to the *highest occupied molecular orbital* (HOMO) – indicative of E_F pinning by the same chemically-induced defect states.

A counter example is the deposition of $F_{16}CuPc$ (Pc, phthalocyanine) on Au, which produces no chemical reaction and does not induce gap states versus Au deposition on $F_{16}CuPc$, which produces doping or a reaction with the organic and E_F shifted $0.5 - 0.6$ eV closer to the organic's HOMO level. [See Track II T15.5. Organic Semiconductor Band Lineups and Dipole Formation.]

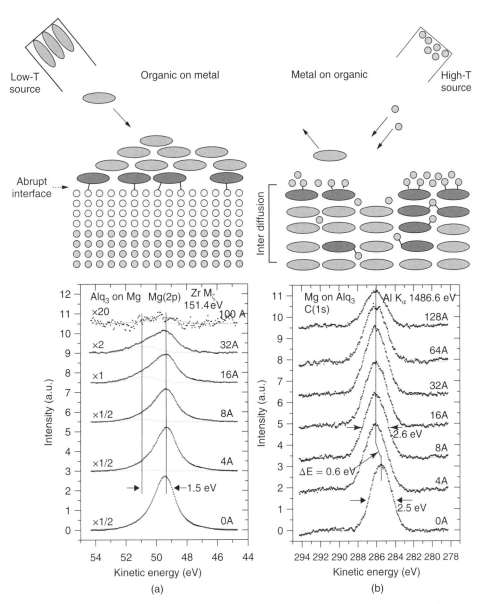

Figure 15.5 *(a) Low-temperature deposition of* Alq_3 *on Mg (top) and Mg 1s core level versus* Alq_3 *coverage. Note the reacted Mg component (dashed) localized at the buried interface. (b) High-temperature deposition on Mg on* Alq_3 *(top) and C 1s core level versus Mg coverage. (Shen, C. and Kahn, A. 2001 [9]. Reproduced with permission of American Institute of Physics.)*

15.4   Self-Consistent Electrostatic Calculations

Independent of the physical mechanism to account for E_F stabilization or pinning in a narrow range of energies, self-consistent electrostatic calculations can gauge the effect of charge states on band bending [10]. In general, both acceptor and donor states can be present in the semiconductor band gap. Their charge states will depend on the E_F position relative to their energy level positions. Hence their total charge density will depend on: (i) the donors and acceptors present, (ii) their densities, and (iii) their relative positions. [See Track II T15.6 Self-Consistent Electrostatic Calculations: Metals on GaAs.] The charges present below the interface, for example, 10 nm, and the countercharges in the metal set up a dipole that alters Φ_{SB} otherwise determined by Φ_M and χ_{SC}.

The self-consistent process involves calculating: first, the charge on the donors/acceptors for E_F at the position determined by Φ_M, then the dipole induced by this charge, the new E_F position relative to the donors/acceptors and, after iterating, the final E_F position with no further changes in donor/acceptor occupancy. One can then construct plots of Φ_{SB} versus Φ_M for n- and p-type semiconductors of different densities along with negative and positive *exchange-correlation interaction potential U* ordering (positive for acceptor above donor, negative for donor above acceptor). The resultant Φ_{SB} versus Φ_M plots exhibit characteristic dependences on the detailed energies, densities, and ordering of the electrically-active defects. These "fingerprints" of dipole formation are independent of the various physical mechanisms proposed to account for E_F stabilization.

Case Study: Barrier Height versus Defect Density

Consider an abrupt metal–semiconductor junction with metal work function $\Phi_M = 5.4$ eV, semiconductor bulk Fermi level position $E_C - E_F = 0.1$ eV, electron affinity $\chi = 4.2$ eV, and relative dielectric constant $\varepsilon = 10$. A donor level density $\sigma = 10^{13}$ cm^{-2} resides 0.9 eV below E_C. Assume that $E_C - E_F = \Phi_M - \chi = 5.4 - 4.2 = 1.2$ eV at the interface. The density of positively charged donors is given by $\sigma^+ = \sigma_0/[1 + \exp[(E_F - E_d)/k_B T]]$ (assuming no degeneracy for simplicity) $= 10^{13}/[1 + \exp[(0.9 - 1.1)/0.0259$ eV]] $\simeq 10^{13}$ cm^{-2}, that is, the donors are fully charged at room temperature. Assuming that the positively charged donor and its image charge on the metal side are 1.0 nm apart, calculate the potential drop across the interface due to the interface dipole and the corresponding interface Fermi level.

Answer:

$$q\Delta V = q^2 \frac{\sigma_a}{\varepsilon} d = \frac{1.6 \times 10^{-19} \times 10^{13}}{10 \times 8.85 \times 10^{-14}} \times 1 \times 10^{-7} = 0.18 \text{ eV} \qquad (15.13)$$

The interface Fermi level is then raised by 0.18 eV toward E_C. $E_C - E_F = 1.2 - 0.18 = 1.02$ eV and $E_F - E_d = 1.02 - 0.90 = 0.12$ eV $>> k_B T$ so that the donors remain nearly fully charged. Thus E_F stabilizes in the vicinity of the defect level.

For work functions $\Phi_M < \chi + 0.9$, the donors are not charged, the dipole contribution is zero, and the Schottky barrier behaves classically. For $\Phi_M - \chi \sim E_C - E_d$, one must take the occupancy of the donor level into account. For Φ_M even higher than 5.4 eV, dipole contribution ΔV remains fixed and $E_C - E_F = \Phi_M - \chi + \Delta V$ increases proportionally to Φ_M but offset by ΔV. See, for example, Figure T15.6(c).

Similarly, increasing defect density increases the dipole contribution. Figure 15.6 illustrates the Fermi level dependence on defect density for an acceptor level located 2.0 eV below E_C for $\Phi_M = 5.1$ eV, $E_C - E_F = 0.15$ eV in the bulk, and $\varepsilon = 8.2$. See Problem 15.8.

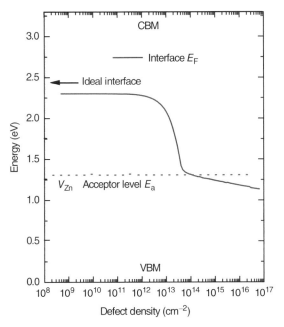

Figure 15.6 E_F position in the band gap versus acceptor defect density. E_F is "pinned" near E_a for defect densities above 10^{12} cm^{-2}. (Brillson 2010. Reproduced with permission of Wiley.)

The self-consistent formalism enables several Schottky barrier calculations. For a given Φ_{SB} and Φ_M, one can obtain $q\Delta V$, then use E_F to solve for E_a and σ. For a given E_A and σ, one can iterate E_F to obtain Φ_{SB}. In general, the Fermi level position, defect charge occupancy, and dipole contribution must be established self-consistently, for example, by computer iteration.

15.5   Experimental Schottky Barriers

Since the early 1950s, researchers have compiled an enormous array of Schottky barrier measurements for most common semiconductors and prepared under many different conditions. These results provide a resource for researchers and technologists to select particular combinations of metals and semiconductors.

15.5.1   Metals on Si and Ge

Initially, Schottky barrier studies focused on the elemental semiconductor Si, given its importance for microelectronics.

15.5.1.1   Clean-Cleaved Si

For metals on clean-cleaved Si, there is good agreement between Φ_{SB} measured by $I - V$ and IPS values, while $C - V$ measurements yielded values consistently $0.1 - 0.3$ eV higher. [See Track II Table T15.7A Barrier Heights: Clean, Cleaved Si.] Higher $C - V$ versus $I - V$ values are common due to inhomogeneities within individual diodes. Nevertheless, Φ_{SB} values extended only ~0.2 eV compared with Φ_M values ranging over ~1 eV.

15.5.1.2   Etched and Oxidized Si

For chemically-etched and oxidized Si as well as transition metal silicides on Si, the Φ_{SB} range increases to ~$0.4 - 0.5$ eV. As expected, higher Φ_M yield larger Φ_{SB} values. There is a wide Φ_{SB} variation for the same metal on Si with different surface treatments, for example, $\Phi_{SB} = 0.5 - 0.7$ eV for Al on etched and oxidized Si. Furthermore, these contacts exhibit *aging* effects, that is, Φ_{SB} changes with annealing or with exposure to different gas ambients. Such Φ_{SB} *aging* can be due to interface chemical interdiffusion or reaction, diffusion of O or H through the metal film or into their periphery, or chemical reaction involving adsorbates prior to metallization. These aging effects are likely to introduce inhomogeneity and multiple barrier heights within the same contact, leading to a lack of reproducibility and depending sensitively on how the interface was prepared.

N-type and p-type Si Φ_{SB} exhibit complementary values within the Si band gap; that is, $\phi_B{}^n - \phi_B{}^p \sim E_G$ (1.1 eV), which indicates a common E_F stabilization energy. [See Track II Table T15.7B.] (Note that $\phi_B{}^n$ and $\phi_B{}^p$ by definition have opposite signs.) In contrast to Si, there are very few results for contacts to the elemental semiconductor Ge. For example, $\Phi_{SB} = 0.45$ eV for Au and 0.48 eV for Al, nearly the same despite the large Φ_M difference.

15.5.2   Metals on III-V Compound Semiconductors

Researchers have studied Schottky barrier formation on III-V compound semiconductors extensively since the mid-1970s, to understand not only E_F pinning, commonly observed for GaAs, but also the nature of localized states at the GaAs–oxide interface. A primary driver of this work was to develop GaAs and InP for gating current in MOS devices as higher mobility alternatives to Si as well as for optoelectronics.

To obtain clean metal–III-V compound interfaces, researchers typically cleaved semiconductor surfaces in UHV before depositing metal and monitored the interface properties using the techniques described in earlier chapters. Chemical, ion bombardment, and plasma etching can also produce clean surfaces but usually also introduce new lattice defects.

15.5.2.1 GaAs(110) Pinned Surfaces

GaAs single crystals grown by the Czochralski crucible method can be cleaved along the (110) direction to expose clean surfaces. XPS band bending studies showed that (110) surfaces without cleavage steps are free of surface states in the GaAs band gap. Metal deposition on this surface produces rapid, that is, submonolayer coverage, E_F movement from E_C or E_V to characteristic energy levels near mid-gap. [See Track II T15.3. Unified Defect Model.] Transport measurements for metals on clean GaAs (110) and (100) surfaces show only a weak dependence of Φ_{SB} on Φ_M, with Φ_{SB} changes of only $0.2 - 0.25$ eV over a wide range of Φ_M, regardless of orientation. [See Track II Table T15.8. I-V and C-V Barrier Heights: GaAs (110) and (100) Surfaces.]

15.5.2.2 Oxidized GaAs(110) Surfaces

Metal contacts to oxidized GaAs surfaces prepared by air-cleavage or chemical etching exhibit a larger Φ_{SB} range, extending between ~0.65 and 1 eV. This larger range could be due to an oxidized intermediate layer, whose composition would depend on the nature of any chemical reaction that occurred between metal and oxide. For example, Al, Ti, and Cr react with oxides whereas Au produces little or no chemical interaction.

15.5.2.3 InP(110) Unpinned Schottky Barriers

Clean, cleaved InP (110) surfaces are also free of surface states in the band gap. Unlike the bulk GaAs(110) surface, InP (110) exhibits monolayer-scale E_F movement and a wide range of energy positions in the band gap. Figure 15.7 illustrates this E_F movement for

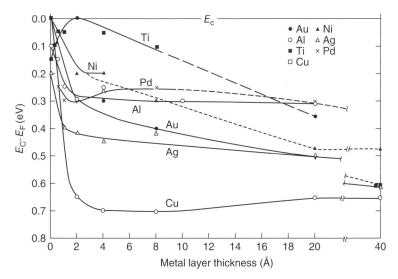

Figure 15.7 *SXPS measurements of E_F versus metal coverage on UHV-cleaved n-type InP (110). E_F moves to a 0.5 eV range of energies with over 20 Å thicknesses of Au, Al, Ti, Cu, Ni, Ag, and Pd. (Brillson 1982 [11]. Reproduced with permission of American Institute of Physics.)*

various metals on n-type InP (110) surfaces. SXPS measurements show different E_F movements and stabilization energies for each metal, extending over more than a few monolayer thicknesses and over $\sim$0.5 eV. Transport measurements of n-InP (110)–metal junctions by multiple groups show a similar $\sim$0.5 eV range of stabilization energies, ranging from 0.6 to 0.22 or ohmic, reflecting the difficulty of measuring barriers of only a few hundred meV. As with the semiconductors in Figure 4.6, there is a transition in interface heat of reaction between reactive versus unreactive metals on InP [12].

15.5.2.4   *GaN Schottky Barriers*

A leading III-V compound semiconductor for optoelectronics is the wide band gap semiconductor GaN with a Φ_{SB} range from 0.58 to 1.65 eV over a Φ_M range of 1.4 eV, increasing with increasing Φ_M. [See Track II Table T15.9. Schottky Barrier Heights for Elemental Metals on GaN.] As with other semiconductors, there is considerable scatter for the same metal on GaN prepared by different methods and measured by different methods. In addition, chemical treatments that reduce or eliminate native oxides on GaN can reduce Φ_{SB} with otherwise inert metals such as Au/Pt.

Point defects near dislocations can act as donors that decrease depletion widths, substantially reducing GaN Φ_{SB}. These barriers also exhibit an orientation dependence, increasing for (0001) surfaces due to piezoelectric polarization-induced surface charges.

15.5.2.5   *Other III-V Binary and Ternary Semiconductors*

Φ_{SB} results for other binary III-V compound semiconductors include: (i) Φ^n_{SB} ranging from 1.0 to 1.4 eV for GaP, (ii) Φ^n_{SB} limited to $\sim$0.55 eV for GaSb, (iii) $a \sim 0.6$ eV Φ^n_{SB} range for InAs with a dependence on air exposure, and (iv) $\Phi^n_{SB}(Au) = 0.17$ eV and $\Phi^n_{SB}(Al) = 0.18$ eV for InSb.

For ternary alloys such as $Al_xGa_{1-x}As(100)$, some evidence exists for a *Common Anion Rule* such that the common anion As determines Φ^p_{SB} regardless of the alloy x. Also $E_G - \Phi^n_{SB}$ for $Au/GaAs_{1-x}P_x = 0.55$ eV, independent of As content. In contrast, UHV results for $Ga_xIn_{1-x}As(100)$ are in direct contradiction with the earlier air-exposed results. Likewise, $E_G - \Phi^n_{SB}$ for $Au - Al_xGa_{1-x}Sb$ contacts increase with Al content, and Φ^p_{SB} increases monotonically with x for $Au/p-Ga_yIn_{1-y}As_{1-x}P_x$ contacts such that Φ^p_{SB}/E_G is approximately constant.

15.5.3   Metals on II-VI Compound Semiconductors

Metals on II-VI compounds exhibit large barrier height ranges with n-type barriers increasing with Φ_M. II-VI oxide surfaces can react with metals with Φ_{SB} dependent on the nature of the reaction. Very limited data for III-VI compounds, such as PbTe displays Schottky-like behavior, whereas GaSe exhibits Schottky-like barriers for unreactive metals and "pinned" E_F positions for reactive metals.

15.5.3.1   *ZnO Schottky Barriers*

As with GaN, the wide band gap semiconductor ZnO is a prime candidate for next generation opto- and microelectronics, including blue and UV LEDs, lasers, transparent conducting oxide electronics, spintronics, biosensors, and nanodevices. Among its attributes, ZnO has a large exciton binding energy (60 meV) that inhibits thermal activation and enhances light emission at room temperature. It readily forms nanostructures that can emit light and sense charge transfer efficiently. It is also radiation hard, relatively abundant in nature, easy to etch chemically, and biocompatible. As with other semiconductors, a major challenge to fabricating high quality ZnO devices is Schottky barrier control.

Besides adsorbates and interface chemical bonding, surface and interface native point defects can influence Φ_{SB} strongly [13]. Furthermore, the ability to dope ZnO p-type remains a significant challenge. The wide range of Φ_{SB} enables ZnO to serve as a test bed for probing physical phenomena of semiconductors in general. [See Track II. Table T15.10. Schottky Barriers for Metals on ZnO.] For example, Au Φ_{SB} on ZnO can range from 0 to 1.2 eV, depending on the crystal, the surface preparation, and the conditions under which the contact is formed.

15.5.3.2   *Effect of Native Defects*

Besides crystalline quality and impurity content, native point defects distributed near metal–semiconductor interfaces can have a major impact on Φ_{SB} for ZnO and other semiconductors. Such native points can be electrically-active, serving to change effective carrier densities locally, particularly since they can segregate to surfaces and interfaces. See Section 11.10.6. Techniques such as *remote oxygen plasma* (ROP) to reduce this near-surface segregation have a pronounced effect on Φ_{SB}. Thus the commonly observed "green" luminescence defects that act as donors and segregate to free ZnO surfaces decrease with ROP treatment, the effect of which is to change the contact properties from ohmic to strongly rectifying in proportion to the extent that these defects are removed [14]. Similarly, crystals with decreasing bulk defect densities exhibit corresponding increases in Au $I - V$ Φ_{SB}. The direct correlation between reverse current leakage and defect density emphasizes the importance of low bulk defect crystals for Schottky barrier studies. Furthermore, the ability to reduce or remove defects that segregate to surfaces provides a tool for interface engineering.

15.5.3.3   *Effect of Polarity*

Schottky barriers on clean Zn- versus O-polar surfaces exhibit a pronounced difference in Φ_{SB} for the same metal. Contacts to Zn-face ZnO have higher Φ_{SB}, lower subsurface deep level defects, and lower subsurface free carrier concentrations than O-face diodes of the same ZnO crystal [15]. Correlation of these Φ_{SB} with both carrier densities and densities within the semiconductor's depletion region indicates that segregation of native point defects to metal–semiconductor interfaces can have major interface electronic effects.

15.5.4   Other Compound Semiconductors

IV-IV semiconductors represent another important class of electronic materials. The wide bandgap semiconductor SiC has been studied extensively due to its high breakdown voltage and high temperature performance for electric vehicles and energy transmission. Φ_{SB} can vary widely for this material, not only due to different surface preparation but also because SiC possesses multiple crystal polytypes with different band gaps as well as orientation-dependent polarities. Most commonly used is the 4H-SiC polytype with either n- and p-type 4H-SiC. [See Track II Table T15.11. SiC Schottky Barriers.] Φ_{SB} variations of over 1 eV are measured, depending on the metal Φ_M and the polarity.

Considerably fewer results are available for the IV-IV semiconductor diamond, in large measure because of the difficulty in doping to obtain conductive material. Surface chemical termination, orientation, and interface chemical bonding can all influence diamond Schottky barriers.

IV-VI compound semiconductors, such as PbTe and PbSe, are useful as near-IR photodetectors and sensors. Low Φ_M contacts produce ohmic contacts to n-PbTe and rectifying contacts to p-type, consistent with a Schottky–Mott model. The III-VI compound Ga_2O_3 shows significant promise for high voltage, high power and high speed applications. UHV Schottky barrier studies are now just beginning.

15.5.5   Compound Semiconductor Summary

For most if not all semiconductors, localized states clearly affect Schottky barrier heights. Specific surfaces, surface treatments, chemical reactions, and annealing all affect Φ_{SB}. Furthermore, there is no simple theoretical model that predicts Schottky barrier heights *a prio*ri.

15.6   Interface Barrier Height Engineering

Schottky barrier heights can be controlled on both a macroscopic and an atomic scale. Since no predictive model for Schottky barrier formation is yet available, technologists have taken a pragmatic approach, developing mesoscopic and microscopic techniques to obtain desirable electric properties in real devices. Now, with the refinement of epitaxial growth techniques to fabricate microelectronic and optoelectronic devices, new interface passivation and control methods have emerged to engineer metal–semiconductor contacts on an atomic scale.

15.6.1   Macroscopic Methods

Figure 15.8 illustrates four macroscopic approaches to barrier control. Figure 15.8a shows the commonly observed $\Phi_{SB} = 0.8$ eV thermionic emission barrier for metals on GaAs. Figure 15.8b illustrates increased field emission due to increased dopant concentration at the GaAs interface. The enhanced tunneling lowers the effective barrier height. With graded composition from n-GaAs to n-InAs, Figure 15.8c shows how E_F at the interface now rises above E_C, producing a low resistance barrier. Instead of a graded interface, Figure 15.8d illustrates an interface between metal and GaAs through an abrupt intermediate $In_xGa_{1-x}As$ layer. This intermediate layer results in two small barriers between metal and GaAs, substantially reducing the effective barrier overall, given the exponential dependence of current on barrier height. Finally, Figure 15.8e illustrates how graded composition from GaAs to AlGaAs at the interface results in a larger Φ_{SB} due to the larger AlGaAs versus GaAs band gap.

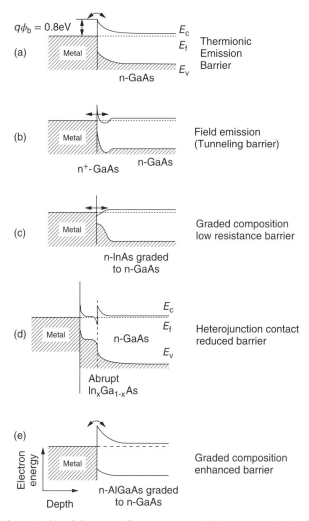

Figure 15.8 *Schematic band diagrams illustrating a metal–GaAs junction before (a) and after the use of macroscopic methods to decrease (b–d) or increase (e) the effective barrier height [16,17]. (Sebestyen 1982 [23]. Reproduced with permission of Elsevier.)*

15.6.2 Defect Formation

The introduction of new localized states in the semiconductor's near-surface region can produce disorder-induced conduction. [See Track II T15.12 Localized State Transport.] Bombardment with ionized species or, on a much coarser scale, mechanical scraping can produce such disordered regions. These disordered regions can be abrupt or graded such that disorder decreases with depth into the semiconductor. The additional localized states at such disordered semiconductor contacts provide pathways for charge to tunnel or hop between sites to reach from the metal to the disorder-free semiconductor bulk. The defects produced at metal–ZnO interfaces, for example, lower barrier heights due to localized states created by chemical reaction.

15.6.3   Thermally-Induced Phase Formation

On a microscopic scale, there has been considerable research on the metallurgy of elemental metals with GaAs for ohmic or Schottky contacts. Annealing these contacts at selected temperatures leads to a variety of thermally-induced phases that depend on the thermodynamics of the specific metal. [See Track II Table T15.13. Barrier Heights and Chemical Interactions for Elemental Metal Contacts to GaAs.] In general, new phases form with increasing temperatures and annealing time, Φ_{SB} changes with the phases that form, different phases can have different stoichiometries and morphologies, and particular morphologies can produce ohmic or rectifying contacts in addition to macroscopic decomposition of the metallization.

Multiple metal contacts to semiconductors are useful to control near-interface diffusion and alloy formation. These contacts can also serve to: (i) increase metal–semiconductor adhesion, (ii) maintain chemical stability at elevated temperatures during device operation, and (iii) minimize interdiffusion of constituents that might introduce unwanted defects or dopants. [See Track II Table T15.14 Barrier Heights and Chemical Interactions for Multi-Metal Contacts to GaAs.]

15.6.4   Interdiffused Ohmic Contacts

Metallizations for producing ohmic contacts to III-V and mixed III-V compound semiconductors have been a major focus of semiconductor device research. The contact resistance inside high mobility devices contributes to the *RC time constant* that limits the *high frequency cutoff* f_T of the device. Contact resistance becomes ever more critical as circuit features shrink in size to increase device speeds. Hence, the sizes of contact features decrease proportionally, increasing resistances in general.

Thermal annealing is used to produce interdiffusion that increases adhesion and produces high densities of dopant atoms or native point defects, which in turn promote tunneling and hopping through initially rectifying or blocking contacts. Metallic contacts with more than one element are used to lower the melting temperature and limit the extent of diffusion.[See Track II Table T15.15 Ohmic Contact Technology for III-V and Mixed III-V Compound Semiconductors and Track II Table T15.16 Very Low Resistance Contacts to GaAs for examples of specific composition, temperature, and time recipes to achieve contact resistivities ρ_C that extend down to the 10^{-7} $\Omega\,cm^2$ range.]

15.7   Atomic-Scale Control

Just as interlayers can modify band offsets at heterojunctions, interface layers at the atomic scale can modify Schottky barriers. These atomic-scale techniques include: (i) reactive metals, (ii) less-reactive buffer layers, (iii) wet-chemical cleaning or bonding, and (iv) gas–surface interactions.

15.7.1 Reactive Metal Interlayers

Reactive metal interlayers less than a monolayer thick between inert metals and either III-V or II-VI compound semiconductors can alter Φ_{SB} dramatically. Figure 15.9 illustrates how increasing Al interlayers below a single monolayer in thickness can convert a 150 Å Au/CdS $(10\bar{1}0)$ rectifying contact with $\Phi_{SB} = 0.8$ eV into an ohmic contact with only 2 Å of Al. A similar effect is observed for Al interlayers at other II-V interfaces. At the Au/II-VI interface, reactive metals retard Cd outdiffusion, leading to excess cations at the interface, increasingly Cd-rich interfaces, and tunneling through the much narrower depletion region.

For III-V compounds, an Al interlayer acts to trap outdiffusing anions at melt-grown GaAs, changing the sub-surface stoichiometry and decreasing Φ_{SB} by $0.1 - 0.2$ eV. Similarly Yb, Sm, and Sm-Al interlayers on MBE-grown GaAs (100) also change Φ_{SB} by nearly 0.2 eV. Note the high precision of Au deposition using a quartz crystal oscillator (QCO) with a wire evaporation source in Figure 15.9.

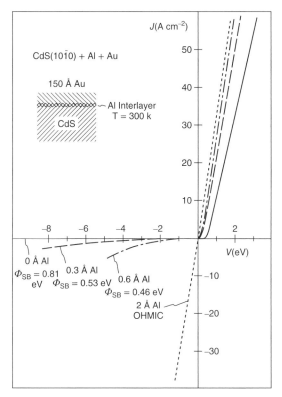

Figure 15.9 *Atomic-scale control of macroscopic device characteristics. Room temperature I − V characteristics of 150 Å thick Au diodes on UHV-cleaved CdS $(10\bar{1}0)$. With increasing submonolayer Al interlayer thickness, Φ_{SB} decreases steadily from 0.8 eV to ohmic. The inset shows a cross-sectional schematic diagram of the interlayer structure. (Brucker and Brillson 1981 [18]. Reproduced with permission of American Institute of Physics.)*

15.7.2   Molecular Buffer Layers

Less reactive interlayers include: H_2S on InP(110) that lower Φ_{SB} from 0.5 to <0.3 eV, believed due to S indiffusion and surface doping that increases tunneling through the barrier. Xe buffer layers deposited at cryogenic temperatures can reduce interface reaction and diffusion associated with metal condensation and clustering on semiconductor surfaces. For cleaved GaAs, the resultant Φ_{SB} vary over several tenths of a volt compared to "pinned" E_F positions without the Xe buffer layers. These layers produce even larger effects at metal–InP and ZnSe interfaces.

15.7.3   Semiconductor Interlayers

Nanometer thicknesses of S, Se, and Te on MBE-grown GaAs (100) can produce >0.6 eV changes. Metals that react strongly with these chalcogens exhibit low Φ_{SB} compared with metals that do not react. Doped Ge or Si interlayers can alter E_F in the MBE-grown GaAs (100) band gap. Depending on whether they are n- or p-doped, these interlayers stabilize E_F at gap levels corresponding to their semiconductor work function either near the semiconductors E_C or E_V, respectively. In other words, the Ge or Si interface layers move E_F according to their effective work functions, which decrease or increase according to whether they are doped n- or p-type. In general, these results show that the work functions of the Si or Ge deposited layers determine the E_F stabilization energies and that E_F is not "pinned" at these MBE-grown GaAs interfaces. [See Track II T15.17. Band Bending Control with Semiconductor Interlayers.]

15.7.4   Wet Chemical Treatments

Wet chemical treatments can passivate semiconductor surfaces to reduce surface and interface state densities.

15.7.4.1   *Photochemical Washing*

Perhaps the gentlest surface treatment introduced to date is photochemical washing with H_2O. Bathing n- or p-type GaAs in 18 MΩ deionized H_2O while under band gap illumination is believed to dissolve and wash away surface As and As oxides. The resultant increase in band gap luminescence corresponds to decreasing surface state density by orders of magnitude. [See Track II T15.18 GaAs Photochemical Washing.]

15.7.4.2   *Inorganic Sulfides, Thermal Oxides, and Hydrogen*

Inorganic sulfides such as NaS and NaOH can reduce *surface recombination velocity* (SRV) by more than six orders of magnitude while passivating surfaces to prevent re-oxidation. Orders of magnitude SRV reductions are found for Si, GaAs, and InGaAs. [See Track II T15.19 Si, GaAs, and $In_{0.53}Ga_{0.47}As$ Recombination Velocities versus Surface Treatment.] For GaAs, the formation of Ga chalcogenides is believed to rehybridize the otherwise dangling bonds associated with mid-gap states so that their energies are no longer within the band gap. In addition, thermal oxides and hydrogen etch treatments with such as HF eliminate surface roughness and produce a monohydride passivation layer.

15.7.5   Crystal Growth

Progress in semiconductor crystal growth has enabled several new approaches to controlling Schottky barriers. Research tools now available include the control of stoichiometry, crystal orientation, surface vicinal misorientation, and the epitaxical growth of binary
alloys. These atomic scale techniques also allow researchers to demonstrate the dominant
effect of extrinsic structural features on interface electronic properties.

15.7.5.1   Stoichiometry and Defect Control

Epitaxial layer-by-layer crystal growth permits control of stoichiometry, purity, point defect
and dislocation density, precipitates, and their associated features. In contrast to GaAs
grown As-rich from the melt to reduce dislocations, MBE-grown GaAs is stoichiometric.
This stoichiometric GaAs exhibits greatly enhanced Φ_{SB} control. Figure 15.10 shows that
the range of E_F stabilization is much wider, with Φ_{SB} extending from 0.18 to 0.92 eV – over
half the GaAs bandgap. Internal photoemission spectroscopy measurements confirm this
wide range. By contrast, the range of Φ_{SB} control for melt-grown GaAs is less than 0.2 eV,
apparently limited by defect levels within the band gap. [See Track II T15.3.]

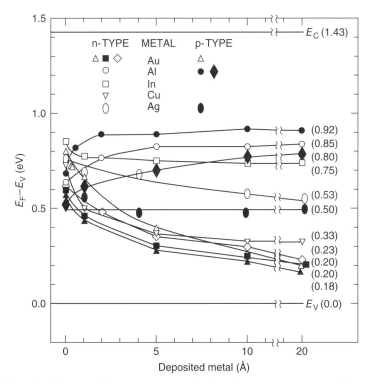

Figure 15.10   *Qualitative difference in metal-induced E_F movements within the GaAs
bandgap as measured by SXPS. Unlike metals on UHV-cleaved GaAs(110), metals on As-de-
capped, MBE-grown GaAs (100) move E_F over several monolayers to a wide range of energies
with matching n- and p-type values. (Brillson 1988 [19]. Reproduced with permission of American Institute of Physics.)*

15.7.5.2   Vicinality

Since defect densities in MBE-grown GaAs are very low, E_F can move over a wide energy range. As a result, metal interfaces with MBE-grown GaAs are a model system to test the effect of different interface states.

Vicinal surfaces with well-defined misorientations can provide controlled densities of interface states. Step edges have dangling bonds extending out from the plane with densities determined by the average distance between steps, which are in turn determined by the angle of misorientation. See, for example, Figure 12.3. Based on the self-consistent electrostatic model presented in Section 15.4, Figure 15.11a shows Φ_{SB} versus Φ_M for mid-gap state densities of defects at the energies shown in the insert. Also shown are experimental Φ_{SB} values for two metals, Al and Au, obtained from SXPS $E_F - E_V$ measurements for a GaAs (100) surface tilted 2° toward different misorientations. The vicinal angle and the direction of misorientation determine the density of As dangling bond sites. CLS measurements provided the gap states at the energies pictured in the inserts.

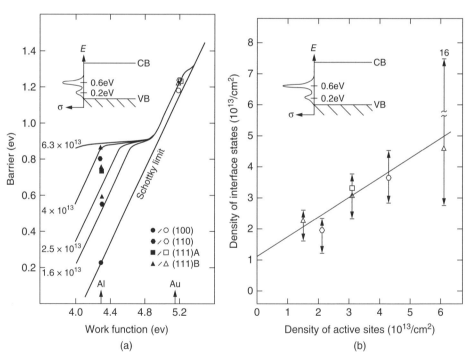

Figure 15.11   (a) Self-consistent analysis of Φ_{SB} versus Φ_M and (b) interface state densities versus active structural sites for Al (closed symbols) and Au (open symbols) on vicinal GaAs (100) surfaces. Φ_{SB} increases monotonically with misorientation angle and direction. The family of density curves for the 0.6 eV and (constant density) 0.2 eV states pictured in the insets yield the interface state density values shown in (b) [21]. (Chang et al. 1990 [20]. Reproduced with permission of American Physical Society.)

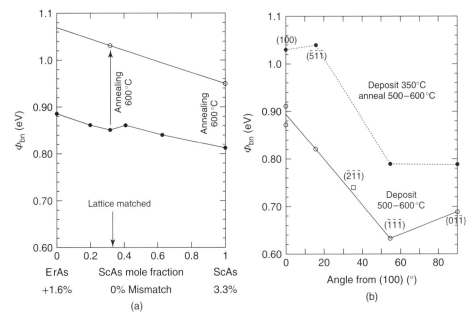

Figure 15.12 Φ_{SB} *of* $Sc_xEr_{1-x}As$ *metal alloys on GaAs (100) versus (a) lattice match and annealing and (b) interface misorientation and annealing. The weak dependence on lattice mismatch contrasts with the strong dependence on misorientation and annealing. [22]. (Palmstrøm 1990 [22]. Reproduced with permission of American Institute of Physics.)*

Figure 15.11b shows the density of interface states determined theoretically for the gap state energies in the insert versus active structural sites due to misorientation angle. The linear correspondence indicates a one-to-one dependence of defect states on active step sites and a direct correlation between dangling bond state densities and energies with Schottky barrier heights.

15.7.5.3   *Metal Epitaxy and Strain*

Epitaxial growth of binary alloys on compound semiconductors can distinguish the effects on Schottky barriers of strain due to lattice mismatch from those due to misorientation. The semi-metallic binary alloy $Sc_xEr_{1-x}As$ on GaAs ($0 \leq x \leq 1$) provides a system to test how Φ_{SB} changes with strain due to lattice mismatch from +1.6% compressive (ErAs) to −3.3% tensile (ScAs). Figure 15.12a shows relatively small Φ_{SB} changes over this large range of strain, even after annealing.

In contrast, a 90° range of orientations includes different densities of {100} and {111} ledges and steps. For these interfaces, Figure 15.12b exhibits much larger Φ_{SB} changes with crystal orientation. Similar large variations are also evident after annealing. Hence active electronic sites rather than strain are the dominating factor. As with vicinal surfaces, one can vary Φ_{SB} over relatively large energy ranges for the same metal–semiconductor contact by controlling interface structure at the atomic scale.

15.8   Summary

Key results presented in this chapter are: (i) there are several intrinsic and extrinsic mechanisms capable of producing localized electronic states and interface dipoles at metal–semiconductor interfaces. In turn, these can strongly affect Schottky barrier formation. (ii) Several different physical mechanisms permit substantial control over Schottky barriers. (iii) Considerable evidence now exists for the role of localized bonding and composition in determining electronic structure. (iv) Even wider Schottky barrier variations can be anticipated at the microscopic domain level compared with their macroscopic averages. (v) Combinations of UHV and epitaxial techniques are now available to attain high reproducibility and control of Schottky barriers at the atomic scale.

15.9   Problems

1. Studies of metal–semiconductor interfaces require clean semiconductor surfaces for metal deposition. Describe four methods of obtaining clean semiconductor surfaces prior to metal deposition and the possible effects these methods could have on the mechanisms of Fermi level stabilization.

2. Calculate S and CNL for n-type InP given $\Phi_{SB}^{N}(Au) = 0.6$ eV, and $\Phi_{SB}^{N}(Al) = 0.3$ eV.

3. According to the S and CNL in Problem 2, what would be the n-type barrier height for Cu?

4. For a Si(111) surface and a density of interface states equal to one electron per atom distributed across the band gap, calculate S. Assume depth $\delta = 1.5$ Å.

5. For Au and In on melt-grown GaAs with barriers $\Phi_{SB}^{N}(Au) = 0.9$ eV and $\Phi_{SB}^{N}(In) = 0.65$ eV, calculate their respective dipoles with the CNL given in Table 15.1 and $S = 0.1$.

6. Line charges within misfit dislocations at a semiconductor heterojunction have surrounding cylindrically symmetric carrier depletion regions. These depletion regions may limit transport through a metal contact to this heterojunction, depending on their spacing. Poisson's equation in cylindrical coordinates yields (Woodall *et al.* (1983) *Phys. Rev. Lett.*, **51**. 1783) $V_D = q\Phi(r_0) = qNr_s^2/2\varepsilon[\ln(r_s/r_0) - 0.5]$ where V_D is the potential at the dislocation core, N is the semiconductor carrier concentration, r_s is the radius of the depletion boundary around the dislocation, and r_0 is the effective radius of the dislocation.

   (a) Calculate r_s for GaAs with $N = 2 \times 10^{16}$ cm^{-3} and $V_D = 0.8$ eV.

   (b) Calculate r_0 for this potential and doping.

   (c) How important is this effect for a dislocation density of 6×10^{10} cm^{-2}? For 10^8 cm^{-2}?

   (d) Is it possible to increase the bulk carrier density to obtain barriers equivalent to reducing dislocation density of 6×10^{10} cm^{-2} to 10^8 cm^{-2}?

7. Defects at metal–semiconductor interfaces can introduce excess free charge that can change the effective barrier height by tunneling through a narrowed depletion region. (a) For an Au-InP diode with $\Phi_{SB} = 0.5$ eV, calculate the probability for electrons at E_{FM}, the Fermi level of the metal, to tunnel under reverse bias $V_R = 1$ V.

(b) How much does this tunneling probability increase due to a change in n-type carrier concentration from $N_D = 10^{15}$ cm^{-3} to $N_D = 5 \times 10^{17}$ cm^{-3}? The InP electron effective mass $m^*/m_0 = 0.077$. Hint: Approximate the reverse-biased depletion region by a triangular barrier, for which the transmission probability is given by $T = \exp(4E_B^{3/2}\sqrt{2m^*}/3q\hbar\,\mathscr{E})$, where $E_B = E_C - E_{FM}$ at the interface and the base width is $\sim \frac{1}{3}W$, the depletion width.

8. Calculate the dipole contribution to the Schottky barrier height of GaAs for a filled surface acceptor density $\sigma_a^- = 10^{13}$ cm^{-2} located 10 Å below the surface.

9. At abrupt Au/n-ZnO ($N_D = 10^{16}$ cm^{-3}) interfaces, there are Zn vacancies, which act as compensating acceptors. The conduction band density of states $N_C = 4 \times 10^{18}$ cm^{-3} and $\varepsilon_{average}/\varepsilon_0 = 8.2$. The density of charged acceptors is: $\sigma^-(E_F) = \frac{\sigma_a}{1+2\exp[(E_F-E_a)/k_BT]}$, where E_a is the energy level of the acceptor (2.0 eV below CBM), E_F is the Fermi level, k_B is Boltzmann's constant, and T is the absolute temperature. Assuming the negatively charged acceptor and its image charge on the metal side are 1.0 nm apart, calculate the potential drop across the interface due to interface dipole and the corresponding interface Fermi level with (a) $\sigma_a = 10^{12}$ cm^{-2} and (b) $\sigma_a = 10^{13}$ cm^{-2}, respectively. Will the Fermi level be pinned around $\sigma_a = 10^{14}$ cm^{-2}? Why?

10. Calculate and plot barrier heights versus metal work function from $\Phi_M = 3$ to 6 eV for ZnO with $E_C - E_F = 0.15$ eV in the bulk and a donor level located 0.9 eV below the conduction band edge with densities 10^{12}, 10^{13}, and 10^{14} cm^{-2}.

11. Calculate and plot the interface Fermi level position in the ZnO band gap as a function of acceptor density for the Au–ZnO interface with $E_a = 2.0$ eV as shown in Figure 15.6.

12. Consider an intrinsic n-type ZnO crystal with defect levels due to oxygen vacancy donors at $E_V + 2.5$ eV and zinc vacancy acceptors at $E_C - 2.1$ eV. (a) Calculate the classical Schottky barrier height for Pt, Ir and Ta assuming no interface states. (b) Calculate the corresponding Schottky barriers assuming interface chemical reactions can create additional native point defects and/or interfacial layers. Use $\chi(\text{ZnO}) = 4.2$ eV.

13. Nanoscale contacts to nanostructures can be made with an electron microscope by focused ion beam or electron beam-induced deposition. Here the incident beam decomposes metal-bearing molecules at the surface. Pt is a commonly deposited metal decomposed on surfaces. For Pt deposited on ZnO in high vacuum, describe whether the expected Pt–ZnO junction is a Schottky barrier or an ohmic contact. Does this electrical behavior depend on the method of deposition?

14. Sketch a band diagram to illustrate an ohmic contact between a metal and n-type GaAs using InAs and a graded InGaAs intermediate layer.

15. For an ohmic contact to an AlGaN high electron mobility transistor, calculate the maximum contact resistivity to permit a cut-off frequency of 50 GHz. Assume an AlGaN thickness of 23 nm and a 30% Al composition. Neglect channel resistance.

References

1. Duke, C.B. and Mailhiiot, C. (1985) A microscopic model of metal-semiconductor contacts. *J. Vac. Sci. Technol. B*, **3**, 1170.

2. (a) Tejedor, C., Flores, F., and Louie, S.G. (1977) The metal-semiconductor interface: Si (111) and zincblende (110) junctions. *J. Phys. C*, **10**, 2163; (b) Flores, F. and Tejedor, C. (1977) On the formation of semiconductor interfaces. *J. Phys. C: Solid State Phys.*, **20**, 145.

3. (a) Tersoff, J. (1984) Schottky barrier heights and the continuum of gap states. *Phys. Rev. Lett.*, **52**, 465; (b) Tersoff, J. (1984) Theory of semiconductor heterojunctions: The role of quantum dipoles. *Phys. Rev. B*, **30**, 4875; (c) *Heterojunction Band Discontinuities: Physics and Device Applications*, Capasso, F. and G. Margaritondo, G. (eds) (1987) North-Holland, Amsterdam, p. **44**.

4. Kahn, A., Stiles, K., Mao, D., Horng, S.F. *et al.* (1989) in *Metallization and Metal-Semiconductor Interfaces*, NATO ASI Series B, Vol. 195 (ed. I.P. Batra) Plenum, New York, p. 163.

5. Spicer, W.E., Linda, I., Skeath, P., and Su, C.Y. (1980) Unified mechanism for Schottky-barrier and III-V oxide interface states. *Phys. Rev. Lett.*, **44**, 420.

6. Weber, E.R., Ennen, H., Kaufmann, V. *et al.* (1982) Identification of As_{Ga} antisites in plastically deformed GaAs. *J. Appl. Phys.*, **53**, 6140.

7. Brillson, L.J., Brucker, C.F., Katnani, A.D. *et al.* (1981) Chemical basis for InP-metal Schottky-barrier formation. *Appl. Phys. Lett.*, **38**, 784.

8. Kahn, A., Koch, N., and Gao, W. (2003) Electronic structure and electrical properties of interfaces between metals and π-conjugated molecular films. *J. Polym. Sci.*, **41**, 2529.

9. Shen, C. and Kahn, A. (2001) Electronic structure, diffusion, and p-doping at the $Au/F_{16}CuPc$ interface. *J. Appl. Phys.*, **90**, 4549.

10. Duke, C.B. and Mailhiot, C. (1985) A microscopic model of metal-semiconductor contacts. *J. Vac. Sci. Technol. B*, **3**, 1170.

11. Brillson, L.J., Brucker, C.F., Katnani, A.D. *et al.* (1982) Fermi-level pinning and chemical structure of InP-metal interfaces. *J. Vac. Sci. Technol.*, **21**, 564.

12. Williams, R.H., Montgomery, V., and Varma, R.R. (1978) Chemical effects in Schottky barrier formation. *J. Phys. C*, **11**, L735.

13. Brillson, L.J. and Lu, Y. (2011) ZnO Schottky barriers and ohmic contacts. *J. Appl. Phys./Appl. Phys. Rev*, **109**, 121301.

14. Brillson, L.J., Mosbacker, H.L., Hetzer, M.J. *et al.* (2007) Dominant effect of near-interface native point defects on ZnO Schottky barriers. *Appl. Phys. Lett.*, **90**, 102116.

15. Dong, Y., Fang, Z.-Q., Look, D.C. *et al.* (2008) Zn- and O-face polarity effects at ZnO surfaces and metal interfaces. *Appl. Phys. Lett.*, **93**, 072111.

16. Murakami, M., Kim, H.J., Shih, Y.-C. *et al.* (1989) Chemical effects in Schottky barrier formation. *Appl. Surf. Sci*, **41/42**, 195.

17. Sands, T., Palmstrøm, C.J., Harbison, J.P. *et al.* (1990) Stable and epitaxial metal/III-V semiconductor heterostructures. *Mater. Sci. Rep.*, **5**, 99.

18. Brucker, C.F. and Brillson, L.J. (1981) New method for control of Schottky-barrier height. *Appl. Phys. Lett.*, **39**, 67.

19. Brillson, L.J., Viturro, R.E., Shaw, J.L. *et al.* (1988) Unpinned Schottky barrier formation at metal-GaAs interfaces. *J. Vac. Sci. Technol.B*, **6**, 1263.

20. Chang, S., Brillson, L.J., Kime *et al* (1990) Orientation-dependent chemistry and Schottky-barrier formation at metal-GaAs interfaces. *Phys. Rev. Lett.*, **64**, 2551.

21. Chang, S., Vitomirov, I.M., Brillson, L.J. *et al.* (1991) Correlation of deep-level and chemically-active densities at vicinal GaAs(100)-Al interfaces. *Phys. Rev. B*, **44**, 1391.
22. Palmstrøm, C.J, Cheeks, T.L., Gilchrist, H.L., et al. (1990) in *Electronic, Optical and Device Properties of Layered Structures*, (eds J.R. Hayes, M.S. Hybertson, and E.R. Weber) Materials Research Society, Pittsburgh, p. 63.
23. Sebestyen, T. (1982) Models for ohmic contacts on graded crystalline or amorphous heterojunctions. *Solid-State Electron.*, **25**, 543.

Further Reading

Brillson, L.J. (1982) The structure and properties of metal–semiconductor interfaces. *Surf. Sci.Rep.* **2**, 123 (1982).
Brillson, L.J. (1992) Surfaces and interfaces: Atomic-scale structure, band bending, and band offsets. in *Handbook on Semiconductors, Vol. 1, Basic Properties of Semiconductors*, ed. P.T. Landsberg, Elsevier, New York, ch.7.
Brillson, L.J. (ed) (1993) *Contacts to Semiconductors*, Noyes Publications, Park Ridge, NJ.
Mönch, W. (2001) *Semiconductor Surfaces and Interfaces*, Springer-Verlag, Berlin.
Rhoderick, E.H. and Williams, R.H. (1988) *Metal-Semiconductor Contacts*, Oxford Science, Oxford.
Tung, R.T. (1992) Electron Transport at Metal-Semiconductor Interfaces: General Theory. *Phys. Rev.*, **45** (23), 13509–13523.

16

Next Generation Surfaces and Interfaces

16.1   Current Status

Over the past few decades, researchers have used the essential techniques and methods described in this book to develop our present understanding of electronic material surfaces and interfaces. Initial models of an atomically abrupt boundary have evolved into a more complex picture involving quantum physics, solid state chemistry, and materials science. Figure 16.1 illustrates this evolution, starting with (a) the abrupt metal–semiconductor interface commonly found in textbooks, (b) the presence of an additional dipole layer between metal and semiconductor that accommodates part or all of the potential difference between the two materials, and (c) the possibility of an extended interface on a scale of single atomic layers to tens of nm or more.

Both chemical reaction and interdiffusion can occur at this extended interface, resulting in physical properties unlike those of either metal or semiconductor. Chemical reactions can produce new interfacial phases with new dielectric properties or metallic alloys whose work function or electronegativity differs from those of the original metal. Interdiffusion of metals into the semiconductor can introduce impurities that act as dopants or recombination centers that can compensate carrier densities. Diffusion out of the semiconductor can introduce lattice defects, such as vacancies, that are electrically active and able to segregate near surfaces and interfaces, changing the carrier densities within the semiconductor's surface space charge region.

For a metal–semiconductor junction, all of these extended interface effects can alter the band bending and Schottky barrier formation. For a semiconductor heterojunction, similar dipole formation and interface chemical effects can occur, again introducing potential changes that can alter heterojunction band offsets. Even for the atomically-abrupt metal semiconductor interface, several mechanisms can contribute to charge transfer across the contact. Figure 16.2 illustrates three major mechanisms: (i) thermionic emission over the barrier set up by the semiconductor space charge region, (ii) tunneling through the barrier,

An Essential Guide to Electronic Material Surfaces and Interfaces, First Edition. Leonard J. Brillson.
© 2016 John Wiley & Sons, Ltd. Published 2016 by John Wiley & Sons, Ltd.
Companion Website: www.wiley.com/go/Brillson/

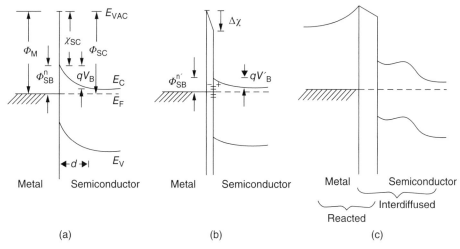

Figure 16.1 *Conceptual evolution of the metal–semiconductor interface from (a) abrupt junction to (b) atomically thin interface dipole plus band bending to (c) extended interface. (Brillson 2010. Reproduced with permission of Wiley.)*

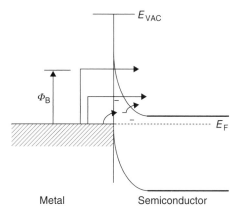

Figure 16.2 *Competing charge transfer mechanisms: thermionic emission, tunneling, and hopping transport through defects levels in the band gap. (Brillson 2010. Reproduced with permission of Wiley.)*

and (iii) hopping across the barrier through deep level defect states within the band bending region.

Besides the intrinsic properties of the metal (*e.g.*, Φ_M or X_M) and semiconductor (*e.g.*, E_G, χ, ε, n or p), the competition between these mechanisms depends on the temperature, the band bending profile, and any densities of deep level defects within the band bending region. With decreasing temperature, thermionic emission decreases. With decreasing barrier width, tunneling across the barrier increases. With increasing deep level densities, trap-assisted hopping through the barrier increases. With nm-scale barrier widths or high deep level densities, charge transport is high enough for the contact to become ohmic.

While the extended interface pictures in Figure 16.1c add complexity, the physical mechanisms that determine its structure and properties comprise a platform on which to develop new solid state electronic technology. We now examine several current device challenges that can be addressed with our current understanding. Beyond these are many emerging directions that depend strongly on electronic material surfaces and interfaces.

16.2 Current Device Challenges

Device research continues to focus both on improving the crystal perfection, doping, carrier mobility, and optical response of semiconductors, and designing new component architectures and their integration. As device dimensions shrink even further into the nanometer range, challenges involving electronic material surfaces and interfaces continue to increase.

For metal contacts to semiconductors, basic challenges are to achieve either: (i) low or ohmic barriers to minimize current loss, voltage drop, or heating at their interface or (ii) high barriers to increase carrier confinement or rectification. Low barrier examples include ohmic contacts for: (i) current injection to lasers, light emitting diodes, or solid-state displays, especially powered by low voltage sources, for example, batteries and (ii) low heat generation inside microelectronics such as computers or laptops. High barrier examples include: (i) large internal potential regions for solar cells to increase open circuit voltage, (ii) diode switches to improve high voltage rectification, especially for automotive applications and power transmission, (iii) transistor gate electrodes to suppress current leakage, and (iv) field effect transistors to improve detection efficiency.

For semiconductor heterojunctions, basic challenges are to design and grow lattice-matched combinations of semiconductors with desirable optical, electronic, and magnetic properties. These heterojunctions require band offsets to: (i) improve carrier confinement and radiative recombination efficiency at quantum well lasers, (ii) increase emitter efficiency in bipolar junction transistors, and (iii) tune confinement energies within quantum-confined devices such as tunnel diodes and high electron mobility transistors. Heterojunctions also require atomically abrupt, low defect interfaces to minimize carrier scattering and parasitic recombination. Examples that require low bulk and interface carrier scattering include high peak-to-valley negative resistance and high frequency oscillations of *resonant tunnel* diodes (RTDs), the high efficiency and low noise of avalanche detectors and quantum cascade lasers, and the step grading across interfaces with large band offsets or barrier heights.

Large band offsets have presented considerable challenges with the advent of very large band gap semiconductors for deep UV light emitter and detector applications. Nevertheless, innovative design of the lattice-matched sequence of materials can address these challenges. For example, growth of p-on-n-type AlGaN on n-type SiC presents a large ΔE_C that retards electron injection into the n-AlGaN. Instead, n-on-p-type AlGaN on p-SiC presents a low ΔE_V that permits easy hole flow while rendering ΔE_C no longer a factor. On the other hand, injection of holes into p-SiC is a challenge due to the lack of high work function metals. As with the semiconductor interlayers discussed in Section 15.7.3, recent work with heavily p-doped wide band gap oxides represents a new avenue for lowering hole barriers to wide band gap semiconductors.

Electron injection into wide band gap semiconductors is also a challenge. For example, AlN is believed to be a negative electron affinity emitter, where p-type band bending at its surface lowers E_C below E_F. However, its low χ makes ohmic n-type contacts difficult. Graded AlGaN interfaces, such as the reduced barrier contact in Figure 15.8, but here between n-SiC and n-AlGaN, can produce a sequence of lower barriers to AlGaN with increasing Al content. With an ohmic contact to the n-type SiC substrate, the graded AlGaN barriers provide more efficient electron injection into AlN to serve as an electron emitter.

16.3   Emerging Directions

The development of new electronic materials and device architectures has opened many new avenues for surface and interface research. New studies of Schottky barrier can benefit from the much wider range of E_F movement possible with semiconductors such as ZnO and other wide gap materials. While intrinsic wave function tunneling models can account for Schottky barriers involving only non-reactive metals with 0.1–0.2 eV precision, the prediction of Schottky barriers requires a more comprehensive framework to comprehend the vast majority of metal–semiconductor interfaces that involve chemical reactions, interdiffusion, and defect formation to a greater or lesser degree. Similarly, the prediction and control of heterojunction band offsets can depend on both intrinsic and extrinsic mechanisms involving altered interface stoichiometry bonding, and dipoles that depend sensitively on the sequence and manner of growing one semiconductor on another. This sensitivity is both a challenge and an opportunity since these extrinsic effects can begin to control band offsets in systematic ways. Nevertheless, such band offset engineering requires epitaxial growth plus structural, chemical, and electronic characterization with monolayer scale precision.

16.3.1   High-K Dielectrics

Surface and interface techniques have a major role to play in the development of high permittivity (high-K) dielectrics. The progression of complementary metal-oxide-semiconductor (CMOS) transistors to ever smaller dimensions presents a challenge for conventional gate oxides. For design rules of only a few tens of nm, maintaining constant gate dielectric capacitance $C = \varepsilon A/d$ for dielectric thickness d and area A, requires d to scale well below 20 Å, thin enough for significant tunneling to occur in conventional SiO_2 dielectric films. New dielectrics, such as HfO_2, ZrO_2 and their alloys, with order-of-magnitude higher ε increase C for the same thickness. However, these new dielectrics have many more bulk and interface defects than SiO_2, whose low 10^{10} cm^{-2} state densities required over half a century to achieve.

Rapid improvement of high-K dielectrics and their interfaces require *in situ* analysis techniques that measure electronic properties of ultrathin films and their interfaces without the need for electrical testing of finished device structures. For example, [see Track II T16.1 Depth-resolved cathodoluminescence spectroscopy: Mo/HfO$_2$/Si.] Here, DRCLS is able to distinguish electronic features of an ultrathin (4 nm) dielectric as well as its buried interfaces with metal gate and Si substrate. The dramatic changes at the HfO$_2$ interface with annealing under different ambient anneals highlights the sensitivity of specific gate dielectric interfaces to processing changes. The ability to probe defects in such ultrathin and buried interface structures, combined with a theoretical analysis of their nature, can contribute to the refinement of new materials, device architectures, and process steps that achieve low film and interface state densities.

16.3.2   Complex Oxides

Complex oxides have attracted considerable interest because of their ferroelectric, ferro-magnetic, and multiferroic properties, which are useful for spintronic and superconducting applications. The properties of these materials and their composite structures depend sensitively on their interface composition, bonding, and lattice structure. Furthermore, their interfaces can exhibit remarkable new properties with conducting or even superconducting layers between otherwise insulating oxides.

Native point defects such as oxygen vacancies that act as donors may also contribute to such effects since they can segregate to complex oxide interfaces. [See, for example, Track II T16.2. Complex Oxides: $(Ba,Sr) TiO_3/SrTiO_3$ Interface Defects.] Here, DRCLS and positron annihilation spectroscopy (PAS) measure the segregation of these defects to the interface between a $(Ba,Sr) TiO_3$ epilayer and its $SrTiO_3$ substrate. Without sufficient oxygen pressure during growth, the growing $(Ba,Sr) TiO_3$ film draws O from the substrate, leaving behind oxygen vacancy-related defects. Transmission electron microscopy and electron loss spectroscopy have provided evidence for interdiffusion, oxygen vacancy segregation, and altered bonding at related complex oxide interfaces. See, for example, Figure 9.15. The dependence of ε on electric field in these materials presents a particular challenge for understanding rectification at interfaces and inside surface space charge regions of these materials, where potentials and dielectric screening both change spatially on a nm scale. Likewise, measuring how external electric fields move electrically-charged defects inside these materials can provide new understanding of memristor mechanisms. Finally, mechanical strain can strongly affect the electronic properties in these materials and at their interfaces due to the orbital-derived nature of their energy levels.

16.3.3   Spintronics and Topological Insulators

Magnetic material architectures that use spin rather than electrons to carry information are the basis for the new technology of *spintronics*. Advantages over conventional charge-based electronic devices include increased speed, reduced energy consumption, and device integration density. However, the transport of spin-polarized carriers across heterojunctions requires atomically abrupt interfaces and low defect densities in order to avoid scattering and loss of polarization.

Extrinsic effects, such as charge native point defects, extended defects, interface roughness, multiple phases, and interface compound formation, can increase charge scattering. Common spintronic devices include spin valves and magnetic tunnel junctions, both of which incorporate multiple heterointerfaces. [See Track II T16.3. Spin-Dependent Transport Architectures.] These devices must maximize charge transport of spin across heterointerfaces. Hence, such interfaces are integral to advances in spintronics. Surface analytic techniques such as XPS and AES can provide information on the chemical composition of these layers either during growth or combined with depth profiling on the structures after growth. Structural techniques, such as TEM and EELS, can evaluate local atomic configurations while DRCLS can detect new interface states at the magnetic material interfaces (see, e.g., Figures 9.15–9.17). Furthermore, the high surface sensitivity of DRCLS at low voltages ($E_B = 0.1$–1 keV) can, in principle, enable detection of electronic states in surface layers of topological insulators.

16.3.4 Nanostructures

The drive to miniaturize electronics includes the creation of nanoscale circuits using novel material structures such as nanowires, nanotubes, and molecules. Ohmic contacts to these materials are difficult to achieve with $I - V$ characteristics that may be space charge-limited, as well as rectifying or blocking due to intervening layers. Dominant factors at the nanometer scale are: (i) three-dimensional electrostatic screening in high aspect ratio structures, (ii) dopant incorporation difficulty, and (iii) surface and bulk trap states that further deplete free carriers. For example, hyperspectral DRCLS reveals that electrically-active defects extend tens of nanometers or more into nanowires, with the potential to alter charge densities and transport along these wires. [See Track II T16.4 Hyperspectral Imaging of Nanostructures.]

Defects that act as majority carrier dopants can increase charge densities, while defects that act as compensating or recombination centers can decrease them. Injection at nanowires will be governed by band alignment, reduced dimension electrostatics, and also surface state recombination. Spatially varying defects and charge densities introduce size-dependent effects that are a challenge to nanoscale circuit development. Again, surface and interface research can contribute significantly to this field's development.

16.3.5 Two-Dimensional Materials

Two-dimensional (2D) materials composed of naturally occurring layered materials have attracted considerable interest due to their band structures indicating high carrier mobilities, new optoelectronic and spintronic features, and the potential to pattern advanced planar heterostructures. Chemical synthesis can produce new Van der Waals solids, such as silicene and germanene, and their covalent surface modification or substitution can alter band structures of these 2D materials for high performance field effect and *valleytronic* devices.

Surface-sensitive excitation techniques, such as XPS and DRCLS, can distinguish 2D from substrate features and provide optical information to test theory calculations of band structure. [See Track II T16.5 Ge-CH_3 Inter-Orbital DRCLS and Band Theory.] Advances in this field can contribute to basic knowledge of 2D physics and chemistry in general. Surface science techniques, such as XPS, UPS, ARPES, SPS, and CLS, are well-suited to explore this emerging field.

16.3.6 Quantum-Scale Interfaces

Quantum-scale interfaces, such as the self-assembled quantum dots formed at III-V compound heterostructures, are another material system with a rich scientific landscape to explore. Such structures can form carrier confinement regions, such as quantum wells, whose electronic properties depend on factors such as: (i) exciton localization, (ii) interface abruptness, and (iii) impurity states within the wells. STM and STS can probe quantum dot states that contribute to tunneling current and can spatially resolve differences in electronic state energies within individual quantum dots. [See Track II T16.6. Quantum Dot Tunneling.] Similar spatially-resolved studies could resolve the influence of free surfaces on deep level energies due to impurities versus depth. Such measurements could draw distinctions between different model calculations of local electronic structure.

16.4   The Essential Guide Conclusions

The aim of this book has been to introduce surfaces and interfaces of electronic materials to students and researchers new to this field. Three primary areas comprised this introductory text: semiconductor surfaces, overlayer–semiconductor interfaces, and semiconductor–semiconductor heterojunctions.

For semiconductor surfaces, major conclusions are:

1. Surface geometric structure differs in often complex ways from that of the bulk lattice.
2. New growth and characterization techniques show the importance of thermodynamics and kinetics, and
3. Surface electronic structure depends sensitively on specifics of surface atom reconstruction with localized states in the band gap related to atomic bonding.

For overlayer–semiconductor interfaces, major conclusions are:

1. Adsorption can induce surface geometric changes.
2. Chemical reaction, diffusion, and new phase formation can occur, depending on local atomic bonding and processing.
3. Metal–semiconductor localized states and band bending are influenced by extrinsic phenomena, and
4. Schottky barriers are controllable by atomic-scale techniques.

For semiconductor–semiconductor heterojunctions, major conclusions are:

1. Geometric structure is defined by the match between crystal lattice parameters, and lattice mismatch can generate local and extended lattice imperfections.
2. Diffusion and reaction can occur, depending on surface bonding and stoichiometry.
3. Atomic-layer template structures can control the nature of epitaxy and growth, and
4. Techniques are now available for nanoscale state-of-the-art analysis.

In general, the field of surfaces and interfaces provides fertile ground for future research. On a fundamental level, research can lead to new understanding of electronic material structure and charge transfer on an atomic scale. On an applied level, the tools and techniques now available are able to give an understanding of the physical properties of electronic device structures on the smallest scale possible.

Appendix A

Glossary of Commonly Used Symbols

A, B	primitive vectors
a_0	lattice constant (Å)
b	impact parameter (Å, cm)
C	capacitance (F), concentration
cpd	contact potential difference (V)
d	depletion width, crystal thickness (Å, μm, cm))
D	grid spacing (Å, μm), induced dipole, overlayer metal thickness (Å, μm)
d_{opt}	optical density (dB)
D_q	diffusion coefficient
$d\Omega$	differential solid angle (steradian)
E	energy (J, eV)
E_B	core level binding energy (eV)
E_F	Fermi level (eV)
E_G	band gap energy level (eV)
E_K	kinetic energy (J)
f	vibration frequency (Hz), lattice mismatch
F	flux of photons (s^{-1} cm^{-2}), diffusion flux (mol s^{-1} cm^{-2}), free energy (J), unit area (cm^2)
$f(E)$	Fermi function
f_C	cut-off frequency (Hz)
f_{dipol}	fluctuating dipole field
f_p	probability of electron emission
f_q	ratio of total J with versus without tunneling
g	vector of the Bravais lattice

An Essential Guide to Electronic Material Surfaces and Interfaces, First Edition. Leonard J. Brillson.
© 2016 John Wiley & Sons, Ltd. Published 2016 by John Wiley & Sons, Ltd.
Companion Website: www.wiley.com/go/Brillson/

G	Gibbs free energy (J), shear modulus
$g_{\square}$	sheet conductance (Ω^{-1})
g_i	statistical weight
H_F	Heat of formation (J mol^{-1})
H_R	Heat of reaction (J mol^{-1})
h	film thickness (Å, μm)
h_c	critical thickness at a heterojunction (Å, μm)
I	ionization energy, ion intensity, mean excitation energy loss (eV)
I_D	source-drain current (A)
I_T	tunneling current (A)
J	current density (A cm^{-2})
J_R	photocurrent per absorbed photon per unit area (A cm^{-2})
k	wave vector (rad μm^{-1})
K	kinematic factor
k	vector momentum (g m s^{-1})
L	scattering length (Å, μm), channel length (Å, μm)
l	layer thickness (Å, μm)
m	molecular mass (g)
M	tunneling matrix element
m^*	carrier effective mass (g)
M_i	incident particle mass (g)
M_{if}	matrix element between initial and final states
n	ideality factor
n	number of molecules, refractive index of a solid, bulk charge density (C cm^{-2})
N	atomic density, semiconductor carrier concentration (cm^{-3}),
n_0	equilibrium carrier concentration (cm^{-3})
N_C	conduction band density of states (cm^{-3})
p	momentum (g m s^{-1})
P	dipole moment (V cm)
P	pressure (Pa)
P_0	incident power (W)
R	rate of photoexcitation, transition probability per unit time (s^{-1})
r	rate of charge flow
R_B	Bohr–Bethe range (nm, μm)
Δr_c	coherence length (Å, μm)
r_{min}	closest approach of particle-solid scattering (Å)
S	index of interface behavior, enthalpy (J), internal potential (V)
S_{eff}	effective pumping speed (l s^{-1})
S_j	scattering efficiency
S_P	pumping speed (l s^{-1})
$[S]$	backscattering energy loss factor
T	temperature (K), energy transfer (J)
U_k	Bloch function
V	applied voltage (V), volume (m^3), potential (V), Coulomb repulsion energy (J)

v	velocity (cm s^{-1})
$<v>$	average thermal velocity (cm s^{-1})
ΔV	band offset (eV)
$V(z)$	lattice potential (eV)
V_B	semiconductor band bending (eV)
v_D	effective diffusion velocity (m s^{-1})
v_R	recombination velocity (m s^{-1})
v_{TR}	transverse acoustic mode velocity (m s^{-1})
W	depletion layer width (Å, μm)
W_A	transition probability
$\boldsymbol{x}$	imaginary refractive index
X	electronegativity
Y	Young's modulus
Z	atomic number, channel width (Å, μm)
α	optical absorption coefficient
$\varepsilon, \varepsilon_0, \varepsilon_S$	permittivity, free space permittivity, static permittivity (F cm^{-1})
ε_i	effective dielectric constant ($\varepsilon = \varepsilon_i \varepsilon_0$)
κ	imaginary part of complex refractive index
η	real part of complex refractive index
λ	wavelength (μm)
μ	electron mobility (cm^2 V^{-1} s^{-1}), Poisson ratio
υ	photon frequency (Hz)
$\mathscr{E}$	applied electric field (V m^{-1})
ρ	density of states (cm^{-3}), density (kg m^{-3})
$\rho(x)$	charge density of a region (C cm^{-2})
σ	conductivity (S m^{-1}), photoelectron cross section (cm^{-2}), strain/dislocation energy (J mol^{-1})
σ_e	ionization cross section, Auger electron yield
τ	lifetime (s), reacted layer thickness (Å, μm)
Φ, Φ_M, Φ_{SB}	potential, metal work function, Schottky barrier (eV)
χ	electron affinity (J mol^{-1}), dielectric susceptibility
ψ	wave function
ω	frequency (rad s^{-1})
ω_X	fluorescence yield
Γ_e	energy of a grid of edge dislocations (J, eV)

Appendix B

Table of Acronyms

AES	Auger electron spectroscopy
AFM	atomic force microscopy
AG	as-grown surface treatment
ARPES	angle-resolved photoelectron spectroscopy
ATMA	acetone, trichloroethylene, methanol and air-exposed
BEEM	ballistic electron energy microscopy
BEP	beam equivalent pressure
BPR	beam pressure ratio
BZ	Brillouin-zone
CNL	charge neutrality level
CBE	chemical beam epitaxy
CCA	chemically-cleaned air-exposed
CFS	constant final state spectroscopy
CIS	constant initial state spectroscopy
CLS	cathodoluminescence spectroscopy
$C-V$	Capacitance–voltage
CRS	confocal Raman spectroscopy
CZ	Czochralski
DIGS	disorder-induced gap state
DLTS	deep level transient spectroscopy
DOS	density of states
DRCLS	depth-resolved cathodoluminescence spectroscopy
EBIC	electron beam-induced current
EDEP	electrochemical deposition
EDC	energy distribution curve
EDXS	energy dispersive X-ray spectroscopy

An Essential Guide to Electronic Material Surfaces and Interfaces, First Edition. Leonard J. Brillson.
© 2016 John Wiley & Sons, Ltd. Published 2016 by John Wiley & Sons, Ltd.
Companion Website: www.wiley.com/go/Brillson/

EELS	electron energy loss spectroscopy
ELS	electron loss spectroscopy
ESCA	electron spectroscopy for chemical analysis
ESR	electron spin resonance
FCC	face centered cubic
FGA	forming gas anneal
FIM	field ion microscopy
FWHM	full width at half maximum
FZ	float-zone
GFA	Ga flux and anneal
HB	horizontal Bridgman
HD	high defect
HEMT	high electron mobility transistor
HOMO	highest occupied molecular orbit
HPA	hydrogen peroxide and air exposed
HREELS	high resolution electron energy loss spectroscopy
HRTEM	high resolution transmission electron microscopy
HT	hydrothermal
IP	in-phase
IPES	inverse photoemission spectroscopy
IPS	internal photoemission spectroscopy
ISA	Ion sputter and anneal
ISS	ion beam scattering spectrometry
I–V	Current–voltage
KPFM	Kelvin force probe microscopy
LA	laser annealed
LAPS	laser-excited photoemission spectroscopy
LEC	liquid encapsulated Czochralski
LD	low defect
LDOS	local density of states
LEED	low energy electron diffraction
LEEM	low energy electron microscopy
LEELS	low energy electron loss spectroscopy
LEPD	low energy positron diffraction
LFM	lateral force microscopy
LUMO	lowest unoccupied molecular orbit
MBE	molecular beam epitaxy
MFM	magnetic force microscopy
MG	melt-grown
MIGS	metal-induced gap states
MIS	metal-insulator-semiconductor
MISFET	metal-insulator field effect transistor
MOCVD	molecular organic chemical vapor deposition
MOSFET	metal-oxide semiconductor field effect transistor
NBOHC	non-bonding oxygen hole centers

NEA	negative electron affinity
NVC	NH_3-based vapor clean
OCA	organic clean and air exposed
OLED	organic light emitting diode
OPS	oxygen-plasma sputtered
PAS	positron annihilation spectroscopy
PLD	pulsed laser deposited
PLS	photoluminescence spectroscopy
PMMA	polymethylmethacrylate
PVD	plasma vapor deposition
QCM	quartz crystal monitors
QD	quantum dot
RBS	Rutherford backscattering spectroscopy
RF	radio frequency
RGA	residual gas analyzer
RHEED	Reflection high-energy electron diffraction
ROP	Remote oxygen plasma
ROP1A	room temperature remote oxygen plasma, one hour and air exposure
ROP2	room temperature remote oxygen plasma for two hours
ROPHT	remote oxygen plasma cleaned at high temperature
ROPRT	remote oxygen plasma cleaned plus room temperature re-exposure
RPAN	remote plasma assisted nitride
RPAO	remote plasma assisted oxide
RSS	Raman scattering spectroscopy
RT	room temperature
RTD	resonant tunnel diode
SCLC	space charge-limited current
SE	spectroscopic ellipsometry
SEM	Scanning electron microscope
SET	secondary electron threshold
SEXAFS	surface extended X-ray absorption fine spectroscopy
SIMS	secondary ion mass spectroscopy
SPC	surface photoconductivity spectroscopy
SPS, SPV	surface photovoltage spectroscopy
SRS	surface reflectance spectroscopy
SRV	surface recombination velocity
SSIMS	static secondary ion mass spectrometry
STM	scanning tunneling microscopy
STS	scanning tunneling spectroscopy
SXAS	soft X-ray absorption spectroscopy
SXPS	soft X-ray photoemission spectroscopy
TEM	transmission electron microscopy
TEXRD	total external X-ray diffraction
TLM	Transmission line measurement
TM	transition metal

TOF-SIMS	Time of flight secondary ion mass spectrometry
TRCI	time-resolved charge injection
TSCAP	thermally stimulated capacitance
TSP	Titanium sublimation pump
UHV	ultra high vacuum
UPS	ultraviolet photoemission spectroscopy
UV	ultraviolet
UVOA	UV ozone and air-exposed
VPE	vapor phase epitaxy
VS	valence states
XCLS	cross-sectional cathodoluminescence spectroscopy
XKPFM	cross-sectional Kelvin probe force microscopy
XPS	X-ray photoemission spectroscopy
XRD	X-ray diffraction

Appendix C

Table of Physical Constants and Conversion Factors

Avogadro's Number, N_A	6.023×10^{23} atoms mole^{-1}
Bohr radius, a_0	5.29×10^{-9} cm
Boltzmann constant, k_B	1.38×10^{-23} J molecule^{-1} K^{-1}
	8.62×10^{-5} eV molecule^{-1} K^{-1}
$k_B T$ at room temperature	0.0259 eV
Electron charge, e	1.602×10^{-19} C
Fine structure constant, α	$\sim 1/137$
Planck's constant, h	6.626×10^{-34} J s
	4.136×10^{-15} eV s
Reduced Planck's constant, $\hbar$	1.054×10^{-34} J s
Rest electron mass, m_e	$9.10938215 \times 10^{-31}$ kg
Speed of light	2.998×10^{10} cm s^{-1}
Vacuum permittivity, ε_0	8.854×10^{-14} F cm^{-1}
	8.854×10^{-12} F m^{-1}
1 Å (Angstrom)	10^{-8} cm
1 nm (nanometer)	10^{-7} cm
1 μ (micron)	10^{-4} cm
1 eV	1.602×10^{-19} J
e^2	14.4 eV Å
$e^2/2a_0$	13.58 eV

An Essential Guide to Electronic Material Surfaces and Interfaces, First Edition. Leonard J. Brillson.
© 2016 John Wiley & Sons, Ltd. Published 2016 by John Wiley & Sons, Ltd.
Companion Website: www.wiley.com/go/Brillson/

Appendix D

Semiconductor Properties

	E_G (eV)		χ (eV)[a]	Structure	a_0(Å)	$\varepsilon_s{}^b$	Density (g cm^{-3})	Melting Point (°C)	Enthalpy (kJ mol^{-1})[c]
Si	1.11	i	4.05	D	5.43	11.8	2.33	1415	0.0
Ge	0.67	i	4.0	D	5.65	16	5.32	936	0.0
C	5.5	i	0.3	D	3.5597	5.70	3.515	3850	0.0
SiC (3C)	2.36	i	4.12	ZB	4.3596	9.72	3.21	~3100 @ 35atm	66.9
SiC (6H)	3.0	i	3.7	W	$a = 3.0806$ $c = 15.11733$	9.66 10.03	3.21	~3100 @ 35atm	66.9
SiC (4H)	3.2	i	3.2	W	$a = 3.0730$ $c = 10.053$	9.66 10.03	3.21	~3100 @ 35atm	66.9
BN	4.5–5.5	i	4.5	W	$a = 2.55$ $c = 4.17$	6.8 5.06	3.48	1400	252.3
BN	6.1–6.4	i	4.5	ZB	3.615	7.1	3.45	1400 @4GPa	252.3
BN	4.0–5.8	d	4.5	H	$a = 2.5–2.9$ $c = 6.66$	6.85 5.06	2.0–2.28	1400	252.3
AlN	6.2	d	0.6	W	$a = 3.112$ $c = 4.982$	8.5	3.23	2750 @100 atm N$_2$	318.4
AlP	2.45	i	3.98	ZB	5.46	9.8	2.4	2000	164.4
AlAs	2.16	i	3.5	ZB	5.66	10.9	3.60	1740	116.3
AlSb	1.6	i	3.6	ZB	6.14	11	4.26	1080	50.2
GaN	3.39	d	3.5	W	$a = 3.189$ $c = 5.186$	8.9	6.15	2530	109.6
GaN	3.2	d	4.1	ZB	4.52	9.7	6.15	2530	109.6
GaP	2.26	i	3.65	ZB	5.45	11.1	4.13	1467	102.5

(continued overleaf)

An Essential Guide to Electronic Material Surfaces and Interfaces, First Edition. Leonard J. Brillson.
© 2016 John Wiley & Sons, Ltd. Published 2016 by John Wiley & Sons, Ltd.
Companion Website: www.wiley.com/go/Brillson/

	E_G (eV)		χ (eV)[a]	Structure	a_0(Å)	$\varepsilon_s{}^b$	Density (g cm^{-3})	Melting Point (°C)	Enthalpy (kJ mol^{-1})[c]
GaAs	1.43	d	4.07	ZB	5.65	13.2	5.31	1238	74.1
GaSb	0.7	d	4.06	ZB	6.09	15.7	5.61	712	43.9
InN	0.7[d]	d		ZB	$a = 3.533$ $c = 5.693$	15.3	6.81	~750 @10^3 atm N$_2$	
InP	1.35	d	4.35	ZB	5.87	12.4	4.79	1070	75.3
InAs	0.36	d	4.90	ZB	6.06	14.6	5.67	943	57.7
InSb	0.18	d	4.72	ZB	6.48	17.7	5.78	525	30.5
CdS	2.42	d	4.79	W	4.137	8.9	4.82	1475	149.4
CdSe	1.73	d		W	4.30	10.2	5.81	1258	144.8
CdTe	1.58	d	4.28	ZB	6.482	10.2	6.20	1098	100.8
ZnS	3.6	d	3.82	W	5.409	8.9	4.09	1650	205.2
ZnSe	2.7	d	4.09	ZB	5.671	9.2	5.65	1100	163.2
ZnTe	2.25	d	3.53	ZB	6.101	10.4	5.51	1238	119.2
ZnO	3.39	d	4.2-4.57	W	$a = 3.253$ $c = 5.213$	7.8 8.75	5.67	~2000	350.5
PbS	0.37	i	3.3	R	5.936	17.0	7.6	1119	98.3
PbSe	0.27	i		R	6.147	23.6	8.73	1081	100.0
PbTe	0.29	i		R	6.452	30	8.16	925	68.6
MgO	7.8	d	0.85	R	4.2112	9.83	3.58	2800	601.6

All values at 300 K

Structure: D = diamond, ZB = zinc blende (T$_d$^{2}-F43m), W = wurtzite (C$_{6v}$^{4}P6$_3$mc), R = rocksalt (Oh5-Fm3m), H = hexagonal (D$_{6v}$P6$_3$mmc)

[a]Preferred values of electron affinity from among literature values.

GaSb χ from: http://www.ioffe.ru/SVA/NSM/Semicond/GaSb/basic.html

SiC ZB and 6H χ from Internet

AlP χ from Internet: http://www.rpi.edu/~Schubert/Educational-resources/Materials-Semiconductors-III-V-phosphides.pdf

AlAs χ from Internet: http://www.worldscibooks.com/phy_etextbook/2046/2046_chap1_1.pdf

AlSb χ from Internet: http://adsabs.harvard.edu/abs/1965PhRv..139.1228F

PbS χ from Internet: http://arxiv.org/ftp/cond-mat/papers/0412/0412307.pdf

InAs χ: Takahashi, T., Kawamukai, T., Ono, S., *et al.* (2000) *Jpn. J. Appl. Phys.*, **39**, 3721.

InSb χ: S. Adachi, (2005) *Properties of Group-IV, III-V and II-VI Semiconductors,* John Wiley & Sons, p. 196.

MgO values from: http://www.oxmat.co.uk/Crysdata/mgo.htm, also de Boer, P.K. and de Groot, R.A. (1998) *J. Phys.: Condens. Matter*, **10**, 10241.

MgO χ: Abarenkov, I.V and Antonova, I.M. (1979) Electron affinities of the alkaline earth chalcogenides, *J. Appl. Phys.*, **45**, 47.

diamond χ: http://pubs.acs.org/doi/full/10.1021/cm801752j?cookieSet=1; Qi, D., Gao, X,Wang, L. *et al.* (2008) *Chem. Mater.*, **20** (21), 6871–6879.

[b]ε_r = relative static dielectric constant, $\varepsilon = \varepsilon_r \varepsilon_0$; for W or H, top value = $\perp c$ axis, bottom value = $\| c$ axis

[c]Enthalpy values $-\Delta H_{298}$ from Kubachewski, O., Alcock, C.B. and Spencer, P.J. (1993) *Materials Thermochemistry*, 6th edn, Pergamon Press, New York.

SiC, BN, GaN, and InN values from Levinshtein, M., Rumyantsev, S., and Shur, M. (eds) (2001) *Properties of Advanced Semiconductor Materials*, John Wiley & Sons, Inc., New York.

[d]Wu, J., Walukiewicz, W., Yu, K.M. *et al.* (2002) *Appl. Phys. Lett.*, **80**, 3967.

BN gap parameters from: http://www.ioffe.ru/SVA/NSM/Semicond/BN/bandstr.html

Index
